T0310692

Integrating Cognitive Architectures into Virtual Character Design

Jeremy Owen Turner
Simon Fraser University, Canada

Michael Nixon
Simon Fraser University, Canada

Ulysses Bernardet
Simon Fraser University, Canada

Steve DiPaola
Simon Fraser University, Canada

A volume in the Advances in Computational Intelligence and Robotics (ACIR) Book Series

Published in the United States of America by
Information Science Reference (an imprint of IGI Global)
701 E. Chocolate Avenue
Hershey PA 17033
Tel: 717-533-8845
Fax: 717-533-8661
E-mail: cust@igi-global.com
Web site: http://www.igi-global.com

Library of Congress Cataloging-in-Publication Data

Names: Turner, Jeremy Owen, 1974- editor. | Nixon, Michael, 1981- | Bernardet, Ulysses, 1969- editor. | DiPaola, Steve, editor.
Title: Integrating cognitive architectures into virtual character design / Jeremy Turner, Michael Nixon, Ulysses Bernardet, and Steve DiPaola, editors.
Description: Hershey, PA : Information Science Reference, [2016] | Includes bibliographical references and index.
Identifiers: LCCN 2016011642| ISBN 9781522504542 (hardcover) | ISBN 9781522504559 (ebook)
Subjects: LCSH: Avatars (Virtual reality)--Design. | Intelligent agents (Computer software)
Classification: LCC QA76.9.C65 .I498 2106 | DDC 006.3--dc23 LC record available at http://lccn.loc.gov/2016011642

This book is published in the IGI Global book series Advances in Computational Intelligence and Robotics (ACIR) (ISSN: 2327-0411; eISSN: 2327-042X)

British Cataloguing in Publication Data
A Cataloguing in Publication record for this book is available from the British Library.

Advances in Computational Intelligence and Robotics (ACIR) Book Series

ISSN: 2327-0411
EISSN: 2327-042X

Mission

While intelligence is traditionally a term applied to humans and human cognition, technology has progressed in such a way to allow for the development of intelligent systems able to simulate many human traits. With this new era of simulated and artificial intelligence, much research is needed in order to continue to advance the field and also to evaluate the ethical and societal concerns of the existence of artificial life and machine learning.

The **Advances in Computational Intelligence and Robotics (ACIR) Book Series** encourages scholarly discourse on all topics pertaining to evolutionary computing, artificial life, computational intelligence, machine learning, and robotics. ACIR presents the latest research being conducted on diverse topics in intelligence technologies with the goal of advancing knowledge and applications in this rapidly evolving field.

Coverage

- Automated Reasoning
- Cyborgs
- Machine Learning
- Artificial Intelligence
- Fuzzy Systems
- Natural Language Processing
- Pattern Recognition
- Heuristics
- Cognitive Informatics
- Algorithmic Learning

IGI Global is currently accepting manuscripts for publication within this series. To submit a proposal for a volume in this series, please contact our Acquisition Editors at Acquisitions@igi-global.com or visit: http://www.igi-global.com/publish/.

The Advances in Computational Intelligence and Robotics (ACIR) Book Series (ISSN 2327-0411) is published by IGI Global, 701 E. Chocolate Avenue, Hershey, PA 17033-1240, USA, www.igi-global.com. This series is composed of titles available for purchase individually; each title is edited to be contextually exclusive from any other title within the series. For pricing and ordering information please visit http://www.igi-global.com/book-series/advances-computational-intelligence-robotics/73674. Postmaster: Send all address changes to above address.

Titles in this Series

For a list of additional titles in this series, please visit: www.igi-global.com

Applied Artificial Higher Order Neural Networks for Control and Recognition
Ming Zhang (Christopher Newport University, USA)
Information Science Reference • copyright 2016 • 511pp • H/C (ISBN: 9781522500636) • US $215.00 (our price)

Handbook of Research on Generalized and Hybrid Set Structures and Applications for Soft Computing
Sunil Jacob John (National Institute of Technology Calicut, India)
Information Science Reference • copyright 2016 • 607pp • H/C (ISBN: 9781466697980) • US $375.00 (our price)

Handbook of Research on Modern Optimization Algorithms and Applications in Engineering and Economics
Pandian Vasant (Universiti Teknologi Petronas, Malaysia) Gerhard-Wilhelm Weber (Middle East Technical University, Turkey) and Vo Ngoc Dieu (Ho Chi Minh City University of Technology, Vietnam)
Engineering Science Reference • copyright 2016 • 960pp • H/C (ISBN: 9781466696440) • US $325.00 (our price)

Problem Solving and Uncertainty Modeling through Optimization and Soft Computing Applications
Pratiksha Saxena (Gautam Buddha University, India) Dipti Singh (Gautam Buddha University, India) and Millie Pant (Indian Institute of Technology - Roorkee, India)
Information Science Reference • copyright 2016 • 403pp • H/C (ISBN: 9781466698857) • US $225.00 (our price)

Emerging Technologies in Intelligent Applications for Image and Video Processing
V. Santhi (VIT University, India) D. P. Acharjya (VIT University, India) and M. Ezhilarasan (Pondichery Engineering College, India)
Information Science Reference • copyright 2016 • 518pp • H/C (ISBN: 9781466696853) • US $235.00 (our price)

Handbook of Research on Design, Control, and Modeling of Swarm Robotics
Ying Tan (Peking University, China)
Information Science Reference • copyright 2016 • 854pp • H/C (ISBN: 9781466695726) • US $465.00 (our price)

Editorial Advisory Board

List of Reviewers

Table of Contents

Foreword xiii

Preface xvii

Acknowledgment xxiii

Chapter 1
On Vision-Based Human-Centric Virtual Character Design: A Closer Look at the Real World from a Virtual One 1
Eugene Borovikov, PercepReal, USA
Ilya Zavorin, PercepReal, USA
Sergey Yershov, PercepReal, USA

Chapter 2
Integrating ACT-R Cognitive Models with the Unity Game Engine 35
Paul Richard Smart, University of Southampton, UK
Tom Scutt, Mudlark, UK
Katia Sycara, Carnegie Mellon University, USA
Nigel R. Shadbolt, University of Oxford, UK

Chapter 3
A Graphical Tool for the Creation of Behaviors in Virtual Worlds 65
Andrea Corradini, Design School Kolding, Denmark
Manish Mehta, Accenture Technology Lab, USA

Chapter 4
Learned Behavior: Enabling Believable Virtual Characters through Reinforcement 94
Jacquelyne Forgette, Western University, Canada
Michael Katchabaw, Western University, Canada

Detailed Table of Contents

Foreword xiii

Preface xvii

Acknowledgment xxiii

Chapter 1
On Vision-Based Human-Centric Virtual Character Design: A Closer Look at the Real World from a Virtual One 1
Eugene Borovikov, PercepReal, USA
Ilya Zavorin, PercepReal, USA
Sergey Yershov, PercepReal, USA

Enabling cognition in a Virtual Character (VC) may be an exciting endeavor for its designer and for the character. A typical VC interacts primarily with its virtual world, but given some sensory capabilities (vision or hearing), it would be expected to explore some of the real world and interact with the intelligent beings there. Thus a virtual character should be equipped with some algorithms to localize and track humans (e.g. via 2D or 3D models), recognize them (e.g. by their faces) and communicate with them. Such perceptual capabilities prompt a sophisticated Cognitive Architecture (CA) to be integrated into the design of a virtual character, which should enable a VC to learn from intelligent beings and reason like one. To seem natural, this CA needs to be fairly seamless, reliable and adaptive. Hence a vision-based human-centric approach to the VC design is explored here.

Chapter 2
Integrating ACT-R Cognitive Models with the Unity Game Engine 35
Paul Richard Smart, University of Southampton, UK
Tom Scutt, Mudlark, UK
Katia Sycara, Carnegie Mellon University, USA
Nigel R. Shadbolt, University of Oxford, UK

The main aim of the chapter is to describe how cognitive models, developed using the ACT-R cognitive architecture, can be integrated with the Unity game engine

in order to support the intelligent control of virtual characters in both 2D and 3D virtual environments. ACT-R is a cognitive architecture that has been widely used to model various aspects of human cognition, such as learning, memory, problem-solving, reasoning and so on. Unity, on the other hand, is a very popular game engine that can be used to develop 2D and 3D environments for both game and non-game purposes. The ability to integrate ACT-R cognitive models with the Unity game engine thus supports the effort to create virtual characters that incorporate at least some of the capabilities and constraints of the human cognitive system.

Typically, the creation of AI-based behaviors for Non-Playing Characters (NPC) in computer games is carried out by people with specialized skills in the broad areas of design and programming. This book chapter presents Second Mind (SM), a digital solution that makes it possible for novice users to successfully author behaviors. In SM, authors can use a graphical interface to define the behaviors of virtual characters in response to interactions with players. Authors can easily endow NPCs with behavior capabilities that make them act in different possible roles such as e.g. shopkeepers, museum hosts, etc. A series of user tests with human subjects to evaluate the behavior authoring process shows that Second Mind is easy to understand and simplifies the process of behavior production.

A key challenge in programming virtual environments is to produce virtual characters that are autonomous and capable of action selections that appear believable. In this chapter, motivations are used as a basis for learning using reinforcements. With motives driving the decisions of characters, their actions will appear less structured and repetitious, and more human in nature. This will also allow developers to easily create virtual characters with specific motivations, based mostly on their narrative purposes or roles in the virtual world. With minimum and maximum desirable motive values, the characters use reinforcement learning to drive action selection to maximize their rewards across all motives. Experimental results show that a character can learn to satisfy as many as four motives, even with significantly delayed rewards,

Chapter 7
Virtual Soar-Agent Implementations: Examples, Issues, and Speculations 181
Jeremy Owen Turner, Simon Fraser University, Canada

This chapter provides a brief overview of those virtual agent implementations directly inspired by the cognitive architecture: Soar. This chapter will take a qualitative approach to discussing examples of virtual Soar-agents. Finally, this chapter will speculate on the future of Soar virtual characters. The goals of this chapter are sixfold. The first goal is to explain why cognitive architectures are becoming increasingly important to virtual agent design(s). The second goal is to convey why this chapter focuses exclusively on virtual agents that utilize the Soar architecture. The third goal is to explore some of Soar's technical details. The fourth goal is to showcase a few diverse examples where Soar is beginning to have a design impact on virtual agents. The fifth goal addresses Soar's limitations – when applied to agent design in virtual environments. The final goal speculates on ways Soar can be expanded for virtual agent design(s) in the future.

Chapter 8
Towards Truly Autonomous Synthetic Characters with the Sigma Cognitive Architecture 213
Volkan Ustun, University of Southern California, USA
Paul S. Rosenbloom, University of Southern California, USA

Realism is required not only for how synthetic characters look but also for how they behave. Many applications, such as simulations, virtual worlds, and video games, require computational models of intelligence that generate realistic and credible behavior for the participating synthetic characters. Sigma (Σ) is being built as a computational model of general intelligence with a long-term goal of understanding and replicating the architecture of the mind; i.e., the fixed structure underlying intelligent behavior. Sigma leverages probabilistic graphical models towards a uniform grand unification of not only traditional cognitive capabilities but also key non-cognitive aspects, creating unique opportunities for the construction of new kinds of non-modular behavioral models. These ambitions strive for the complete control of synthetic characters that behave as humanly as possible. In this paper, Sigma is introduced along with two disparate proof-of-concept virtual humans – one conversational and the other a pair of ambulatory agents – that demonstrate its diverse capabilities.

In this chapter, the characteristics of a cognitive architecture that can migrate among various embodiments are discussed and the feasibility of designing such architecture is investigated. The migration refers to the ability of an agent to transfer its internal state among different embodiments without altering its underlying cognitive processes. Designing such architecture will address both weak and strong aspects of AI. The authors propose a Universal Migrating Cognitive Agent (UMCA) inspired by onboard autonomous frameworks utilized in interplanetary missions in which the embodiment can be tailored by defining a set of possible actions and perceptions associated with the new body. The proposed architecture is then evaluated within a few virtual environments to analyze the consistency between its deliberative and reactive behaviors. Finally, UMCA is tailored to automatically create computer animations using a natural language interface.

Artificial General Intelligence has traditionally used games as a testbed to develop domain-agnostic game playing techniques. Yet games are about more than winning. This chapter reviews recent efforts that have broadened the ways Artificial Intelligence (AI) is used in games, covering: modeling and managing player experiences, creating novel game structures based in interacting with AI, and enabling AI agents to make games. Many of the techniques used to address these challenges have been ad hoc approaches to solving specific problems. This chapter discusses open challenges in each of these areas and the potential for cognitive architectures to provide unified techniques that address these challenges.

Foreword

This book explores the coming together of two different, related research and development areas, virtual characters and cognitive architectures – both of which are currently experiencing phases of exciting and unprecedentedly rapid progress.

Virtual characters of various sorts have become an everyday reality for numerous technology users. Smartphones contain text and voice based virtual characters integrated with their operating systems; video games contain animated virtual characters as routine aspects of gameplay. Customer support systems blur the lines between virtual humans and actual humans, via delivering customers with chat interactions that interleave human and auto-generated responses. Healthcare systems aim to provide low-cost emotional support to patients by having them interact with animated virtual medical staff displayed on tablet screens.

In short, virtual characters are now a well accepted part of life in the developed world. But the fundamental limitations of understanding and intelligence displayed by current virtual characters are also well understood. The average person who interacts with a virtual character doesn't know what a "cognitive architecture" is – but they can appreciate when one character is more sensibly and appropriately responsive than another. And more often than not, this sort of behavioral difference between two characters boils down, behind the scenes, to a difference in the sophistication of the cognitive architecture behind the two characters.

The critical value of cognitive architectures for artificial intelligence was apparent to savvy researchers very shortly after the inception of the AI field in the late 1950s. During subsequent decades, as both the cross-disciplinary field of cognitive science and the technology of AI algorithms and representations advanced, a variety of increasingly sophisticated cognitive architectures emerged. Today's cognitive science incorporates data and ideas from neuroscience, psychology, computer science, linguistics, philosophy, engineering, mathematics and other disciplines to give an impressively detailed and intricate understanding of how the different structures and processes in the human mind interact to produce human intelligent behavior. Today's cognitive architectures incorporate more and more of this rich, multidimensional understanding in their designs and dynamics.

A decade or two ago, most cognitive architectures explored in academic research had the appearance of "toy models" – useful for research exploration of certain core aspects of human cognition, and for building practical systems in appropriately limited domains, but with numerous huge and obvious gaps. Today things are different, and now it is common for the same cognitive architecture to incorporate multiple kinds of memory, knowledge, reasoning, learning, perception and action into the same framework. Now it is understood that, to make real progress on cognitive science and Artificial General Intelligence (AGI), it is necessary to integrate these various aspects into unified information processing models embodying plausible hypotheses regarding the nature of human cognition.

In short, due to the dramatic recent progress on both sides, virtual characters and cognitive architectures are now at a stage of development where there is tremendous mutual benefit in focusing on the intersection of the two domains.

Virtual environments are now sufficiently sophisticated to serve as interesting playgrounds for characters controlled by cognitive architectures. Furthermore, the overall technology ecosystem now brings a wide variety of people in contact with virtual characters in various aspects of their lives, thus enabling virtual characters to serve as windows to the world for information and interaction hungry artificial cognitive systems.

Cognitive architectures are now sufficiently sophisticated to usefully control virtual characters. Via integrating perception, action, learning, memory and knowledge in the same framework, a modern cognitive architecture can provide a viable framework for controlling a character – a practical and in many cases demonstrably better alternative to controlling a virtual character using a conventional, non-cognitively-structured software program.

Indeed, it would be reasonable to contend that virtual characters and virtual environments constitute the best available path for the refinement of cognitive architectures toward human-level general intelligence – as well as for the advancement of virtual characters toward massive practical utility and commercial profitability. An argument for the first half of this contention would go as follows: Intelligence is not just about representations and reasoning, learning and perception and action algorithms, it's also about environments and embodiment and experience. If we want to create a human-like cognitive system, we will likely need to embody this system in an at least, a vaguely human-like body and environment, and let it learn from experience.

While this might seem to argue in favor of robotics as the preferred route for cognitive architecture development and experimentation, the fact remains that, at the present time, robotics technology is still quite primitive and difficult to work with compared to human bodies. Most research robots are turned on for brief intervals for the running of specific experiments, and then shut down till the next experiments; they don't get the opportunity for lifelong learning. Sensors and actuators fall far short of human level in multiple regards, so that most robotics research ends up focusing on perception and action rather than the more critical cognitive issues. Virtual characters, on the other hand, combine all the different aspects of human-level intelligence in a single compact package. Perception, action, language, learning, reasoning, memory, social interaction, and more – it's all there, and without the need to mess with finicky hardware sensors and manipulators and quickly-draining batteries. Robotics has a lot to offer, but virtual characters and environments also have dramatic advantages, and at very least deserve a prominent role in the research and practical deployment of contemporary cognitive architectures.

The multidimensional aspect of virtual character control also has the benefit of highlighting the complexity and diversity of human intelligence. As an example of why this is important, consider deep learning networks, which are lately – and justly – receiving great attention in the AI community and tech media. The achievements of deep learning networks in visual and auditory perception, and other limited domains like game-playing, are impressive and exciting. But focusing on the problem of controlling virtual characters in complex environments brings home some of the significant shortcomings of the current batch of deep learning algorithms, as compared to the requirements of human-like intelligence. Recognizing hierarchical patterns in high-dimensional data is a critical part of intelligence, but it's not the whole thing by any means. Cognitive architecture researchers have been thinking deeply about the "whole thing" for decades, and virtual character control is a domain that truly requires the whole thing, not just hierarchical perception and not just any one sort of cognitive capability.

From the practical and commercial rather than theoretical point of view, the synergy between cognitive architectures and virtual agents is apparent to everyone who owns a smartphone and tries interacting with the simple text and voice based virtual characters contained therein – Siri, Google Now, Cortana, and so forth. Such systems, like the current crop of game AI characters, are useful and interesting, but there is a fundamental sense in which they don't really understand what's going on. This lack of understanding, readily observable to even the most naïve end user, is rooted in the fact that cognitive systems research has not yet progressed to the point where we can put a complex cognitive architecture behind a user-facing virtual character and have it supply this character with genuine, embodied, context-aware common sense understanding. In contrast to decades past, however, this sort of achievement no longer seems purely science fictional or incomprehensibly remote. Indeed, many of the papers in this volume are palpably taking steps in this direction. Also, several of the papers report practical systems which, via implementing sophisticated cognitive architectures behind virtual characters, supply these characters with impressive degrees of appropriate real-world responsiveness.

While recent progress has been tremendous and currently ongoing work – including the work reported here – is very exciting, the challenges required to bring cognitive architectures and virtual characters together in a compelling way remain significant. We need to improve our cognitive architectures, our learning and reasoning algorithms and knowledge representations, our virtual environments, our theoretical concepts and our tools for measurement and evaluation of our systems. These improvements are happening, with very real progress year by year, and this book reports some of the intellectual and practical movement occurring in these directions.

Ben Goertzel
Artificial General Intelligence Society, USA & OpenCog Foundation, USA & Hanson Robotics, Hong Kong & Aidyia Limited, Hong Kong

Ben Goertzel *is the Chairman of the Artificial General Intelligence Society, which hosts the annual AGI research conference series, and of the OpenCog Foundation. He is Chief Scientist of Hanson Robotics, a Hong Kong robotics company that creates the world's most advanced humanoid robots; and is also Chief Scientist of Aidyia Limited, a Hong Kong investment management firm using advanced AI for financial prediction. Before relocating to Hong Kong in 2011, Dr. Goertzel held executive roles at AI consulting and product development firms in Washington DC (CEO, Chairman and Chief Scientist at Novamente LLC and Biomind LLC) and New York City (CTO at Webmind Inc.). Prior to that, Dr. Goertzel served as faculty in mathematics at the University of Nevada Las Vegas, in cognitive science at the University of Western Australia, and in computer science at Waikato University in New Zealand, at the City University of New York and at the University of New Mexico in Albuquerque. He has published nearly 20 scientific books and well over 100 scientific research papers. Dr. Goertzel holds a PhD degree in mathematics from Temple University in Philadelphia, USA.*

Preface

"Integrating Cognitive Architectures into Virtual Character Design" is a book addressing a disciplinary merger between cognitive science, artificial intelligence and character design.

INTRODUCTION TO THE SUBJECT AREA

This book is interdisciplinary in nature as it combines cognitive architectures with virtual character design. The various contributor disciplines and sub-disciplines include*: Artificial Intelligence (AI), Artificial General Intelligence (AGI), Cognitive Science, Virtual Worlds, Video Games, Communications/Critical Theory, and New Media Studies*.

Cognitive architectures represent an umbrella term to describe ways in which the flow of thought can be engineered towards cerebral and behavioral outcomes. There are many varieties of cognitive architectures outlined in the available literature and they range from purely reactive systems (e.g. YMIR; Thórisson, 1999) to those with very deliberate decision cycles (e.g. SOAR; Laird, 2012). Some architectures are composed out of thought modules and sub-systems (e.g. CLARION; Sun, 2006). Other architectures use non-modular instantiations such as the deployment of factor graphs (e.g. Sigma; Rosenbloom, 2013). In fact, some so-called "architectures" are barely architectural and resemble models (e.g. ACT-*; Anderson, 1996). The most casually represented cognitive representations are little more than a collection of behavioral heuristics and algorithms.

Virtual characters may or may not have their mind enhanced by a cognitive architecture. Virtual characters are agents (i.e. humanoid and non-human types) that are embedded within a virtual environment (world) and are known by many domain-specific categories. In video-games, these agents are known as Non-Player Characters (NPCs). In chat based virtual worlds, these same agents are known as "bots" (short for "robots").

SYNOPSIS

This book ultimately showcases the contributions of Artificial (General) Intelligence scholars from various disciplines. Predictably, many of these chapters were from the Computer Science, Information Science, and Cognitive Science fields. This book expands the disciplinary discourse by adding some contributions from the Humanities disciplines as well as the virtual entertainment industries. This book also focuses on the explicit and implicit connections between higher-level cognitive architectures and characters situated in virtual environments (e.g. video games and virtual worlds).

Cognitive architectures can be sub-categorized into the following:

1. Production or rule-based systems – (e.g. SOAR)
2. Biologically Inspired Cognitive Architecture (BICA)
3. Hybrid Systems (e.g. CLARION) that employ both biological metaphors (e.g. Artificial Neural Nets) and production systems.

Virtual (aka. synthetic) characters, on the other hand, are often semantically categorized by the type of environment they are situated within and are known within each discipline respectively as:

1. Non-Player Characters (NPCs) in Video Games
2. Virtual Agents in generic AI or AGI domains
3. Bots as Characters in chat-based virtual worlds.

Ultimately, this book explicitly shows the variety of ways a cognitive architecture can deliberately guide an intelligent virtual character in a virtual environment.

OVERALL OBJECTIVES AND MISSION OF THIS BOOK

The overall objectives and mission of this book is to gather a diverse range of virtual agent cognitive implementations into a single print publication. The goals of this book are three-fold.

1. To inspire virtual character designers to consider cognitive architectures and thereby, considering the ideal of Artificial General Intelligence (AGI) as a potential alternative to current industry-approved character-AI systems and procedures.

2. To show that cognitive architectures are no longer restricted to military, scientific, philosophical, cognitive science and medical applications.
3. To provide state-of-the-art examples of the explicit usage of cognitive architectures (rather than the conventional hierarchical planning structures, e.g. HTNs, hierarchical planners) in video-games.

Our mission is to inspire the proliferation of cognitive architectures – as compared to mere algorithms, heuristics or industry "tricks" reactively simulating intelligence – for the next-generation of artificially intelligent virtual characters.

This book was published, in part, to make AGI more accessible to the character designers and character-design hobbyists, and academics working within Video-Games and Virtual Worlds industries. This book also exists to explicitly show that cognitive architectures are becoming more common in virtual/synthetic environments.

The theme of this book contributes new scholarly value by advocating the explicit usage of cognitive architectures for complex deliberations and interactions in virtual environments. In many previous publications, cognitive architectures served as theoretical and technical introductions or elaborations. As such, they focused more of the core discussions on general higher-level applications rather than specifically commenting on how the cognitive architectures could act as intellectual shapers of idiosyncratic personality traits and personified behaviours. Although some scholars (e.g. Magerko et al.) have focused their discussions about cognitive architectures around virtual character implementations, there has not yet been an academic book handing the specific thematic correlations between cognitive architectures and character-driven interactions in virtual environments.

This book serves to expand the perspectives of scholars from the information and engineering sciences by bridging their disciplinary paradigm to both scholarly contributions from the Humanities and practical considerations from corporate industries (e.g. video-games, virtual worlds). Most importantly, some chapters of this book have illustrated explicit cross-disciplinary connections that unite both academic theory and praxis as well as tractable correlations uniting the AI and AGI sub-disciplinary discourse.

We are aware that the current competition includes previously published books showcasing particular cognitive architectures (e.g. Laird's SOAR book, MIT 2012), online tutorials (videos/courseware), an official AGI Journal (e.g. Wang & Goertzel, Ed.), conference proceedings and other decentralized cognitive architecture and virtual character papers discovered using Google Scholar. However, this book has showcased and discussed updated and recently created versions of cognitive architectures for video games. Some cognitive architectures (e.g. SOAR) are frequently updated with new features. The creators of these cognitive architectures are constantly seeking new publications to promote their new features and therefore, previous publications are not really in direct competition with our book. Further, it is becoming increasingly common for brand new cognitive architectures to be invented from scratch by emerging AGI/AI scholars/engineers. Because of the increasing popularity of cognitive architectures in general, this book will likely inspire brand new cognitive architectures specifically tailored to virtual characters and personalities. This book has essentially centralized all previous publications on cognitive architectures tailored for virtual character implementation(s).

Due to the rapid iteration and implementation of cognitive architectures and the growing popularity of virtual environments (video games, virtual worlds), a small amount of non-canonical cognitive architectures are likely to become outdated shortly after the time of this publication's release. However, even academic theories about outmoded cognitive architectures and virtual environments can still provide historical inspiration for new architectures and implementations.

INTENDED AUDIENCE

This book will be most beneficial to AGI scholars, AGI/AI engineers and character-AI designers with computer science, virtual worlds industry, and/or humanities (arts) backgrounds. Usually, this demographic has to mine for papers regarding this specialized topic and application-space. Until the publication of this book, there was no centralized resource for this precise topic as-of-yet. Usually, this mixed demographic usually locates such papers from numerous disparate sources such as: AGI anthologies on cognitive architectures in general, conference proceedings, books dealing with only one particular cognitive architecture, or from public and proprietary papers within the video-games industry. The book aggregates the state-of-the-art and historically relevant cognitive architectures and their relationship to heuristics and implementations of characters in virtual environments (video-games, virtual worlds).

CHAPTER RECOMMENDATIONS

This book contains valuable chapters dealing with higher-level issues regarding agent-embodied cognition alongside more practical issues relating to the technical interface that connects a virtual agent with a cognitive architecture.

For those interested in the cognitive affordances unique to virtual characters as humanoids, we recommend reading "On Vision-Based Human-Centric Virtual Character Design: A Closer Look at the Real World from a Virtual One" by Eugene Borovikov, Ilya Zavorin and Sergey Yershov. In addition, the subject of SmartBody humanoid implementations using mixed-reality scenarios has been addressed in the chapter "Personality-based Cognitive Design of Characters in Virtual Environments" (Ch. 5) by Maryam Saberi.

For those yearning for a character who is capable of parsing higher-level logic structures, we recommend the chapter "The Contemporary Craft of Creating Characters Meets Today's Cognitive Architectures: A Case Study in Expressivity" (Ch. 6) by Selmer Bringsjord, John Licato and Alexander Bringsjord from the Rensselaer Polytechnic Institute (RPI).

For those who still value the utility of reinforcement learning techniques when designing compelling virtual characters, we recommend the chapter, "Learned Behavior: Enabling Believable Virtual Characters Through Reinforcement" by Jacquelyne Forgette and Michael James Katchabaw (Ch. 4).

For those who may have designed their own character already but are having issues interfacing a particular cognitive architecture with their preferred implementation (gaming) engine, it would best to consult the case study about ACT-R and Unity set forth by Paul Richard Smart, Tom Scutt, Katia Sycara, and Nigel Shadbolt. This chapter has been titled, "Integrating ACT-R Cognitive Models with the Unity Game Engine" (Ch. 2).

For those who simply wish to explore a historical overview of previous character implementation examples, one of the editors provided a chapter on "Virtual SOAR-Agent Implementations: Examples, Issues and Speculations" (Jeremy Owen Turner, Ch. 7).

For those who are getting impatient reading academic theories about cognitive architectures interfacing with virtual agents and would actually like to begin making their own cognitively-enabled agent in the social virtual world, Second Life; Andrea Corradini and Manish Mehta have created a personality authoring system for non-programmers, called "Second Mind". The name of this chapter is "A Graphical Tool for the Creation of Behaviors in Virtual Worlds" (Ch. 3).

For those who would prefer to read about more abstract higher-level concepts related to AGI and their relationship to more speculative character designs, we recommend three chapters from this book. The first chapter to recommend is "Towards Truly Autonomous Synthetic Characters with the Sigma Cognitive Architecture" by Volkan Ustan and Paul S. Rosenbloom (one of the creators of the SOAR cognitive architecture). The second chapter that corresponds to the more general theme of universality is, "A Universal Architecture for Migrating Cognitive Agents" by Kaveh Hassani and Won-Sook Lee. Overall, speculations on AGI and virtual characters concludes this book in "Game AGI Beyond Characters" by Alexander Zook.

These chapters cover a broad range of virtual agents and address the degree to not only what kinds of cognitive architectures are most useful for human-level believability and beyond, but whether it is useful to even implement a cognitive architecture at all.

REFERENCES

Anderson, J. R. (1996). ACT: A simple theory of complex cognition. *The American Psychologist*, *51*(4), 355–365. doi:10.1037/0003-066X.51.4.355

Laird, J. (2012). *The Soar cognitive architecture*. MIT Press.

Magerko, B., Laird, J., Assanie, M., Kerfoot, A., & Stokes, D. (2004). AI characters and directors for interactive computer games. Ann Arbor, 1001(48), 109-2110.

Rosenbloom, P. S. (2013). The Sigma cognitive architecture and system. *AISB Quarterly*, *136*, 4–13.

Sun, R. (2006). The CLARION cognitive architecture: Extending cognitive modeling to social simulation. *Cognition and multi-agent interaction*, 79-99.

Thórisson, K. R. (1999). A Mind Model for Multimodal Communicative Creatures and Humanoids. *International Journal of Applied Artificial Intelligence*, *13*(4-5), 4. doi:10.1080/088395199117342

Wang, P., & Goertzel, B. (2007, June). Introduction: Aspects of artificial general intelligence. In *Proceedings of the 2007 conference on Advances in Artificial General Intelligence: Concepts, Architectures and Algorithms: Proceedings of the AGI Workshop 2006* (pp. 1-16). IOS Press.

Acknowledgment

The editors would like to thank the chapter authors, the peer reviewers, and the publisher for their invaluable support.

Jeremy would like to additionally thank his fellow editors. Jeremy would like to acknowledge his COGS 100 students and colleagues from each semester, including his Teaching Assistant, Meehae Song. On that note, Jeremy would like to thank Linda Kaastra for encouraging him to lecture on cognitive architectures while he was her Teaching Assistant. Presently, Jeremy is also grateful to his PhD supervisory committee for allowing him to pursue research into cognitive architectures as part of his inter-disciplinary dissertation.

Jeremy would like to thank the conference organizers at BICA 2014 for the book's widely publicized poster session at MIT. Jeremy would also like to thank Mike van Lent from SoarTech and John Laird of Soar for connecting him with various Soar cognitive architects and scholars. This helped devise the historical content on Soar for his own chapter. Glenn Taylor from SoarTech should also be thanked for spending extra time providing his co-authored publications for the historical overview on Soar agents. Jeremy also enjoyed corresponding with Selmer Bringsjord about cognitive architectures and the speculative possibilities of hyper-computation. Jeremy would like to also give credit to anyone else that he had forgotten to mention in the haste of preparing this acknowledgements list. Personally, Jeremy would like to thank his family and especially his son, Nolan Turner-Skuce. It turns out that so far, Nolan shares his Dad's interest with researching "robot brains".

Chapter 1
On Vision-Based Human-Centric Virtual Character Design:
A Closer Look at the Real World from a Virtual One

Eugene Borovikov
PercepReal, USA

Ilya Zavorin
PercepReal, USA

Sergey Yershov
PercepReal, USA

ABSTRACT

Enabling cognition in a Virtual Character (VC) may be an exciting endeavor for its designer and for the character. A typical VC interacts primarily with its virtual world, but given some sensory capabilities (vision or hearing), it would be expected to explore some of the real world and interact with the intelligent beings there. Thus a virtual character should be equipped with some algorithms to localize and track humans (e.g. via 2D or 3D models), recognize them (e.g. by their faces) and communicate with them. Such perceptual capabilities prompt a sophisticated Cognitive Architecture (CA) to be integrated into the design of a virtual character, which should enable a VC to learn from intelligent beings and reason like one. To seem natural, this CA needs to be fairly seamless, reliable and adaptive. Hence a vision-based human-centric approach to the VC design is explored here.

DOI: 10.4018/978-1-5225-0454-2.ch001

He's as blind as he can be
Just sees what he wants to see
Nowhere Man, can you see me at all?
-John Lennon

INTRODUCTION

A pure virtual character (VC) is typically limited to interactions with and reasoning about its virtual world. However, given certain abilities to perceive and explore some of the real world and interact with the intelligent beings there, can a VC evolve into an intelligent virtual being? Let us equip a VC with visual sensors, include some algorithms for object recognition and tracking, and provide some ability to learn and reason. Then such a virtual character, much like Alice stepping through the looking-glass (as in Figure 1) and becoming aware of the other world, should have a chance to eventually discover some intelligent characters there, observe their traits, and by virtue of interacting with them, learn and reason about that world and its beings. Such perceptual capabilities evidently prompt a sophisticated cognitive architecture (CA) to be integrated into the design of a virtual character, and to seem

Figure 1. Alice Through the Looking Glass sculpture by Jeanne Argent at Guildford Castle, Surrey, UK

natural, this CA needs to be fairly seamless, reliable and adaptive at both sides of the virtual looking glass. Thus, enabling cognition in a virtual character may truly be an exciting endeavor for the VC designers and hopefully for the VCs themselves.

In general, there is a difference between cognitive architecture approaches and Artificial Intelligence (AI) approaches to intelligent agents design. The latter usually are optimized for the maximum task performance, while the former are optimized for a human-like performance. This chapter focuses on the human-centric CA that enable a perception-capable VC to learn and imitate the traits of the intelligent agents it observes and interacts with ultimately striving towards a human-like performance, but also allowing for developing and optimizing certain abilities that may eventually surpass those of the humans, e.g. very fast and accurate content based image retrieval.

Virtual character's perceptual abilities would naturally rely on the given sensory capabilities, e.g. video cameras for its eyes or microphones for its ears. Clearly those sensory streams should be synchronized and carry enough of the signal resolution to distinguish the important features of the objects and beings a VC would need to interact with. Those features would be extracted by various signal and image processing algorithms accompanying the given sensors, and hence be known and referred to as the *basic perceptual abilities* that our virtual character does not need to develop. A perceptually capable VC, however, would need to use its evolving cognitive architecture at deciding on a combination of important features characterizing the real-world objects and beings it needs to reason about and interact with.

Communications between VC and humans would naturally be of the most interest to this study, and thus a virtual character should be able to localize and track humans (e.g. via non-rigid 2D or 3D models), recognize them (e.g. by their faces and/or voices) and communicate with them, preferably via natural (for both parties) interfaces, e.g. a human-like virtual reality (VR) avatar, for the purpose of our *human-centric* approach, which puts the humans at the center of the VC's attention with the intent of learning some of the human behavioral traits via the given senses, especially vision. This means that such a VC needs to work in visually unconstrained environments, perform its perceptual sub-tasks in real time, and constantly learn from its experiences with both virtual and real worlds, interacting with their inhabitants. Such real-time interactions between a VC and the real world should result in a gradual development of that virtual character, ultimately resulting in a highly realistic virtual or mixed reality experience for the humans.

The authors propose a vision-based human-centric approach to the design of a virtual character equipped with visual sensors. A general vision-based solution for the problem of VC design is beyond the scope of this discussion, so we focus on arguably the most visually expressive and natural human real-world manifestations: face and body. The main contribution of this work is a set of methods for several visual perception tasks that we believe are essential for a flexible, real-time and continuously learning human-centric VC development system, namely:

- Adaptive color-plus-texture-based image segmentation,
- Real-world objects detection and classification,
- Hierarchical model reconstruction of non-rigid objects, e.g. the human body,
- Most expressive body part detection and tracking, e.g. face and hand,
- Motion and gesture recognition and interpretation.

The proposed techniques aim to address the required trade-off between robustness, real-time performance and incremental learning of a realistic VC-based system.

BACKGROUND

A cognitive architecture (CA) is a system that usually comprises multiple computational modules that, working as a whole, attempts to approach human-level intelligence (Goertzel, Lian, Arel, De Garis, & Chen, 2010). It is therefore not surprising that the use of cognitive architectures to generate more human-like virtual characters has attracted considerable attention. Recent advances in 3D sampling technology (e.g. multi-view or multi-mode video capture as with Microsoft Kinect) provide more detailed face and body shape and motion analysis, which allows a vision-enabled VC to watch and learn human traits more precisely. We expect that the human interactions with human-centric VCs could benefit both parties.

To observe the real world, reason about its objects, interpret their actions and then transfer this knowledge into the virtual environment for the purpose of emulation, a vision-capable intelligent VC needs to be able to deal with the complexity, uncertainty, and immediacy of the real world calling for moment-to-moment improvisation. Investigation of the dynamics of everyday routine activity reveals important regularities in the interaction of simple machinery with its environment, and the Pengi system uses that, engaging in fairly complex planning activity without requiring explicit models of the world (Agre & Chapman, 1987; Chapman, 1989).

In (McCollum, Barba, Santarelli, & Deaton, 2004), a version of an architecture called iGEN (Zachary, Ryder, Ross, & Weiland, 1992) was used to control virtual characters in an immersive training system developed for the U.S. Army. These characters represented typical members of indigenous population in a target area, with their behavior and affect changing in response to actions of a human trainee. One of the innovations of the resulting VECTOR (Virtual Environment Cultural Training for Operational Readiness) environment was the use of emotion models which guided VC's behavior toward the trainee. However, as is often the case with virtual training systems, it is the human trainee that had to learn based on interactions with the VCs and not the other way around, with the VC models remaining cognitively static.

In (Trafton et al., 2013), the cognitive architecture ACT-R/E was used to develop better human-robot interactions where humans and robots share the same virtual or physical task environment and where actions of each participant affects those of others. The authors note that in HRI an assumption is often made that the human participant of these interactions is in fact more of a perfect machine than a human, i.e. he or she never makes mistakes, is always fully predictable and is never affected by fatigue or negative emotions. They focused on addressing this gap by developing a CA that is capable of deeper modeling of human cognition to enable the robot participant to recognize when its human counterpart does something wrong. The resulting architecture called ACT-R/E is shown in Figure 2. While very interesting in its own right, this study appears to step outside of the VC world, because a robot is a physical (usually active and mobile) entity with its own set of real-world experiences to learn from. However, it is still valuable because the assumption of an *imperfect* human participant is critical to the development of a robust VC cognition system.

In (Choi, Könik, Nejati, Park, & Langley, 2007), the authors address the specific topic of first-person shooter (FPS) video games. They used a CA called ICARUS to develop an agent capable of playing the game Urban Combat. While the learning process of a virtual player agent did not involve observing a human player

Figure 2. ACT-R/E architecture

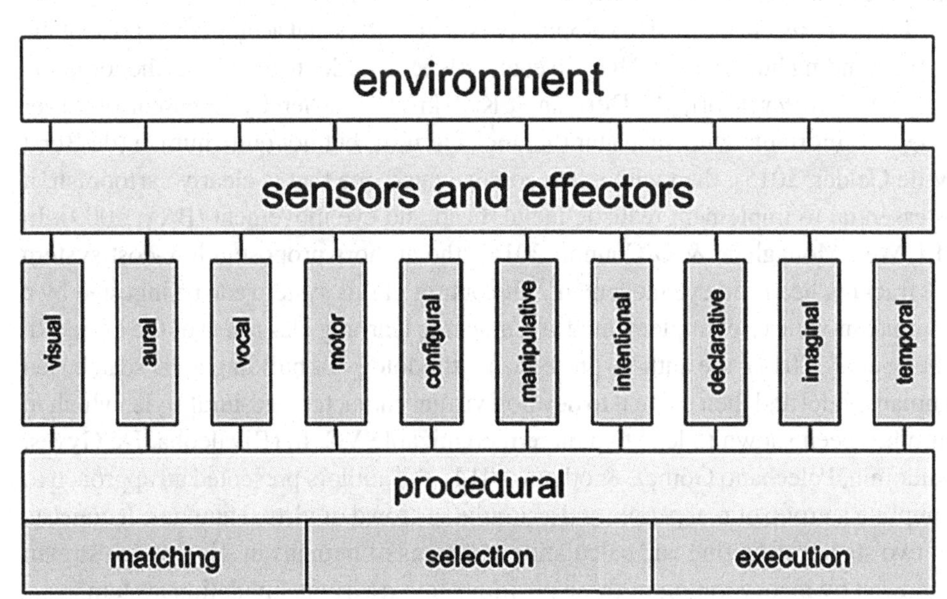

in action, its learning model was goal-oriented (such as overcoming obstacles in order to find and diffuse an IED) and incremental, the latter feature being essential to a robust VC cognition model.

In another work related to FPS (Asensio et al., 2014), different controllers for virtual characters are developed based on fusion of CERA–CRANIUM and SOAR cognitive architectures as well as on a system for the automatic evolution and adaptation of artificial neural networks (ANN). The resulting virtual agents showed improved performance based on BotPrize competition believability assessments. The authors also compare different approaches to assessing believability of virtual characters.

The biologically inspired (Hubel & Wiesel, 1968) convolutional neural networks (CNN) as introduced as *neocognitron* (Fukushima, 1980) and then improved, generalized and simplified (Simard, Steinkraus, & Platt, 2003), have seen a spectacular renaissance in the recent decade (LeCun, Bengio, & Hinton, 2015) due to the emergence of the affordable GPU computing power, which made the non-trivial image processing tractable for many visual tasks (Abdolali & Seyyedsalehi, 2012; Fan, Xu, Wu, & Gong, 2010) that may be considered quite important for visually capable virtual character adaptive cognitive architecture that needs to learn important features straight from visual sensors. Modern deep learning (Yue-Hei Ng, Yang, & Davis, 2015) content based image retrieval (CBIR) techniques (Wan et al., 2014) could also help with the VC's robust long-term memory sub-system development, e.g. by transferring deep networks trained on image classification to image retrieval tasks. Deep learning models are an essential part of the approach we envision for integrating the vision modules described in this work.

There is a recent trend of developing personal avatars that accurately represent the corresponding human user (Nguyen et al., 2005). In order to avoid the phenomenon of *uncanny valley* (Mori, MacDorman, & Kageki, 2012) whereby the human observer is less comfortable with an avatar that looks almost, but not quite human (de Borst & de Gelder, 2015), than with a less sophisticated one that is clearly cartoonish, it is essential to implement realistic facial, head and eye movement (Brey, 2009). In (Li, Wei, Monaghan, & O'Connor, 2014), the authors propose a low-cost system for tracking head and eye movement. The output of this system can be ingested by a training module (e.g. implemented as a machine learning classifier) of the avatar. In (Feng et al., 2014), the authors present a methodology of building a 3D scan based human model and then using it to develop virtual character's gestural style, which in turn has been shown to lead to a more recognizable VC. In (Rivalcoba, De Gyves, Rudomin, Pelechano Gómez, & others, 2014), the authors presented an approach to coupling a group of real people and a simulated crowd of virtual humans. It consists of two stages: detecting and calculating positions of humans in a live video stream followed by incorporation of these positions into a virtual simulation system.

Along with the realistic appearance and individual gesture modeling, for a humanoid VC, it is important to understand the human behavioral traits, capturing different personalities for modeling realistic behavior of a VC as suggested by (Saberi, Bernardet, & DiPaola, 2014), where the authors present a hybrid cognitive architecture that combines the control of discrete behavior of the VC moving through states of the interaction with continuous updates of the emotional state of the VC depending on the feedback from the environment. Testing their approach using turn-taking interaction between a human and a 3D humanoid VC, the authors noticed more individualized and believably humanoid artifacts in their VC's behavior.

One of the natural extensions of the cognitive architecture for virtual character capable of observing the real world would be in relaxing some of the VC's appearance and behavioral constraints, allowing for some unsupervised learning by observation, and discovering important objects (with their prominent features), object relationships and behaviors along with their natural constraints, e.g. based on the real world physics as well as on the living objects physiology and behavior.

VISION BASED EVOLUTION OF A VIRTUAL CHARACTER

A virtual character (VC) with cognitive abilities has a chance to evolve into a sophisticated (virtual) being because of the possibility of the direct communication with the real world intelligent agents. Provided with a sound cognitive architecture (CA), such a VC should be able to utilize its knowledge of the virtual world (e.g. 3D object models and hierarchical infrastructure) for its real-world perception tasks via the given sensors (e.g. cameras). Once aware of the real-world objects (and possibly of their inter-relationships and inter-actions), a VC should be able to bring the learned concepts as models to the virtual world, reason about them, emulate some of them, and possibly share them with other intelligent agents, virtual or human.

The authors argue that the vision-based human-centric approach (Buxton & Fitzmaurice, 1998), involving face and body modeling, can provide the necessary foundation for introducing a sound cognitive architecture to the VC design, especially if it encompasses a human-like avatar because:

- Those are the most visually expressive modes of the human real-world appearance.
- Fusing several observable modalities (texture, color, depth) should result in finer virtual models.
- VC can learn to distinguish between general and person-specific models.
- Real-time interactions promise a natural and incremental VC evolution.
- Evolving mixed reality experience could benefit both VC and humans.

One could measure the accuracy of the VC design objectively by comparing the target appearance or requested behavior to the captured data, but the ultimate judgment of how naturally the evolved VC looks or behaves is likely to be made by a human.

A vision based virtual character is therefore expected to gradually evolve (Boots, Nundy, & Purves, 2007) the abilities to observe and reason about the real-world objects with their relationships, study and emulate them in the virtual environment, eventually becoming more natural in the interactions with the entities of the virtual and real worlds. The envisioned evolutionary steps include (but are not limited to):

1. Distinguish important (statistically significant) visual objects from their surroundings.
2. Track the objects to learn the basic physical concepts (e.g. continuity, gravity, elasticity).
3. Apply the object and motion models in the virtual world to emulate their physical counterparts.
4. Distinguish and track the human users, recognize the most important ones for interaction.
5. Learn detailed traits of natural interaction and attempt to emulate them the best way possible.

Given a rich enough representation and computing resources, the cognitive architecture of such a virtual character should allow it to imitate virtual or human personalities it interacts with, adapt the observed traits to is virtual environment, and then eventually develop a non-trivial virtual personality of its own via exploration and experimentation with different behavior patterns and analysis of the implicit or explicit feedback from the humans.

The study of VC behavioral traits could perhaps lead to a separate research field, e.g. virtual character psychology. It would be interesting to see, if or when such a character becomes self-aware as an intelligent individual, and what positive and negative behavioral traits it may develop and exhibit. Finding an optimal VC behavior trajectory with respect to its environment may be addressed via genetic algorithms run on individual personality traits. If VCs are allowed to be replicated, it would be interesting to watch how the multiple personalities may diverge given their continuing experience in both worlds. An exciting opportunity for the evolutionally algorithms could be found given vision capable VC *society*, whose members could adapt to their dynamic environment within a generation, and then pass their appearance and behavioral traits to the future generations. This could open yet another research field that may be called VC sociology, but that exciting and possibly very controversial study clearly falls outside of the scope of this chapter.

Issues, Controversies, Problems

Have you ever wondered: why your computer literally does what you ask, not what you mean? Even if you have not, you may still observe that there is a place for misunderstanding in any interaction between the intelligent systems, artificial or natural. Such misunderstanding may have many causes (noisy signal, communication errors, ambiguous semantics, etc.) but it is rooted most likely in what we call the *perception gap*, which can be defined broadly as the *difference in interpretation of the perceived reality*.

As an illustration of the perception gap consider some difference in interpretation of the visual query and its expected results, as in Figure 3: for the query image in the top left corner one of the commercially successful search engines produced what it perceived to be most visually similar images from the publicly available ones that it was aware of. Before one could judge how good the results are, let us ask: what was this visual query about? If the query was about the horse that is in the foreground, then the results are clearly not acceptable, as they mostly depict the wonderful autumn forest and park views without any four-legged animals in them that could even remotely resemble a horse. The visual search engine probably guessed that since the yellow-red trees leafy mass occupies most of the picture, the results reflected exactly that, and the user's *information need* may have been left unsatisfied due to the apparent perception gap (PG) between human and machine interpreta-

Figure 3. Visual query results for the query image in the top left corner as an example of a possible visual perception gap

tion of the query picture. Could this PG be narrowed, bridged or even closed? The answer depends on many factors, e.g. the kind of communication the interacting agents are engaged in, their intellectual abilities, their environments, as well as on the information processing and communication tools and that are available to them.

Humans can bridge this perception gap by simply talking to each other, hence bringing more common context into their understanding of the perceived reality. In our visual query example, when asked to pick the most visually similar pictures to the given one, a person would not have much trouble recognizing the horse on the foreground and would have likely preferred the pictures with horses and some trees on the background. All ambiguities would have been naturally resolved via a question-answer exchange.

Computers, on the other hand, while being much more predictable than humans and more capable at processing large arrays of well-structure data, typically lack the natural interaction abilities, human intelligence and experience, thus rendering most of the modern artificial intelligence (AI) systems to be quite imperfect at parsing the real world stimuli in the way humans do, hence increasing the perception gap chances in human-computer interactions.

Modern image retrieval systems (e.g. images.google.com or yandex.com/images) usually do not ask additional questions regarding the query images, but would rather try guessing on the pictorial information need, which sometimes leads to irrelevant results because of the PG. One way to narrow down the visual perception gap in this case is by a human imposing a more specific query via a cropped image clearly depicting the wanted horse, as in Figure 4. This time the results depict mostly horse-

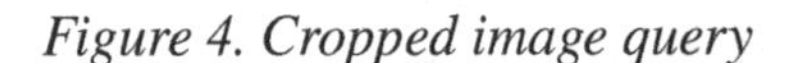

Figure 4. Cropped image query

looking animals (and an occasional cat), as Figure 5 illustrates, but they do not contain any white horse in their top hits. Cropping the query image further brings more house animals (including some goats and rams along with the horses) to the picture, but let us skip the display to save space, and invite our readers to experiment with the on-line image search engines with images of their choice. Hence, one of the primary research and development (R&D) goals when designing intelligent virtual characters would be narrowing and possibly closing the perception gaps in human-computer interactions.

In addition to resolving various interaction semantics ambiguities, designers of vision-capable VCs would also need to provide some low-level computer vision algorithms to enable their virtual characters to reliably solve some basic visual tasks (e.g. image segmentation, object detection) and develop some more intellectually involved visual capabilities for distinguishing individual objects and their classes evolving some detailed models depending on the VC cognitive needs, which may or may not have a quantifiable measure of success because (a) perception is always subjective and (b) virtual characters do not come into direct contact with the real world (not needing to survive in it), hence would not develop human-like reflexes or get similar to human real-world experience. Thus whatever is expected from the human visual system may not be as needed in the VC vision, yet a virtual character being equipped with non-human visual sensors (e.g. a visual array with cameras distributed in some 3D space, possibly moving independently) may develop some visual abilities that are quite different from or (in certain cases) beyond those of human's.

Figure 5. Refined visual query results narrowing the perception gap

SOLUTIONS AND RECOMMENDATIONS

The search for a general vision solution in the context of a virtual character design is beyond the scope of this discussion, but since a vision-capable VC is likely to see and interact with humans, we would like to provide our view of a human-centric approach to the VC design. In particular, the authors would like to explore specifics of visual perception regarding:

- Adaptive color and texture image segmentation, e.g. foreground-background separation.
- Real-world objects detection and classification.
- Non-rigid object (e.g. human body) hierarchical model reconstruction.
- Most expressive body parts (face+landmarks, hand+fingers) detection and tracking.
- Gesture recognition and interpretation.

and discuss their impact on the intelligent virtual character development that needs to smoothly interact with both natural and virtual intelligent agents. Such a cognitive architecture calls for VC to mentally separate the real and virtual worlds, treating the window to the real world (given by its visual sensors) as an exploratory tool for creating abstract object concepts (e.g. rigid vs. articulate), thus employing high-level reasoning about objects and their relationships, which could be applied in both sides of the separating virtual boundary.

For example, a virtual character can learn the specifics of a real human face and body motion and attempt to mimic them in virtual character representation by slightly altering its underlying model, hence enhancing the virtual reality user experience. At the same time, in some of the augmented reality (AR) scenarios, virtual objects and characters can be inserted into the real-world video streams and allowed to interact with some real objects in real time, building-up on the VC physical world experience, and ultimately enhancing the AR user experience.

Adaptive Image Segmentation

One of the important vision tasks for any perceptual system is input image segmentation, that is breaking it into distinct blobs (corresponding to the objects and background elements) using various cues, e.g. color, texture and motion. Let us consider the problem of automatic foreground and background separation in live video streams requiring real-time algorithms that are robust to complex backgrounds, accommodating changes in illumination or location where the application is deployed. A variety of background modeling and foreground separation methods have already been developed (Bouwmans, 2014), but many of them assume a relatively static camera (which may not be the case with the virtual character sensors) and rely on computationally expensive iterative optimization routines, which can hardly be executed in real time. To overcome these difficulties several real-world challenges need to be addressed, specifically:

- Quick and accurate processing of high resolution images.
- Real-time temporal information accommodation.
- Adaptation to changing lighting changes and sensor motion.

One can utilize a Bayesian probabilistic approach to adaptive foreground/background separation, relying on motion, color and texture features, which can bypass the prohibitively expensive computations and leads to the development of a robust and efficient technology for foreground/background separation, resulting in a solution running much faster than the existing methods, hence very suitable for the real-time vision processing. The described here Adaptive Real-Time Object Segmenter (ARTOS) aims at establishing a sound mathematical framework for probabilistic image segmentation to:

- Dynamically model and maintain background information in video feeds,
- Automatically detect and segment foreground blobs in real time,
- Adapt to the current environment and usage patterns to improve segmentation,
- Identify and enforce smoothness constraints to be robust to spatio-temporal noise,
- Use heuristics in place of time-consuming optimization routines.

Foreground/background separation is one of the core computer vision (CV) tasks being solved in many practical systems for segmenting the object of interest from its surroundings. Hence, it is a particular case of a more general multi-class image segmentation problem. Classification is binary, because there are two mutually exclusive classes (foreground and background) to consider for the picture elements, e.g. (super-)pixels or pixel blocks in the image pyramid, but called (generalized) pixels further on.

Given a live video stream from a video camera, let us use the motion cues to identify the foreground object candidates, and approach the problem of real-time foreground/background separation from the machine learning perspective, by estimating each picture element's probability of belonging to foreground (versus background) using its observable features (e.g. color and texture), using the non-parametric Bayesian approach (Peter Müller, 2004)

$$P(X|O)=\frac{P(O|X)P(X)}{\left(P(O|X)P(X)+P(O|\bar{X})P(\bar{X})\right)}$$

where X models foreground/background pixel, $\bar{X}$ models its complement, and O are its observations or computed features. Note that all of the right-hand side quantities can be

- Efficiently estimated/pre-computed a-priori from previously labeled data, and
- Updated incrementally from frame to frame, using more of high confidence pixels.

resulting in an efficient and incrementally updated estimate of the resulting pixel class probability with the a-priori known theoretical threshold of 0.5, indeed:

$P(O\mid X)$ and $P(O\mid \bar{X})$ are the probabilities of observation given the pixel kind or its complement can be estimated via the observation histograms computed from the labeled data;

$P(X)$ is the ratio of the pixel kind count to the total pixel count (e.g. foreground/all);

$P(\bar{X})=1-P(X)$ as a complement event.

Updates to the above probabilities/histograms can be done by adding the already classified (high confidence) pixel's features to the underlying probability histograms, adaptively re-sampling the scene in real-time. The adaptive re-sampling loop converges when the incremental changes to the foreground/background conditional probabilities become negligible or when it's time to process the next frame.

Based on a solid mathematical framework, the described segmentation approach is mathematically *sound* and *general* because it makes no a-priori assumptions on foreground/background features (observations) or on their distributions. It is data-driven and flexible, as its model can be initialized and incrementally updated, adapting to the current scenery and to the usage patterns, e.g. estimating camera ego-motion and user's tendency to focus on the object of interest.

Given a dynamic scene with a relatively static background (e.g. in surveillance scenarios) a statistical background model can be precomputed, and the foreground statistics would be sampled from the blobs that move in the way that the background model cannot explain, in which case the adaptive re-sampling loop will kick in and converge to some foreground model, performing segmentation in real time.

In a dynamic scene with a predictably moving camera (i.e. fairly smooth translation, rotation or zooming, without any jerky motions), the background model initialization will have to be more elaborate (than with a statistically static background), accounting for the estimated regular motion, but still manageable at the interactive speeds for the statistics to be collected and processed. The foreground blobs will have to be assumed to be the ones that are not explained by the background model, or pointed out by the user. The frames with the unpredictable (jerky) camera motions will have to be identified and discarded as no reliable segmentation can be provided for such frames.

One important practical application of this approach is color-based real-time human skin mapping. If in the Bayesian approach equation, we let X=*skin* and O=*color* (in some color space, e.g. RGB or HSV) we then can compute the conditional probability of a pixel belonging to the *skin* class given its color. The target probability computation requires collecting several histograms from the data, taking the motion cues into consideration as probable a-priori foreground, treating the relatively static pixels as probable background.

For example, Figure 6 represents a 3D histogram (with bins represented by the spheres reflecting individual bins color and size) of color pixels (in RGB) that one is likely to collect from the *skin* class: the color distribution tends to form a nice cluster in this color space (Jones & Rehg, 2002). Figure 7 represents a histogram of color pixels from the *non-skin* class: note that the colors are distributed along the main RGB cube diagonal, do not form a clear single cluster, and do not seem to overlap much with the skin color cluster (considering the bin colors), which indicates that it is possible to build a fairly reliable probabilistic classifier to separate skin from non-skin pixels. The resulting conditional probability histogram for skin given color is depicted in Figure 8: as we can see the distribution is far from being Normal and contains several surprising outliers (e.g. in the blue area of the color cube) probably due to the noise in our input data and some overlap in the distributions.

When tested with a live video of a person in a room, the method successfully detects the exposed skin patches, correctly thresholds them at the theoretical level of 0.5, as shown in Figure 9, and dynamically updates the probability histograms using the motion cues and resampled pixels that are likely to represent the skin (or background) with a high level of confidence.

This adaptive foreground/background mapping using color can be utilized for more advanced vision stages, e.g. face and hand detection. When coupled with the texture information (and possibly other image features) it could be utilized for more general blob/object segmentation in the *early* vision modules of the proposed human-centric cognitive architecture, ultimately enabling our intelligent virtual character to pay attention to the important content in the live video feed, and build the set of robust probabilistic color/texture classifiers based on the important features, forgetting about the non-essential ones.

Figure 6. P(color|skin)

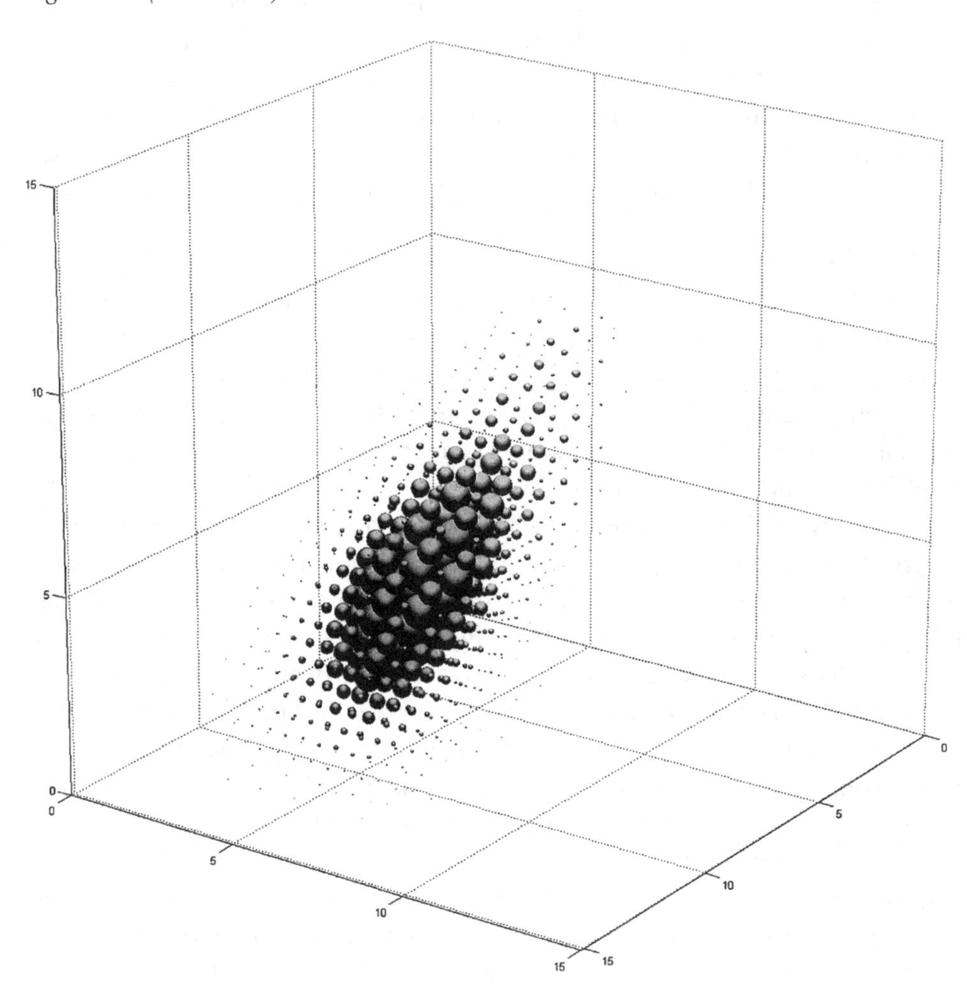

Human Face Detection

Given the human-centric nature of the cognitive architecture for our virtual character design, we must consider the robust face detection stage, which should enable or VC localize and track the very special class of objects (human faces) anywhere in the image or video sequence. Face detection is a critical early stage for any face recognition (FR) module. Face detection problem has been thoroughly studied and many robust on- and off-line solutions have been proposed (Zhang & Zhang, 2010), yet many challenges still remain, especially for the unconstrained images (Eugene Borovikov et al., 2012) and videos, as shown in Figure 10: circles indicate the face

Figure 7. P(color|non-skin)

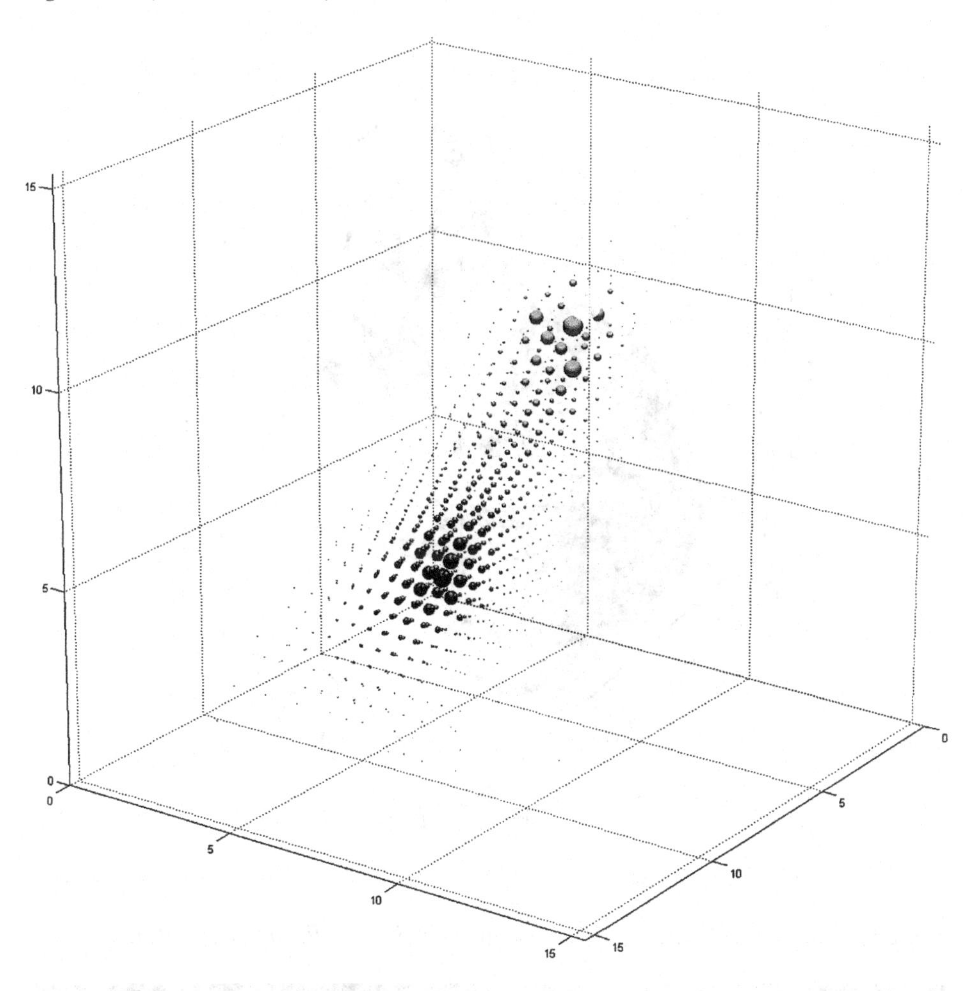

detection candidates, and we can see a number of false positives (FP) and false negatives (FN) that can confuse the face recognition system down the cognitive pipeline.

To equip our virtual character with a more reliable human face localizer, we propose an approach, which takes advantage of multiple color image cues and works with images/videos of virtually any quality detecting faces in real time. The described here method uses an ensemble of three algorithms working together:

- Gray-scale face detector based on (Markus, Frljak, Pandzic, Ahlberg, & Forchheimer, 2013).
- Color/texture based skin mapper based on the adaptive image segmentation (in previous section).

Figure 8. P(skin|color)

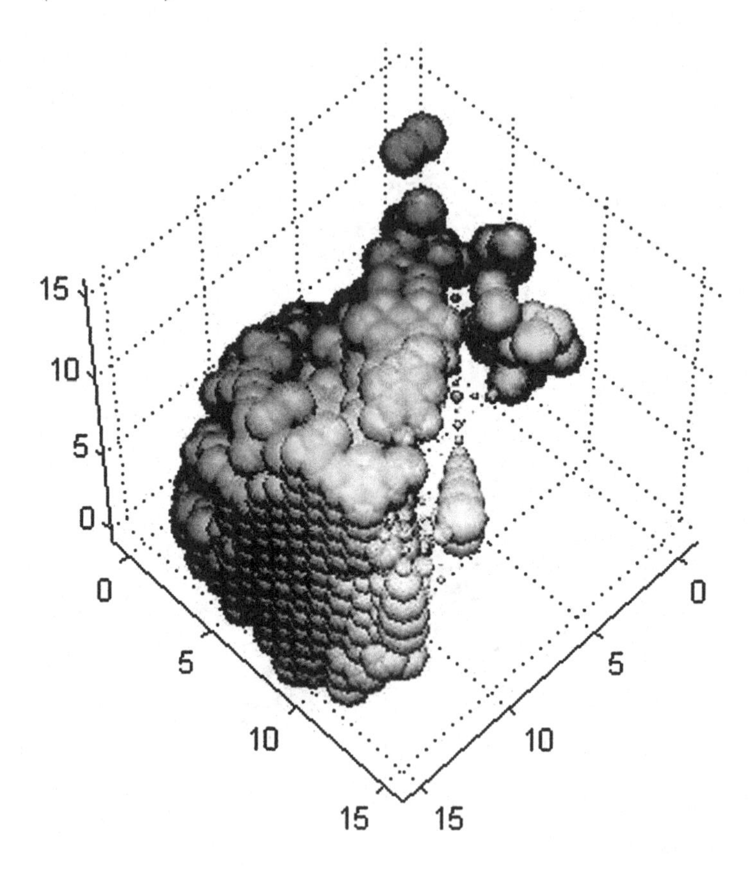

Figure 9. Skin map adaptive estimation through the conditional probabilities

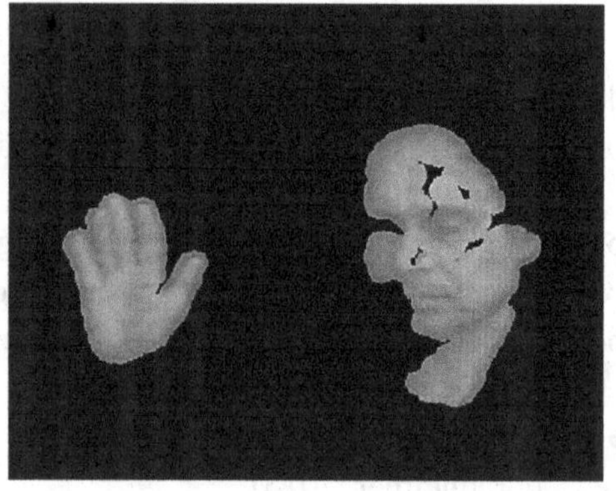

Figure 10. Typical face localization errors by an open source face detector on unconstrained images
(Viola & Jones, 2004)

- Incremental landmark detector/tracker based on (Asthana, Zafeiriou, Cheng, & Pantic, 2014).

Our base gray-scale face detector with classifiers for frontal and profile views can be quite inaccurate on unconstrained images. It is applied it to an image and record the results of both (frontal and profile). When some detected regions overlap, we favor the frontal faces over profiles, unless the profile contains an ear.

In order to increase the accuracy the gray-scale face detector under the unconstrained conditions, we utilize some color and texture information in the images. The real-valued skin mapping module (described earlier) helps diminish the non-skin regions (reducing the false alarms) and enhance the large skin blobs (recovering some missing face candidates). The skin-tone enhanced large blobs are re-inspected by the face localizer for new possible faces not found originally by the gray-scale face detector. In order to better detect the rotated faces, we employ the landmark detection approach based on the incremental face alignment. As a result we get a robust face (and facial landmark) detector that works in real time and resolves many of the challenges presented by the unconstrained images and videos. Refer to Figure 11 to observe major steps of the color-aware face localizer (from left to right): the original image (where gray-scale detector failed), then the real-valued skin likelihood map, then the skin-enhanced face candidates (rotated to align the eye line with the horizon, if eyes are detected), and the corrected output (with all major landmarks - eyes, nose, mouth - correctly identified, for the higher resolution images).

Given the possibility for incremental training in all of the components, one can obtain a highly adaptive face detection module for the cognitive architecture of the human-centric virtual character, which the latter can utilize for the face recognition

Figure 11. Color-aware face and facial landmark localization with face alignment

stage, or evolve into a more generic object detector, e.g. by training its components on different (from faces) objects, and improving them during the interactions with the real world. A human in the loop would certainly help to expedite the training process by hinting what the new objects are and indicating the errors during the learning process.

Human Face Recognition

In a human-centric cognitive architecture of a vision-capable virtual character, it is important to have a reliable face recognition (FR) module to distinguish among the humans, and identify some important people this VC interacts with. This module working in real time, needs be robust to its video signal noise, variations in the visual environment as well as to a potentially open set of humans and their appearances (due to some occlusions, facial hair, jewelry, make-up, etc.) A typical FR system would target the face recognition problem in one of the formulations:

- **Verification:** do the two given photos depict the same person? i.e. verifying the ID by picture, or
- **Identification:** searching a (static) set of individuals for closest in appearance to the query image.

Both formulations would typically imply on some form of model training either for 1-to-1 verification inferring a visual distance function between two appearances, or 1-of-N classification typically on a fixed set of individuals. A more challenging case of matching against open set collections with very few shots per person implies no model training, but rather relying on a robust ensemble of weighted image descriptors, as in Figure 12: same person gets fewer key-spot mismatches (CalTech,

Figure 12. Unconstrained key-spot face image matching: same person, different photos (left pair), different people (right pair)

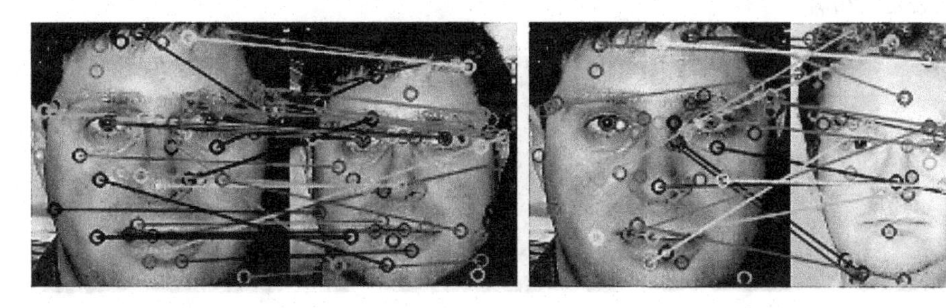

n.d.), for content based face image retrieval (Jain, Klare, & Park, 2012; Eugene Borovikov et al., 2012; Eugene Borovikov, Vajda, Lingappa, Antani, & Thoma, 2013).

The recent decade has seen a considerable progress in the face recognition technology, in some cases approaching the human-level accuracy in FR tasks (Lu & Tang, 2014; Taigman, Yang, Ranzato, & Wolf, 2014), especially in some fairly controlled environments or in the presence of the large amounts of training data (Sharif, Mohsin, & Javed, 2012; Beham & Roomi, 2013; Azeem, Sharif, Raza, & Murtaza, 2014). None of the two conditions can really be satisfied during the real-time interactions of a virtual character with the humans, hence it would be more advisable to rely on the FR techniques that emphasize some incremental (semi-supervised) training (El Gayar, Shaban, & Hamdy, 2006; David Rim & Pal, 2011) and fast unconstrained face matching/retrieval using a dynamically changing photo collection (Eugene Borovikov et al., 2013). This approach may be not very accurate initially, but with time as the VC's face/people photo collection expands, it has a good chance of increasing its matching accuracy without sacrificing the interaction speed.

As the FR subsystem gets more pictures of the same person, it can cluster them by the face ID and learn (from different views of the same face) various facial expressions by localizing the facial landmarks, as shown in Figure 13, and collecting features around them for subsequent more adaptive facial detection and recognition stages (Yun, Deng, & Hiscock, 2009), which can be effectively used in VC facial expression understanding via imitation (Husk, 2012).

When, however, the number of the most frequent users interacting with the virtual character is limited, then one could envision training a classifier and perform 1-of-N face recognition with an occasional enrollment of either a new user or a new appearance of the existing one, followed by a re-training session optimizing the classification boundaries given the new data, hence facilitating the short-to-long term memory information assimilation. Such a limited circle of acquaintances may

Figure 13. Capturing various face expressions by localizing facial landmarks by Chehra-0.1
(Asthana et al., 2014)

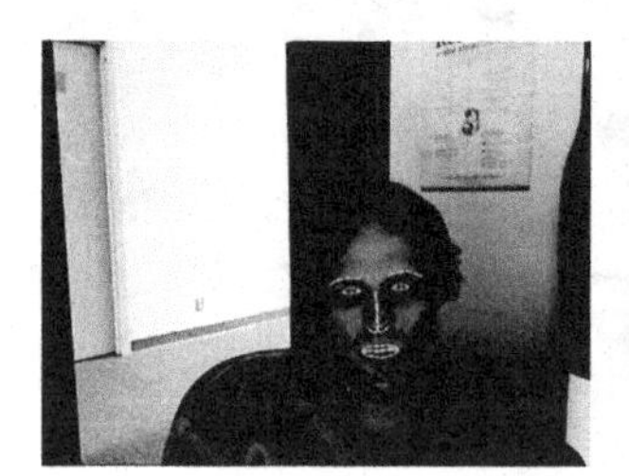
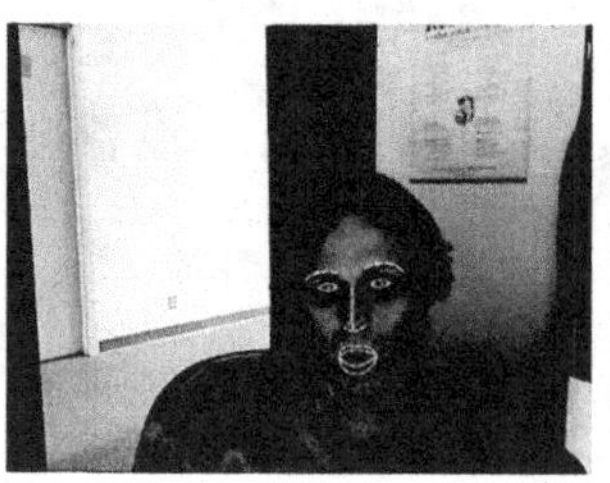
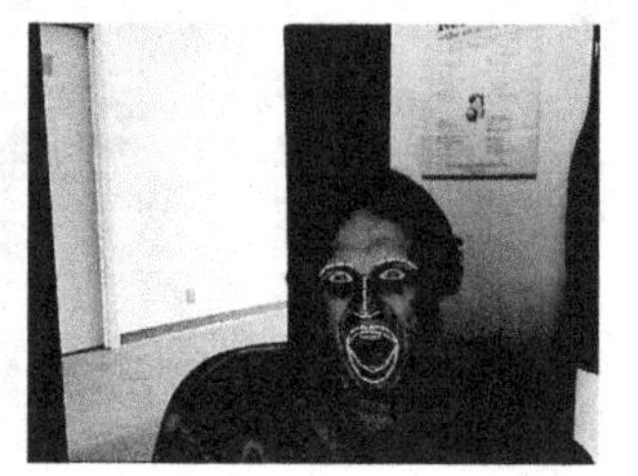

protect the developing VC from being overwhelmed, but this kind of architecture (if not allowed to change dynamically) may considerably limit that VC's cognitive development with respect to recognizing some new faces, when their set expands beyond just a few in a fairly short amount of time.

3D Non-Rigid Shape Reconstruction and Tracking

A human-centric cognitive architecture for a vision-capable virtual character may need to deal with various manifestations of the human appearance accessible non-intrusively by the vision sensors. The most expressive ones (aside from faces) could be human hands and the whole human body. All three categories call for efficient means of representing and handling non-rigid 3D shapes, which may be challenging given just a single camera due to the shape ambiguities that a projective transformation inevitably introduces causing multiple complexities in reconstructing the 3D shapes from a single point of view (Aloimonos & Shulman, 1989; Bülthoff, 1990).

Reconstructing unknown shapes from a single point of view is quite challenging and not always possible because it necessitates solving some ill-posed problems for noisy data. Hence, such solutions even if they exist and are feasible, may not be stable or practical. Solving 3D reconstruction problems for some known (rigid) shapes is more constrained, typically corresponds to the camera calibration problem, and hence is more practical, but those solutions may also be quite ambiguous and unstable due to the same depth-losing projection transform, multiple feasible solutions (from ambiguous correspondences) and noise in the data (Eugene Borovikov, 1998; David, Dementhon, Duraiswami, & Samet, 2004).

Dealing with non-rigid shapes from a single camera with no markers is not feasible in general, but in some particular cases is possible, when such shapes are known, and their motions are fairly constrained, as it may be the case with the marker-less human hand palm detection and tracking (Lee & Hollerer, 2009). This particular system detects hands candidates by the skin color (e.g. as described in the adaptive image segmentation section), computes the hand outline and all its fingers, and fits a five-finger hand model using the constraints on the natural hand to estimate its pose with respect to the camera; it then can place a virtual object/character on the hand palm, and track it with the Kalman filter algorithm (Chan, Hu, & Plant, 1979).

This approach works well when the hand is the largest object in the scene, which does not have too many skin-like blobs, and the fingers are not occluding each other. This hand tracking module can be naturally extended to learn different hand sizes (e.g. adult vs. child), shapes (e.g. left vs. right), and gestures (e.g. finger clicks vs. taps) to be used and/or understood by the vision-capable virtual characters, e.g. (a) humanoid VC trying to mimic some natural human hand shapes and motions, or (b) a pet VC that can be placed on a hand palm in augmented reality worlds for interacting with the humans as shown in Figure 14 (from left to right): virtual 3D axes, virtual pet on a hand, and virtual pet in pajamas pointing towards its owner's face.

One practical application of hand tracking for a humanoid VC could be mastering a sign language, especially in the conjunction with the face tracking and lip motion interpretation.

Human Body Shape Modeling and Tracking

The whole human body shape and motion tracking research and development has been in progress for many decades, and there were many successful solutions for various applications (Lepetit & Fua, 2005) from 2D shape based visual surveillance through skeleton model tracking to 3D body model reconstruction (Perez-Sala, Escalera, Angulo, & Gonzalez, 2014).

Figure 14. Hand tracking and interaction with virtual characters in augmented reality settings (collaboration with 3Bots.com)

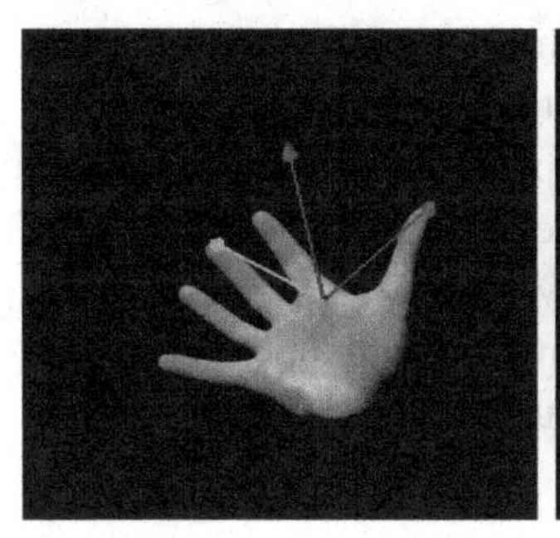

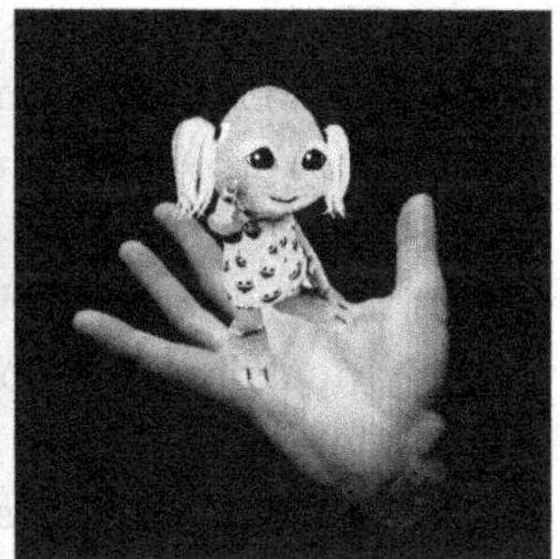

Virtual worlds are typically three-dimensional, and it is natural for a human-centric VC cognitive architecture to operate 3D human body models using 3D data. One can sample the real world in 3D using some active sensors (e.g. Kinect), which can help produce very accurate depth information (RGBD) from a given observation point, and many (especially gaming) applications can take advantage of that extra depth channel, as can our vision-capable virtual character, but those active sensors still may be counted as intrusive (casing IR rays), losing their accuracy in presence of occlusions, and may become unstable when other active sensors are active. Hence a multi-view alternative becomes an attractive possibility for such a cognitive architecture.

Multi-sensor (consisting of two or more cameras) image capturing arrays are not particularly easy to set up and maintain, but multi-view approach can simplify many tasks that are difficult for monocular vision (e.g. volume reconstruction, occlusion handling), but for some tasks, benefits of parallel sequence capturing and processing outweigh difficulties in setting up the image capturing cluster.

The most basic multi-sensor array is a video-pair: it consists of two tightly coupled synchronized vision sensors, often co-planar to simplify stereo depth reconstruction algorithms. A virtual character can benefit from a stereo-pair experience because it mimics human binocular vision in many ways and thus provides some 3D information about the scene (especially in the near vision filed) in an intuitive human way. Our hand tracking module could definitely benefit from utilizing stereo vision, relaxing some of the hand and finger motion constraints by detecting and tracking a hand in 3D and handling finger occlusions more gracefully.

When it comes to whole body shape reconstruction and tracking it in 3D, one would need a more elaborate visual sensors array depicted in Figure 15. When assemble correctly, such an array (along with its computing cluster) allows for an efficient multi-view volumetric non-rigid shape reconstruction that can run in real-time (Eugene Borovikov & Davis, 2000), producing hierarchical (octree based) occupancy maps, that can be efficiently fit with some (also hierarchical) density-based models (E. Borovikov & Davis, 2002) of human figures, as shown in Figure 16.

Such 3D density models can then be tracked over time for the VC to study human body shape and typical movements, trying to imitate them for a more natural VR experience. The 3D vision module can naturally supply data for back-projecting natural colors and textures back to the models and other objects that a human may interact with, as shown in Figure 17 from various view-points.

The virtual character could then learn other object models, attempt to introduce their virtual replicas in its virtual world and attempt to interact with them, again mimicking the human behavior.

Figure 15. A multi-view vision sensor array and computing cluster model

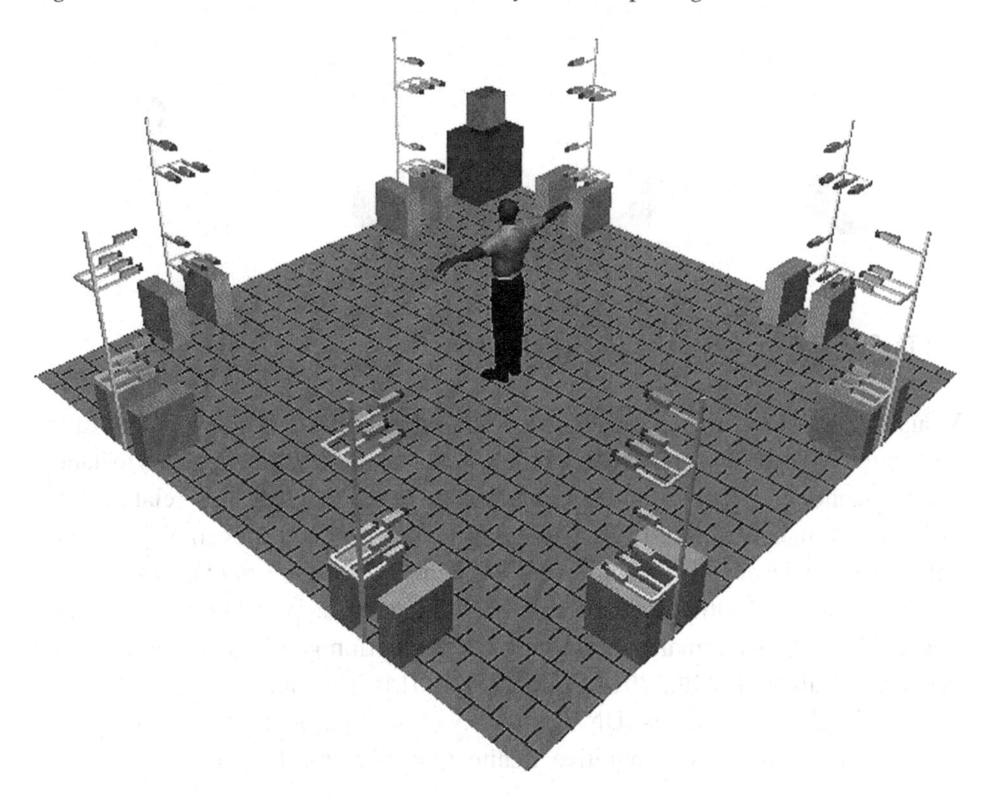

Figure 16. Volumetric 3D shapes (first row) and the corresponding human body 3D density models (second row)

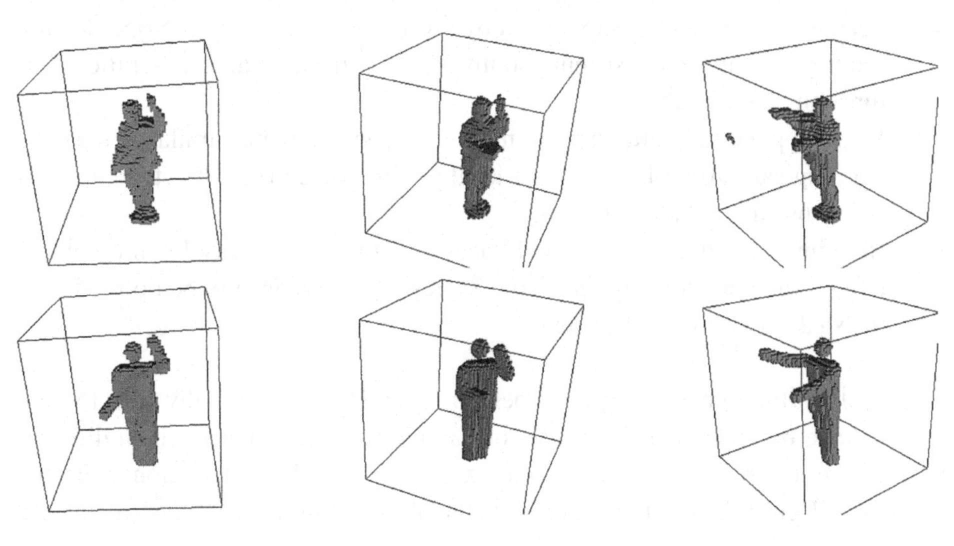

Figure 17. Color and texture back-projection to the recovered 3D shapes

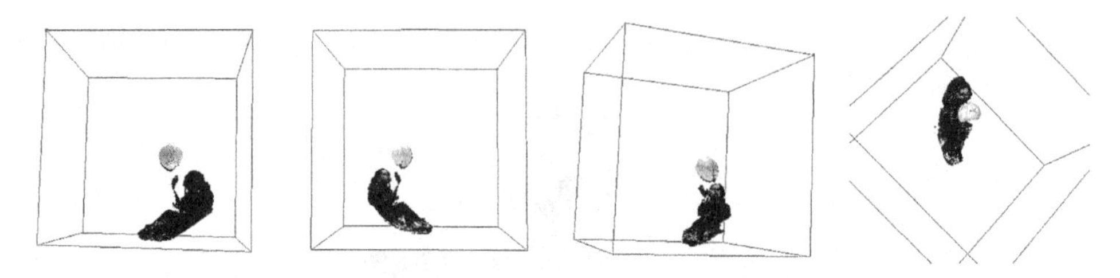

FUTURE RESEARCH DIRECTIONS

Visually capable virtual characters do not have to limit their development to the interactions with humans, as the set of their visual stimuli can be expanded to handle general scenes and the multitude of objects in them, along with their relationships. To help a VC handle this vast variety of imagery, one may consider employing trainable image retrieval sub-system for the VC's short-to-long-term memory transfer, e.g. based on convolutional neural nets (CNN), as they appear to show very good image matching performance narrowing the perception gap between humans and machines (Taigman et al., 2014; Wan et al., 2014; Yue-Hei Ng et al., 2015). In general, the deep neural nets (DNN and CNN as their particular case) appear to be very promising for the VC cognitive architectures because they demonstrate that

- Models pre-trained on large scale dataset can be directly used for features extraction in new visual tasks, capturing high semantic information directly from the raw pixels.
- Features representations extracted by pre-trained model (with proper feature refining schemes) consistently outperform conventional hand-crafted features on all datasets.
- When applied for feature representation in a new domain, similarity measure learning can further boost the retrieval performance of the direct feature output of pre-trained deep models.
- Retraining the deep models with classification or similarity learning objective on the new domain allows the retrieval performance to be boosted w.r.t the shallow similarity learning.

In spite of their good recognition performance, DNNs are usually slow to train and evaluate because of their vast set of parameters, which makes them difficult to deploy in the real-time scenarios that are critical for VC interactions with the real world. Hence the future R&D efforts on DNN should focus on optimizing the

DNN performance by reducing the set of parameters (e.g. as in CNN) and massive parallelization (e.g. via multi-core CPU and GPU).

Cognitive architectures based on perception of the real world do not have to stop at just vision sensors, as elaborate as they may be. Introducing sound sensors, e.g. microphones or even microphone arrays (Duraiswami, Zotkin, & Davis, 2001), as well as touch, taste and smell sensors (as they become available for such applications) should allow to expand the cognitive abilities of a virtual character, making it more aware of the real world, and thus providing a more naturalistic experience for the users. Much like humans gain the real-world experience from their senses and evolve in order to survive in their environments using their cognitive abilities, our sensor equipped VCs with their simulated brains based on DNNs could be expected to strive for the better adaptation to their virtual worlds with the benefit of experiencing the real world and interacting with intelligent beings there (Yang, Li, Fermüller, & Aloimonos, 2015).

In case of vision, the cognition development process could start from learning how to distinguish important foreground blobs from their not so important background. Then some statistically significant blobs could be classified into objects rigid and non-rigid, e.g. by fitting VC's models of the virtual objects to the real world objects. Then figuring out physical object motion by tracking and applying the basic physical concepts of space-time continuity and gravity could lead to deeper understanding of the physical world and the way its intelligent beings move themselves and objects around them in their natural environment, which could then be projected to the virtual shapes and VC motion modes in the virtual environment. Then the human behavior traits could be observed, learned, tested and applied in the virtual environment to be applied by association with the situation. The concepts that are irrelevant (hot vs. cold objects) or not important (e.g. potential damage from a sharp object) in the virtual world in absence of the appropriate sensors probably would not be assimilated by VC, but perhaps they could be labeled as important to the humans, to learn what humans could call empathy in cases of the observed pleasure or discomfort. The evolution of a VC will then very much depend on its experience with the outside world.

CONCLUSION

Vision-based cognitive architectures focusing on humans equip the virtual character designers with an exciting opportunity to enable their VCs experience the real world, meet its intelligent inhabitants, interact with them and become increasingly more intelligent in the course of such interactions. The authors have reviewed several aspects of vision based interfaces, focusing primarily on the most expressive aspects of human appearance: faces, hands and their whole body. The existence of the challenging vision problems that a VC should solve prompted a discussion of several (technical) solutions that may serve as building blocks of its cognitive architecture, whose ultimate goal appears to be in narrowing and bridging the perception gap.

The authors have proposed implementation of several essential components of a flexible, real-time and continuously learning human-centric VC development system. These include a Bayesian framework for adaptive foreground/background separation, an ensemble method combining several heterogeneous algorithms for face detection and matching, and 3D model based non-rigid body (and parts) detection and tracking in multi-view environments. The discussed vision components are expected to compose the core of the human-centric VC cognitive architecture that when combined with the modern machine learning frameworks (e.g. deep neural nets, or DNN) should allow a virtual character to evolve into an intelligent virtual being.

From the practical perspective, the immediate applications of the human-centric virtual character design (even without allowing the full virtual personality development) are quite numerous: VR Q&A kiosks, VR instructors, VR orchestra conductors, VR museum guides, virtual cinema/theatre, video games, etc. The authors also envision the possibility of non-human-like virtual characters (e.g. cartoons in AR world), that would have to learn their own (augmented) reality, interacting with their perceived world. The field of robotics may also benefit from the vision-based human-centric approach by letting robots better imitate human movements and/or face expressions. There is also hope that the proposed approach could help the humans interacting with the human-centric virtual characters to start viewing and treating such VC more as intelligent (although artificial) beings.

REFERENCES

Abdolali, F., & Seyyedsalehi, S. A. (2012). Improving face recognition from a single image per person via virtual images produced by a bidirectional network. *Procedia: Social and Behavioral Sciences*, *32*, 108–116. doi:10.1016/j.sbspro.2012.01.019

Agre, P. E., & Chapman, D. (1987). Pengi: An Implementation of a Theory of Activity. In Proceedings of the Sixth National Conference on Artificial Intelligence (Vol. 1, pp. 268–272). Seattle, WA: AAAI Press. Retrieved from http://dl.acm.org/citation.cfm?id=1856670.1856718

Aloimonos, J. Y., & Shulman, D. (1989). *Integration of Visual Modules: An Extension of the Marr Paradigm*. San Diego, CA: Academic Press Professional, Inc.

Asensio, J. M. L., Peralta, J., Arrabales, R., Bedia, M. G., Cortez, P., & Peña, A. L. (2014). Artificial Intelligence approaches for the generation and assessment of believable human-like behaviour in virtual characters. *Expert Systems with Applications*, *41*(16), 7281–7290. doi:10.1016/j.eswa.2014.05.004

Asthana, A., Zafeiriou, S., Cheng, S., & Pantic, M. (2014). Incremental Face Alignment in the Wild. In *Computer Vision and Pattern Recognition (CVPR), 2014 IEEE Conference on* (pp. 1859–1866). http://doi.org/ doi:10.1109/CVPR.2014.240

Azeem, A., Sharif, M., Raza, M., & Murtaza, M. (2014). A survey: Face recognition techniques under partial occlusion. *Int. Arab J. Inf. Technol.*, *11*(1), 1–10.

Beham, M. P., & Roomi, S. M. M. (2013). A Review Of Face Recognition Methods. *International Journal of Pattern Recognition and Artificial Intelligence*, *27*(04), 1356005. doi:10.1142/S0218001413560053

Boots, B., Nundy, S., & Purves, D. (2007). Evolution of visually guided behavior in artificial agents. *Network (Bristol, England)*, *18*(1), 11–34. doi:10.1080/09548980601113254 PMID:17454680

Borovikov, E. (1998). *Human head pose estimation by facial features location.* University of Maryland Institute for Computer Studies College Park.

Borovikov, E., & Davis, L. (2002). 3D shape estimation based on density driven model fitting. In *First International Symposium on 3D Data Processing Visualization and Transmission* (pp. 116–125). http://doi.org/ doi:10.1109/TDPVT.2002.1024051

Borovikov, E., & Davis, L. S. (2000). A Distributed System for Real-Time Volume Reconstruction. In *Fifth International Workshop on Computer Architectures for Machine Perception (CAMP 2000)* (pp. 183–189). http://doi.org/ doi:10.1109/CAMP.2000.875976

Borovikov, E., Vajda, S., Ghosh, P., Lingappa, G., Antani, S., Gill, M., & Thoma, G. (2012). *FaceMatch: visual search for pictures of missing and found persons during a disaster event* NIH Research Festival.

Borovikov, E., Vajda, S., Lingappa, G., Antani, S., & Thoma, G. (2013). Face Matching for Post-Disaster Family Reunification. In *IEEE International Conference on Healthcare Informatics* (pp. 131–140). http://doi.org/ doi:<ALIGNMENT.qj></ALIGNMENT>10.1109/ICHI.2013.23

Bouwmans, T. (2014). Traditional and Recent Approaches in Background Modeling for Foreground Detection: An Overview. *Computer Science Review*, *11-12*, 31–66. doi:10.1016/j.cosrev.2014.04.001

Brey, P. (2009). Virtual Reality and Computer Simulation. In The Handbook of Information and Computer Ethics (pp. 361–384). John Wiley & Sons, Inc. Retrieved from doi:10.1002/9780470281819.ch15

Bülthoff, H. (1990). Shape-from-X: psychophysics and computation. In P. S. Schenker (Ed.), *Sensor Fusion III: 3-D Perception and Recognition* (pp. 235–246). SPIE; doi:10.1117/12.25259

Buxton, B., & Fitzmaurice, G. W. (1998). HMDs, Caves &Amp; Chameleon: A Human-centric Analysis of Interaction in Virtual Space. *SIGGRAPH Comput. Graph.*, *32*(4), 69–74. doi:10.1145/307710.307732

CalTech. (n.d.). *Caltech Frontal Face Dataset.* Retrieved from http://www.vision.caltech.edu/archive.html/

Chan, Y., Hu, A. G., & Plant, J. (1979). A Kalman filter based tracking scheme with input estimation. *Aerospace and Electronic Systems, IEEE Transactions on*, (2), 237–244.

Chapman, D. (1989). Penguins Can Make Cake. *AI Magazine*, *10*(4), 45–50.

Choi, D., Könik, T., Nejati, N., Park, C., & Langley, P. (2007). A Believable Agent for First-Person Shooter Games (pp. 71–73). AIIDE. Retrieved from http://www.aaai.org/Papers/AIIDE/2007/AIIDE07-013.pdf

David, P., Dementhon, D., Duraiswami, R., & Samet, H. (2004). SoftPOSIT: Simultaneous pose and correspondence determination. *International Journal of Computer Vision*, *59*(3), 259–284. doi:10.1023/B:VISI.0000025800.10423.1f

David Rim, K. H., & Pal, C. (2011). Semi Supervised Learning for Wild Faces and Video. In *Proceedings of the British Machine Vision Conference* (pp. 3.1–3.12). BMVA Press.

de Borst, A. W., & de Gelder, B. (2015). Is it the real deal? Perception of virtual characters versus humans: An affective cognitive neuroscience perspective. *Frontiers in Psychology*, *6*. doi:10.3389/fpsyg.2015.00576 PMID:26029133

Duraiswami, R., Zotkin, D., & Davis, L. S. (2001). Active speech source localization by a dual coarse-to-fine search. In *Acoustics, Speech, and Signal Processing, 2001. Proceedings.(ICASSP'01). 2001 IEEE International Conference on* (Vol. 5, pp. 3309–3312). IEEE. doi:10.1109/ICASSP.2001.940366

El Gayar, N., Shaban, S. A., & Hamdy, S. (2006). Face Recognition with Semi-supervised Learning and Multiple Classifiers. In *Proceedings of the 5th WSEAS International Conference on Computational Intelligence, Man-Machine Systems and Cybernetics* (pp. 296–301). Stevens Point, WI: World Scientific and Engineering Academy and Society (WSEAS). Retrieved from http://dl.acm.org/citation.cfm?id=1374086.1374137

Fan, J., Xu, W., Wu, Y., & Gong, Y. (2010). Human Tracking Using Convolutional Neural Networks. *Neural Networks. IEEE Transactions on*, *21*(10), 1610–1623. doi:10.1109/TNN.2010.2066286 PMID:20805052

Feng, A., Lucas, G., Marsella, S., Suma, E., Chiu, C.-C., Casas, D., & Shapiro, A. (2014). Acting the part: the role of gesture on avatar identity. In Proceedings of the Seventh International Conference on Motion in Games (pp. 49–54). ACM. Retrieved from http://www.arishapiro.com/MIG2014_gestures.pdf doi:10.1145/2668064.2668102

Fukushima, K. (1980). Neocognitron: A Self-Organizing Neural Network Model for a Mechanism of Pattern Recognition Unaffected by Shift in Position. *Biological Cybernetics*, *36*(4), 193–202. doi:10.1007/BF00344251 PMID:7370364

Goertzel, B., Lian, R., Arel, I., De Garis, H., & Chen, S. (2010). A world survey of artificial brain projects, Part II: Biologically inspired cognitive architectures. *Neurocomputing*, *74*(1), 30–49. doi:10.1016/j.neucom.2010.08.012

Hubel, D. H., & Wiesel, T. N. (1968). Receptive fields and functional architecture of monkey striate cortex. *The Journal of Physiology*, *195*(1), 215–243. doi:10.1113/jphysiol.1968.sp008455 PMID:4966457

Husk, E. R. (2012). *Imitating individualized facial expressions in a human-like avatar through a hybrid particle swarm optimization–TABU search algorithm.* University of Central Florida Orlando.

Jain, A. K., Klare, B., & Park, U. (2012). Face Matching and Retrieval in Forensics Applications. *MultiMedia, IEEE*, *19*(1), 20–20. doi:10.1109/MMUL.2012.4

Jones, M., & Rehg, J. M. (2002). Statistical Color Models with Application to Skin Detection. International Journal of Computer Vision.

LeCun, Y., Bengio, Y., & Hinton, G. (2015). Deep learning. *Nature*, *521*(7553), 436–444. doi:10.1038/nature14539 PMID:26017442

Lee, T., & Hollerer, T. (2009). Multithreaded Hybrid Feature Tracking for Markerless Augmented Reality. *IEEE Transactions on Visualization and Computer Graphics, 15*(3), 355–368. doi:10.1109/TVCG.2008.190 PMID:19282544

Lepetit, V., & Fua, P. (2005). Monocular model-based 3D tracking of rigid objects: A survey. *Foundations and Trends in Computer Graphics and Vision, 1*(CVLAB-ARTICLE-2005-002), 1–89.

Li, Y., Wei, H., Monaghan, D. S., & O'Connor, N. E. (2014). A Low-Cost Head and Eye Tracking System for Realistic Eye Movements in Virtual Avatars. In MultiMedia Modeling (pp. 461–472). Springer. Retrieved from http://doras.dcu.ie/19587/1/eyetracking.pdf

Lu, C., & Tang, X. (2014). *Surpassing Human-Level Face Verification Performance on LFW with GaussianFace.* CoRR, abs/1404.3840

Markus, N., Frljak, M., Pandzic, I. S., Ahlberg, J., & Forchheimer, R. (2013). *A method for object detection based on pixel intensity comparisons.* CoRR, abs/1305.4537. Retrieved from http://arxiv.org/abs/1305.4537

McCollum, C., Barba, C., Santarelli, T., & Deaton, J. (2004). Applying a cognitive architecture to control of virtual non-player characters. In *Proceedings of Winter Simulation Conference* (*Vol. 1*). IEEE.

Mori, M., MacDorman, K. F., & Kageki, N. (2012). The Uncanny Valley. *Robotics Automation Magazine, IEEE, 19*(2), 98–100. doi:10.1109/MRA.2012.2192811

Nguyen, T. H. D., Qui, T. C. T., Xu, K., Cheok, A. D., Teo, S. L., Zhou, Z., & Kato, H. et al. (2005). Real-Time 3D Human Capture System for Mixed-Reality Art and Entertainment. *IEEE Transactions on Visualization and Computer Graphics, 11*(6), 706–721. doi:10.1109/TVCG.2005.105 PMID:16270863

Perez-Sala, X., Escalera, S., Angulo, C., & Gonzalez, J. (2014). A survey on model based approaches for 2D and 3D visual human pose recovery. *Sensors (Basel, Switzerland), 14*(3), 4189–4210. doi:10.3390/s140304189 PMID:24594613

Peter Müller, F. A. Q. (2004). Nonparametric Bayesian Data Analysis. *Statistical Science, 19*(1), 95–110. doi:10.1214/088342304000000017

Rivalcoba, J. I., De Gyves, O., Rudomin, I., Pelechano Gómez, N., & Associates. (2014). *Coupling camera-tracked humans with a simulated virtual crowd.* Retrieved from http://upcommons.upc.edu/e-prints/bitstream/2117/24668/1/Pelechano.pdf

Saberi, M., Bernardet, U., & DiPaola, S. (2014). An Architecture for Personality-based, Nonverbal Behavior in Affective Virtual Humanoid Character. *Procedia Computer Science, 41*, 204 – 211.

Sharif, M., Mohsin, S., & Javed, M. Y. (2012). A Survey: Face Recognition Techniques. *Research Journal of Applied Sciences. Engineering and Technology, 4*(23), 4979–4990.

Simard, P. Y., Steinkraus, D., & Platt, J. C. (2003). Best practices for convolutional neural networks applied to visual document analysis. In *Document Analysis and Recognition, 2003. Proceedings. Seventh International Conference on* (pp. 958–963). http://doi.org/ doi:10.1109/ICDAR.2003.1227801

Taigman, Y., Yang, M., Ranzato, M. A., & Wolf, L. (2014). DeepFace: Closing the Gap to Human-Level Performance in Face Verification. In *Proceedings of the IEEE Computer Society Conference on Computer Vision and Pattern Recognition*. doi:10.1109/CVPR.2014.220

Trafton, G., Hiatt, L., Harrison, A., Tamborello, F., Khemlani, S., & Schultz, A. (2013). ACT-R/E: An embodied cognitive architecture for human-robot interaction. *Journal of Human-Robot Interaction, 2*(1), 30–55. doi:10.5898/JHRI.2.1.Trafton

Viola, P., & Jones, M. (2004). Robust real-time face detection. *International Journal of Computer Vision, 57*(2), 137–154. doi:10.1023/B:VISI.0000013087.49260.fb

Wan, J., Wang, D., Hoi, S. C. H., Wu, P., Zhu, J., Zhang, Y., & Li, J. (2014). Deep Learning for Content-Based Image Retrieval: A Comprehensive Study. In *Proceedings of the 22Nd ACM International Conference on Multimedia* (pp. 157–166). New York, NY: ACM. http://doi.org/ doi:10.1145/2647868.2654948

Yang, Y., Li, Y., Fermüller, C., & Aloimonos, Y. (2015). *Neural Self Talk: Image Understanding via Continuous Questioning and Answering*. CoRR, abs/1512.03460. Retrieved from http://arxiv.org/abs/1512.03460

Yue-Hei Ng, J., Yang, F., & Davis, L. S. (2015). Exploiting Local Features From Deep Networks for Image Retrieval. In *The IEEE Conference on Computer Vision and Pattern Recognition (CVPR) Workshops*.

Yun, C., Deng, Z., & Hiscock, M. (2009). Can Local Avatars Satisfy a Global Audience? A Case Study of High-fidelity 3D Facial Avatar Animation in Subject Identification and Emotion Perception by US and International Groups. *Comput. Entertain., 7*(2), 21:1–21:26. http://doi.org/<ALIGNMENT.qj></ALIGNMENT>10.1145/1541895.1541901

Zachary, W., Ryder, J., Ross, L., & Weiland, M. Z. (1992). *Intelligent computer-human interaction in real-time, multi-tasking process control and monitoring systems. In Human Factors in Design for Manufacturability* (pp. 377–402). New York: Taylor and Francis.

Zhang, C., & Zhang, Z. (2010). *A Survey of Recent Advances in Face detection.* Microsoft.

Chapter 2
Integrating ACT-R Cognitive Models with the Unity Game Engine

Paul Richard Smart
University of Southampton, UK

Katia Sycara
Carnegie Mellon University, USA

Tom Scutt
Mudlark, UK

Nigel R. Shadbolt
University of Oxford, UK

ABSTRACT

The main aim of the chapter is to describe how cognitive models, developed using the ACT-R cognitive architecture, can be integrated with the Unity game engine in order to support the intelligent control of virtual characters in both 2D and 3D virtual environments. ACT-R is a cognitive architecture that has been widely used to model various aspects of human cognition, such as learning, memory, problem-solving, reasoning and so on. Unity, on the other hand, is a very popular game engine that can be used to develop 2D and 3D environments for both game and non-game purposes. The ability to integrate ACT-R cognitive models with the Unity game engine thus supports the effort to create virtual characters that incorporate at least some of the capabilities and constraints of the human cognitive system.

INTRODUCTION

Cognitive architectures are computational frameworks that can be used to develop computational models of human cognitive processes (Langley et al., 2009; Taatgen & Anderson, 2010; Thagard, 2012). Cognitive architectures have been useful in terms of advancing our understanding of human cognition in specific task environ-

DOI: 10.4018/978-1-5225-0454-2.ch002

ments, and they have also been used to support the development of a variety of intelligent systems and agents (e.g., cognitive robots). Although a variety of cognitive architectures are available, such as SOAR (Laird, 2012; Laird et al., 1987), ACT-R (Anderson, 2007; Anderson et al., 2004) and CLARION (Sun, 2006a; Sun, 2007), the focus of the current chapter is on ACT-R. ACT-R is a rule-based system that has been widely used by cognitive scientists to model aspects of human cognitive performance. It is also one of the few cognitive architectures that has an explicit link to research in the neurocognitive domain: the structural elements of the core ACT-R architecture (i.e., its modules and buffers) map onto different regions of the human brain (Anderson, 2007), and this enables cognitive modelers to make predictions about the activity of different brain regions at specific junctures in a cognitive task (see Anderson et al., 2007)[1].

Given their role in the computational modeling of cognitive processes, it is perhaps unsurprising that cognitive architectures have been used in the design of intelligent virtual characters. The SOAR architecture, for example, has been used to control a humanoid character that co-exists in a virtual 3D environment alongside a human-controlled avatar (Rickel & Johnson, 2000). The aim, in this case, is to provide a training environment in which the SOAR-controlled character possesses expertise in a particular domain of interest and then mentors human subjects as they progress through the stages of skill acquisition. This is an excellent example of the productive merger of cognitive architectures with virtual environments. As Rickel and Johnson (2000) point out, virtual tutors that cohabit a virtual environment with human players benefit from the ability to communicate nonverbally using gestures, gaze, facial expression and locomotion. In addition, the virtual agents can closely monitor the behavior of human subjects in a way that is not typically possible outside of a virtual environment; for example, a user's actions and field of view can be carefully monitored to determine their likely focus of attention. Rickel and Johnson's (2000) work is also a clear example where a multidisciplinary focus is required to engineer the intelligent virtual agent: the development of an intelligent virtual tutor draws on technical and scientific advances in the fields of knowledge elicitation (Shadbolt & Smart, 2015), knowledge modeling (Schreiber et al., 2000), human expertise development (Chi et al., 1988), and educational psychology.

Another example of cognitive architectures being used in virtual character design is provided by Best and Lebiere (2006). They used ACT-R to control the behavior of synthetic team-mates in a virtual environment as part of military training simulations[2]. As is the case with virtual tutors, the use of a virtual environment is important here because it enables human actions to be closely monitored in a way that is difficult (if not impossible) with real-world environments. As a result of such monitoring, the behavior of ACT-R-controlled virtual agents can be adjusted in ways that respect the norms and conventions of team-based behavior. This is a topic of particular interest in the context of military behavior simulations, where issues of team coordination and the synchronization of team-member responses (often in alignment with doctrinal specifications) are all-important.

In addition to situations where cognitive architectures have been used to control virtual characters as part of training simulations or tutoring applications, there are a number of other research and development contexts where one sees a convergence of issues relating to cognitive architectures, virtual environments and virtual character design. These include the use of cognitive architectures to model the behavior of human game players (Moon & Anderson, 2012), as well as the use of cognitive architectures and virtual environments to study issues in embodied, extended and embedded cognition (Smart & Sycara, 2015b). Other areas that benefit from the integrative use of cognitive architectures and virtual environments include computer simulations of socially-situated behaviors and collaborative problem-solving processes (Smart & Sycara, 2015b; Sun, 2006b), the development of believable game characters (Arrabales et al., 2009), the implementation of virtual coaches and mentors in therapeutic applications (Niehaus, 2013), the creation of game characters with psychologically-realistic properties (Bringsjord et al., 2005), and the digital modeling of human behavior in a variety of occupational and ergonomic settings (Lawson & Burnett, 2015).

No matter what the motivation for integrating cognitive architectures with virtual environments, all integration efforts rely on the existence of mechanisms that support the seamless inter-operation of the cognitive architecture with whatever system is used to implement the virtual environment. In situations where the virtual environment is a 3D environment similar to those encountered in contemporary video games, then the target of the integration solution will typically be a game engine. This can present challenges to integration, since the game engine and the cognitive architecture are typically systems that use different code bases and run in different processes[3]. In addition, both kinds of systems often place significant demands on the computational resources of the host machine. This can make it difficult or impossible to run the cognitive architecture and the game engine on the same machine at the same time.

In the current chapter, we focus on an integration solution for a particular cognitive architecture (ACT-R) and a particular game engine (Unity). We first provide an overview of the ACT-R architecture and the Unity game engine. We describe the key features of both systems and discuss why they provide such a compelling target for integration. We then go on to describe the nature of the integration solution itself. We outline the extensions to the ACT-R architecture that enable ACT-R models to exchange information with Unity. We then go on to present the Unity components that enable virtual characters to be controlled or influenced by ACT-R models. Finally, we provide a concrete example of the use of the integration solution: we show how an ACT-R model can be used to control the behavior of a virtual robotic character that inhabits a Unity-based virtual environment.

THE ACT-R COGNITIVE ARCHITECTURE

ACT-R is one of a number of cognitive architectures that have been used for cognitive modeling (Anderson, 2007; Anderson et al., 2004)[4]. It is primarily a symbolic cognitive architecture in that it features the use of symbolic representations and explicit production rules; however, it also makes use of a number of sub-symbolic processes that contribute to various aspects of performance (Anderson et al., 2004).

ACT-R consists of a number of modules (see Figure 1), each of which is devoted to processing a particular kind of information. Each module is associated with one or more capacity-constrained buffers that can contain a single item of information, called a 'chunk'[5]. The modules access and deposit information in the buffers, and coordination between the modules is achieved by a centralized production system module – the procedural module – that can respond to the contents of the buffers and change buffer contents (via the execution of production rules). Importantly, the procedural module can only respond to the contents of the buffers; it cannot participate in the internal encapsulated activity of the modules, although it can influence module-based processes.

As shown in Figure 1, there are eight core modules in the latest version of ACT-R[6]. These modules assume responsibility for the implementation of specific cognitive functions as part of an integrated cognitive system. The goal module, for example, is a specialized form of 'working memory' that maintains information relevant to task goals. It serves to contextualize the activity of other modules. Another important module is the declarative module. This module is responsible for the mnemonic encoding and retrieval of information. It stores information in the form of chunks, each of which is associated with an activation level that determines its probability of recall. The vision, audio, speech and motor modules function as points of perceptuo-motor contact between an ACT-R agent and the external environment. They provide support for the modeling of agent-world interactions.

The ACT-R architecture has been used to model human cognitive performance in a wide variety of experimental settings[7]. It has generated findings of predictive and explanatory relevance to hundreds of phenomena encountered in the cognitive psychology and human factors literature, and this has earned it a reputation as the cognitive architecture that is probably the "best grounded in the experimental research literature" (Morrison, 2003; p. 24). ACT-R has also been used to model

Figure 1. The core modules of the ACT-R v.6 cognitive architecture

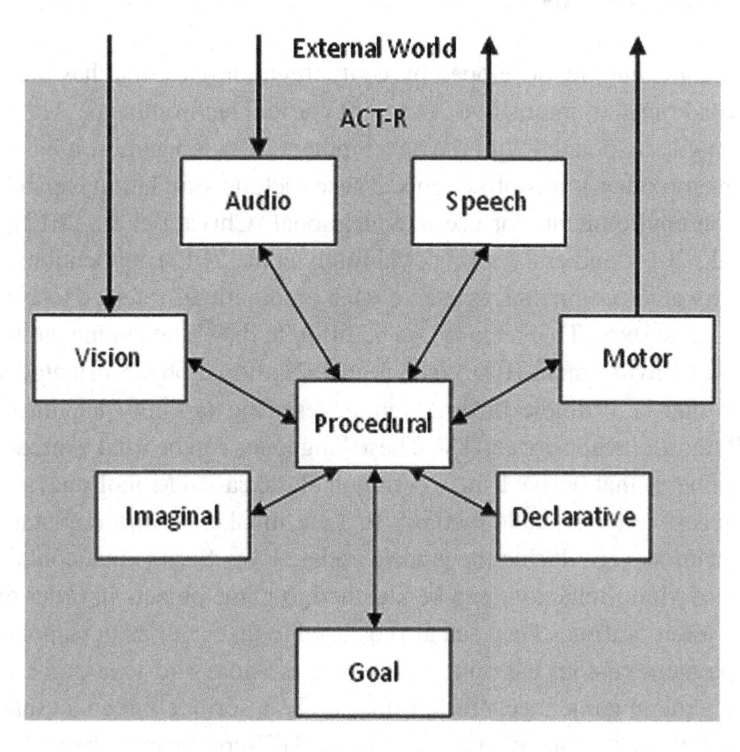

behavior in a range of complex task settings, such as driver behavior (Salvucci, 2006) and collaborative problem-solving (Reitter & Lebiere, 2012). These features make ACT-R a compelling target for research that seeks to model cognitive performance and test assumptions regarding the characteristics of the human cognitive system. In addition, the various cognitive capabilities and features of the ACT-R architecture, especially those relating to learning and memory, make it an interesting choice for developers who are not so much interested in the modeling of human cognition as in the development of agents and systems that exhibit signs of human-level intelligence. This has given rise to studies that have attempted to use ACT-R to control the behavior of real-world robotic systems (Best & Lebiere, 2006; Kurup & Lebiere, 2012; Trafton et al., 2013). It is also the basis of recent work concerning the use of ACT-R models to control the behavior of virtual robots that inhabit virtual 3D environments (Smart & Sycara, 2015a, 2015c). Finally, the rich support that ACT-R provides for the computational modeling of cognitive processes supports its use in controlling the behavior of virtual synthetic agents that engage in interaction with human-controlled characters in the context of (e.g.) training simulations (Best & Lebiere, 2006).

UNITY GAME ENGINE

Unity[8] is a game engine, developed by Unity Technologies, that has been used to create a broad range of interactive 2D and 3D virtual environments. Although, it is most commonly associated with the development of video games, it has also been used to develop other kinds of systems. These include simulation capabilities and visualization environments for use in educational (Christel et al., 2012), medical (Rizzo et al., 2014) and engineering (Mattingly et al., 2012) applications.

The Unity game engine forms part of what is sometimes referred to as the Unity game creation system. This includes, in addition to the game engine, an Integrated Development Environment (IDE) (see Figure 2) and an object-oriented scripting framework that is available in three languages: Boo (a Unity language that resembles Python), JavaScript and C#. These languages can be used to create custom code components that derive from a common class, called MonoBehaviour, which provides access to overridable methods that are invoked at different stages of the game's execution, e.g., during the game's update loop. Scripts that contain classes derived from MonoBehaviour can be attached to game objects in order to control their behavior at runtime. They can also be used to take actions in response to user input, implement custom user interface overlays, store and load game data, and implement general game mechanics. Importantly, a script component that inherits from MonoBehaviour can be attached to several different game objects in order to implement common functionalities, and multiple scripts can be attached to the same object in order to create combined functionalities. Using such features, it is possible to create complex control mechanisms for characters within a game. In the example described later in this chapter, for example, a custom MonoBehaviour component was developed to respond to commands from an ACT-R model in order to control the behavior of a virtual robot.

The Unity IDE (see Figure 2) provides extensive support for the development of 2D and 3D virtual environments. It includes a number of windows, such as the Scene, Project, Hierarchy, Game and Inspector windows. These enable a developer to visualize the current scene (Scene window); adjust the properties of game objects (Inspector window); manage game assets (Project window), such as texture, audio, code and model assets; and play test the current state of the game (Game window). The IDE also integrates seamlessly with external code editors, such as MonoDevelop[9] or Microsoft Visual Studio.

A key function of the Unity IDE is to provide easy access to a number of built-in components that form part of the Unity game engine. One component of particular importance is the Camera component. This provides a view of a 3D or 2D scene from a particular vantage point. While Camera components typically render their view of the scene to the main game window during game play, they can also be

Figure 2. View of the unity IDE showing the game, hierarchy, project and inspector windows

used to render a scene to a special type of 2D image asset, called a RenderTexture. This can then be processed using standard 2D graphic processing routines in order to retrieve information about luminance levels in the red, green and blue (RGB) color channels of the image's pixel data. In the context of our own work, we rely on this technique to implement virtual 'eyes' that generate a series of 2D images as an agent moves through a 3D landscape. Using image processing techniques, it is possible to extract simple visual features from these images in order to implement various forms of visuo-motor control.

One of the features that makes Unity a compelling target for integration efforts is its popularity. Unity is undoubtedly one of the most popular game creation systems currently in use. Underlying this popularity, particularly with the indie game community, is the support it provides for multi-platform development. Unity can thus be used to create applications that are deployable to the Web, game consoles (e.g., Xbox, Wii, and PS3), mobile devices (e.g., iPhone, iPad, and Android devices), and personal computers (e.g., Windows, Mac OS and Linux). In addition to its deployment features, the Unity game engine boasts extensive technical support, a community-based asset store, an impressive list of game development features, and compatibility with Microsoft's .NET Framework. Moreover, the latest version of the game engine (Version 5) is free to use for research and development purposes[10].

INTEGRATING ACT-R WITH UNITY

Integration Challenges

A number of challenges confront the attempt to develop an integration solution for ACT-R and Unity. One of the issues to address concerns the fact that Unity and ACT-R are implemented using different languages: Unity is implemented in C++, while ACT-R is implemented in Lisp. This complicates the attempt to straightforwardly embed ACT-R within Unity.

A second challenge relates to performance. Many 3D environments depend on computational operations that are highly processor-intensive (e.g., the operations associated with graphics rendering or physics simulation). Similarly, the execution of ACT-R models can consume significant computational resources, particularly when multiple ACT-R models are used to perform multi-agent simulations. If the aim of an integration solution is to use ACT-R in conjunction with a virtual environment and have both cognitive and virtual world simulations rely on the same set of processing resources, steps must be taken to ensure that the complexity of the ACT-R model and virtual world simulation are within the performance capabilities of the target deployment platform. This is particularly so if the aim is to run simulations in real-time.

Finally, it should be noted that ACT-R is typically intended to be used as a stand-alone system. By default, the ACT-R system includes very little support for inter-operation with external systems and applications. This underscores the need to implement extensions to ACT-R that support flexible modes of inter-process communication and systemic integration.

A Network-Based Approach to Integration

One approach to integrating ACT-R with Unity is based on a networking solution that supports bidirectional modes of communication between one or more ACT-R models and a single Unity-based virtual environment. This approach has the advantage of dealing with the aforementioned challenges[11]. The issue regarding different code bases is resolved by enabling ACT-R and Unity to run in different processes and communicate using standard network protocols and messaging systems. In addition, the issue of performance overheads can be tackled by having ACT-R run on a different machine to that hosting the Unity-based environment. This enables the cognitive and virtual world simulations to be handled by dedicated processors that run in parallel and synchronize their activity via the exchange of messages. It should also be clear that the strategy of running ACT-R and Unity on different machines has the added bonus of addressing issues associated with compilation and deployment. As mentioned above, the ability to deploy Unity games and applications to multiple platforms is one of the features that motivates its use above other game engines. Although the ACT-R system can be compiled to run on systems, such as Mac OS, Linux and Windows, the use of a networking solution to integrate ACT-R with Unity simplifies the deployment strategy: a developer can deploy the Unity-based application to a target platform using the features available within the Unity editor and then rely on network connections to communicate with an instance of ACT-R that runs on another (perhaps dedicated) machine within the local network environment or the Internet.

The particular integration solution described here relies on the use of an ACT-R component developed by Hope et al (2014) called the JSON Network Interface, or JNI. JSON, in this case, is an acronym that stands for JavaScript Object Notation Format. It is a data interchange format that is commonly encountered in the context of Web-based communications. By using the JNI, it is possible to run ACT-R models that communicate with a target external environment by posting JSON messages. Although, in our case, the external environment of interest is an application or simulation that runs on top of the Unity game engine, it is important to recognize that many different systems could play a similar role. The notion of an external environment could thus be applied to a conventional Windows desktop application that communicates with ACT-R as part of an attempt to evaluate the cognitive ergonomic design of different graphical user interface designs. It could also apply to a real-world robotic platform, where ACT-R is used to process sensor information and control the robot's effector systems. What is required, in all these cases, is a means by which messages sent by ACT-R can be interpreted by the target external environment. Typically, this involves the development of an Application Programming Interface (API) for the target environment. In the present case, we have developed an API for Unity that supports communication with ACT-R via the JNI. We refer to this as the ACT-R Unity Interface (ACT-R UI) Framework.

JSON Network Interface

In order to support the bidirectional exchange of information between ACT-R models and a range of external environments, Hope et al (2014) advocate the use of a network-based approach to integration, in which an external system plays the role of a server and individual ACT-R models play the role of clients. Communication between the two systems is then established via a series of client-server interactions mediated by TCP/IP socket connections.

Hope et al's (2014) solution is encapsulated in a custom ACT-R module – the JSON Network Interface, or JNI. This module can be installed alongside the core modules that form part of the default ACT-R architecture (see Figure 1). Each JNI module implements the functionality needed to establish connections and interact with the external environment. It comes with parameters that identify the IP address of the external environment as well as the number of the TCP port that the external environment is listening on. This is clearly important in terms of enabling a specific ACT-R model to connect to the environment during the initialization process. The JNI module communicates with the external environment using a set of commands that form part of the JNI API. These commands are transmitted as JSON-formatted messages that either inform the external environment about the state of model execution or which instruct the environment to take particular actions based on model outputs. For example, the model-run command is posted by the JNI whenever an ACT-R model has started running (thereby providing state information); the keypress command, in contrast, is posted whenever the ACT-R model has decided to initiate a keypress action via the motor module. In all cases, the messages sent to the external environment include the name of the ACT-R model from which the message originates, the name of the command that is being issued, and any data that is relevant to the interpretation or processing of the command named in the message. The inclusion of the name of the originating model in the message is important here because each ACT-R model has its own instance of the JNI module. This means that multiple ACT-R modules can communicate with the same external environment (e.g., Unity) either as part of a multiplayer online game (perhaps featuring a combination of human-controlled and ACT-R-controlled characters) or as part of an experimental simulation into socially-situated or socially-distributed cognition (see Smart & Sycara, 2015b). The ability to identify the originating ACT-R model, in this case, enables the external environment to delegate the processing of received messages to specific virtual agents within the environment.

In addition to the commands posted by an ACT-R model (known as 'module commands'), the JNI API includes commands that an ACT-R model expects to receive from the external environment. These commands, known as 'environment commands', provide information about the state of the external environment, and they are typically used to update the content of ACT-R's perceptual modules. Figure 3, for example, shows the structure of a sample display-new environment command that is used to update the content of the vision module of an ACT-R model called 'myModel'. Other environment commands enable the external environment to add items to declarative memory, trigger the presentation of rewards and present auditory stimuli.

Figure 3. An example environment command (formatted using JSON) that is used to update the contents of an ACT-R model's vision module

```
{"model": "myModel",
  "method": "update-display",
  "params": {
    "visual-location-chunks": [
      {"isa": "visual-location",
        "slots": {"screen-x": 100,
                  "screen-y": 200}},
      {"isa": "visual-location",
        "slots": {"screen-x": 300,
                  "screen-y": 400}}],
    "visual-object-chunks": [
      [{"isa": "visual-object",
        "slots": {"value": "hello"}},
      {"isa": "visual-object",
        "slots": {"value": ":world"}}]],
  "clear": true}}
```

The set of module and environment commands that comprise the standard JNI API provide the basis for forms of communication in which an ACT-R model is interacting with a particular kind of environment, namely a display screen. Obviously, this is unlike the kind of environment that is encountered in the context of a video game, where a virtual character is required to respond to the perceptual features of a 3D scene[12]. Fortunately, the set of commands that can be exchanged via the JNI module is easy to extend. In terms of extending the range of module commands that are available, the JNI API provides access to a set of Lisp functions that can be called upon to create and format the new command messages. Extending the range of environment commands (commands originating from the external environment) can be accomplished by relying on a particular environment command called trigger-event. This command enables the external environment to trigger the execution of named Lisp functions that execute in the context of the ACT-R environment. Such functions can be used to great effect in situations where the kind of sensory information that needs to be handled by ACT-R is not easily accommodated by the core perceptual modules of the ACT-R architecture. In the case of our own work with virtual robots, for example, we often include tactile information in the package of sensor information that is posted to ACT-R. The core ACT-R architecture has no module that can handle this kind of information, so it is necessary to implement a custom module and rely on the JNI trigger-event command to update the contents of the buffers associated with the new module. Figure 4 provides an example of

Figure 4. Example of a Lisp function that is used to update the contents of a buffer (called robot-sensor) in response to the receipt of a trigger-event command from an external environment

```
(defun update-tactile-sensor (left-whisker  right-whisker)
  "Create a new chunk representing the state of the robots left and right
  whiskers and place it in the 'robot-sensor' buffer, if the buffer is
  currently empty."
  (let ((buffer 'robot-sensor)
    (left-whisker-val (intern (string-upcase left-whisker)))
    (right-whisker-val (intern (string-upcase right-whisker)))
    (chunk (car (define-chunks (isa robot-sensor-state)))))
  (if (query-buffer buffer '((buffer . empty)))
    (progn
      (set-chunk-slot-value-fct chunk 'left-whisker left-whisker-val)
      (set-chunk-slot-value-fct chunk 'right-whisker right-whisker-val)
      (schedule-event-relative 0 (lambda nil (set-buffer-chunk buffer chunk)))))))
)
```

such a function. It shows a Lisp function being used to update a sensor buffer in response to the posting of a trigger-event command by an external environment.

ACT-R Unity Interface Framework

Although the JNI provides the basis for ACT-R integration efforts, additional work needs to be done in order to enable the external environment to communicate with ACT-R. In the case of Unity, for example, it is necessary to develop components that can handle incoming connection requests from ACT-R models, interpret the JSON messages received from ACT-R models and implement whatever commands are contained in the messages. Furthermore, as mentioned above, it is often necessary to extend the functionality of the ACT-R environment in ways that enable an ACT-R model to function within the specific sensorimotor niche provided by the external environment. Often this involves the implementation of additional ACT-R modules that can handle specific kinds of sensory or motor information.

In the context of our own work, we have developed a framework, called the ACT-R Unity Interface (ACT-R UI) Framework, which enables ACT-R models to communicate with applications that run on top of the Unity game engine (Smart et al., forthcoming). The ACT-R UI Framework consists of a number of components, all of which are implemented as Unity-compatible C# scripts. The main components of the framework are the following[13]:

- **ACTRNetworkInterface:** The main function of the ACTRNetworkInterface component is to handle connection requests from ACT-R models. It also implements the functionality to post messages to ACT-R clients on the network.

The component relies on the native support that Unity provides for .NET socket connections.

- **ACTRCommand:** The ACTRCommand class is the base class for all the commands that are recognized by ACT-R and Unity in the context of a particular integration solution. There are two main subclasses of the ACTRCommand class: ModuleCommand and EnvironmentCommand. As is suggested by these names, these abstract classes are intended to represent the two categories of commands that form part of the JNI API. Subclasses of these top-level classes represent all the commands that are included in the JNI API. Additional subclasses can be created as required to extend the range of messages that Unity and ACT-R are able to process.
- **ACTRMessageInterface:** This is a component that inherits from the Unity MonoBehaviour class. Its primary function is to engage in the preliminary processing of ACT-R messages. It processes the raw JSON content of the message and attempts to create appropriate instances of the ACTRCommand class based on the content of the message. Once created, these commands are posted to the relevant ACTRAgent component (see below) according to the name of the originating ACT-R model (which is included in the message). The ACTRMessageInterface component contains a user-editable field (visible in Unity's Inspector window) that specifies the port number that ACT-R clients should use to connect to Unity. The component also references an instance of the ACTRNetworkInterface class, and it initializes this instance whenever the game starts. This results in Unity listening for incoming connection requests from whatever ACT-R models are running on the network.
- **ACTRAgent:** This MonoBehaviour component represents the entity in the virtual environment that is controlled by a particular ACT-R model. It is typically attached to a game object representing a non-player character, such as a virtual robot or humanoid avatar. It exposes a property that enables a developer to specify the name of the ACT-R model that will control the entity. At runtime, the main function of the ACTRAgent component is to engage in sensor processing and implement the motor instructions received from ACT-R. In general, ACTRAgent components can be configured to periodically poll available sensors and post information back to ACT-R as part of what is called the sensor processing cycle. They can also respond to instructions from ACT-R (e.g., to move forward or turn to the right). Unlike the other components that form part of the ACT-R UI Framework, the ACTRAgent component is not intended to be used in specific simulations. This is because the nature of the sensorimotor processing routines for ACT-R-controlled agents are likely to be very different depending on the kind of simulation

that is being run. A simulation involving virtual robots, for example, is likely to feature a different set of sensor and effector components compared to a simulation involving humanoid characters. For this reason, most simulations will need to create components that derive from the ACTRAgent component in order to adapt the capabilities of the virtual character to the specific sensorimotor niche they occupy within the virtual environment.

Using the components of the ACT-R UI Framework, it is possible to create a network-based integration solution in which one or more ACT-R models control the behavior of virtual characters that are embedded within a virtual environment. Given the network-based nature of the integration solution, it is clearly possible to distribute the computational burden associated with the execution of both cognitive models and the virtual environment: multiple cognitive models, each representing a distinct cognitive agent, can be run on different machines and communicate with a Unity-based application that itself runs on a dedicated machine[14]. This solution can be used to integrate ACT-R cognitive models into online multiplayer games that feature the participation of human-controlled characters.

CASE STUDY: MAZE NAVIGATION USING A VIRTUAL ROBOT

In order to demonstrate the use of the ACT-R UI Framework, we describe a simulation capability that focuses on the spatial cognitive capabilities of a virtual robot. The simulation features a robot that is embedded in a virtual maze environment that was created using the Unity IDE. The simulation is organized into two phases. In an initial exploratory phase, the robot is required to explore the maze at random and learn about its structure. Then, in a subsequent navigation phase, a goal location is specified and the robot must use its previous experience of exploring the maze in order to plan a route and navigate to the target location. A video showing the performance of the robot across both phases of the simulation is available for viewing on the YouTube website[15].

Virtual Environment Design

The environment used for the maze navigation task is a simple virtual maze created from a combination of geometric shapes, such as blocks and cylinders. The design of the maze is based on that described by Barrera and Weitzenfeld (2007) who used a similar maze as part of their effort to test bio-inspired spatial cognitive capabilities in a real-world robot. The maze consists of a number of vertically- and horizontally-aligned corridors that are shaped like the letter 'H'. An additional vertically-aligned corridor serves as a common point of departure for the robot during the exploration and navigation phases of the simulation (see Figure 5).

A number of brightly colored blocks and cylinders were placed around the walls of the maze to function as visual landmarks. These objects show up as colored patches in the images that are rendered by each of the robot's eyes, and they can thus be used by the robot to identify its location within the maze.

Robot Design

The virtual robot used in the simulation is a disc-shaped 3D model that is capable of linear and rotational movements. The robot comes equipped with a number of virtual sensors that are capable of processing visual, tactile and directional information. These sensors correspond to the robot's eyes, whiskers, and onboard compass, respectively. The eyes are represented by Unity Camera components that are positioned around the edge of the robot and oriented slightly upwards at an angle of 15 degrees. For convenience, these cameras are referred to as 'eye cameras'. There are four eye cameras in total. They capture views from the left, right, forward and backward directions of the robot. The output of the eye cameras is captured as a RenderTexture asset as described above and is subjected to lightweight image processing techniques in order to extract visual features. The primary aim of the robot's visual processing routines, in the current simulation, is to detect the brightly colored

Figure 5. View of the 'H' Maze from a first-person perspective (left) and from a top-down perspective (right). The robot is located on the right hand side of the maze in both images. The camera that renders the top-down view of the maze (on the right) has been configured to simplify the rendering output. This makes it easier to visualize and analyze simulation results. The white cross in the right hand image represents the starting location for the robot on all training and testing trials. The compass indicator (again in the right-hand image) shows the direction of 'north'.

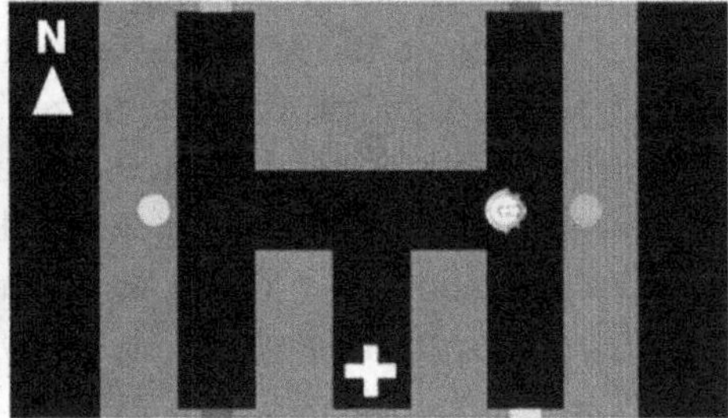

objects arrayed around the walls of the maze. This is accomplished by matching the luminance levels of image pixels in the red, green and blue color channels to a number of target colors corresponding to those of the visual landmarks (i.e., the pink, blue, green and yellow objects). The target colors in this case are specified by the user at design time and can be set to any color. The sensitivity of the robot's eyes is governed by two values: the 'tolerance' and 'threshold' values. The tolerance value represents the range of luminance levels in each color channel that is recognized as a match to the target luminance level. The threshold value, in contrast, specifies the minimum number of matching pixels that must be counted in order for the eye to signal the detection of a particular color. For the purposes of the current study, the tolerance value was set to a value of 0.01 and the threshold value was set to a value of 1500. In addition, each retina was sized to 200 x 200 pixels to give a total of 40,000 pixels per eye camera on each render cycle.

In addition to eyes, the robot comes equipped with 'whiskers'. These function as tactile sensors. Similar to the eye cameras, there are four whiskers situated around the robot's body, and these project outwards in the forward, backward, left and right directions. The whiskers respond to contact with the walls of the maze. This tactile information, in combination with the visual input, serves to identify specific locations in the maze. It also provides affordances for action-related decisions, helping to inform the robot when it needs to turn and what directions it can move in. From an implementation perspective, the whiskers rely on the use of ray casting techniques: each time the robot is required to report sensory information to ACT-R, rays are projected from the vehicle's body and any collisions with the walls of the maze are recorded.

The final sensor used by the robot is a compass. The robot is capable of detecting which direction it is facing based on the compass reading, which is based on the rotation of the vehicle's transform in the world coordinate system. A rotation of zero degrees, for example, corresponds to a heading value of 'NORTH'.

For the purposes of this work, the linear movement of the robot was restricted to the north, south, east and west directions: these are, in fact, the only directions that are needed to fully explore the 'H' Maze environment, and restricting movement in this way reduces the complexity of the simulation. The robot is also capable of making 360 degree turns. This enables the robot to turn around when it reaches the end of one of the arms of the maze. Turning movements are implemented by rotating the robot's transform based on the commands received from ACT-R; linear movements, in contrast, are implemented by specifying the velocity of the robot's Rigidbody component, a component that enables the robot to participate in the physics calculations made by Unity's physics engine.

Cognitive Model

The cognitive modeling effort involved the development of an ACT-R model that could support the initial exploration of the maze and also enable the robot to navigate to target locations. The requirements of the model were the following:

- **Motor Control:** The model needs to issue motor instructions to the robot in response to sensory information in order to orient and move the robot within the maze.
- **Maze Learning:** The model needs to detect novel locations within the maze and memorize the sensory information associated with these locations.
- **Route Planning:** The model needs to use the memorized locations in order to construct a route to a target location.
- **Maze Navigation:** The model needs to use route-related information in conjunction with sensory feedback in order to monitor its progress towards a goal location.

The ACT-R model developed to support the target behavior consists of 110 productions in addition to ancillary functions that control the communication with Unity. A key function of the model is to memorize spatial locations that are individuated with respect to their sensory context (i.e., unique combinations of visual and tactile information). These locations are referred to as 'place fields' in the context of the cognitive model. Each place field is created as a chunk in declarative memory and retrieval operations against declarative memory are subsequently used to recall the information encoded by the place field as the robot moves through the maze. The collection of place fields constitutes the robot's 'cognitive map' of the maze. This map is structured as a directed graph in which the place fields act as nodes and the connections between the nodes are established based on the directional information that is recorded by the robot as it explores the maze[16]. Any two place fields that are created in succession will be connected via ACT-R chunks that record the direction the robot was heading in when the connection was made. For example, if the robot creates a place field (PF1) at the start of the simulation and then creates a second place field (PF2) while heading north from the starting location, a connection is established between PF1 and PF2 that records PF1 as the source of the connection, PF2 as the target of the connection and 'NORTH' as the direction of the connection. The collection of place fields and the connections that join them thus serve to record information about the topological relationships between maze locations based on a combination of sensory experiences and the robot's own movements.

The productions of the ACT-R model were used to realize the motor control, maze learning and maze navigation functions mentioned above; the route planning function, however, was implemented using separate Lisp routines. In order to plan a route, the robot first needs to be given a target location. This was specified at the beginning of trials that tested navigational performance. The robot then needs to identify its current location in the maze. This was achieved by the model comparing current sensory information with all the place fields stored in memory. The result (assuming that the robot's initial navigation of the maze is complete) is a place field that corresponds to the robot's actual position in the maze. Finally, the robot needs to compute a sequence of place fields that encode the path from the start location to the target location. This is achieved using a spreading activation solution that operates over the network of place fields contained in memory. The chain of activated place fields from the start location to the target location specifies the sequence of place fields (identified by combinations of sensory information) that must be detected by the robot as it navigates to the target. In addition, the connections between adjacent place fields along the computed route contain information about the direction the robot needs to move in as each place field is encountered. For example, if the connection between the first and second place fields in the route has an associated value of 'NORTH' and the robot is currently facing north, then the model can simply instruct the robot to move forward. If the robot is facing south, then the robot needs to perform a 180 degree turn before moving forward and the model thus needs to issue instructions for the robot to engage in a turning maneuver.

Sensorimotor Interface Module

The modules that form part of the core ACT-R architecture include modules that deal with the processing of sensory and motor information. The vision module, for example, is designed to handle requests about the presence, location and properties of visual objects within the visual field. These modules were not used as part of the current research effort. Instead, a custom module was used as a single point of sensorimotor contact between ACT-R and Unity. The module was designed to handle all sensory information received from Unity and to also issue motor instructions that controlled the behavior of the virtual robot. One reason why the existing vision module was not used to process visual information relates to the fact that the module was not intended to support the processing of low-level sensor data. While it is not inconceivable that the vision module could be used to handle the kind of visual input seen in the case of the current simulations, we deemed it more straightforward to implement a custom module. Another reason motivating the choice of a custom module to handle sensorimotor processing concerns the fact ACT-R is not designed to accommodate perceptuo-motor processes that derive from the kind of sensory and motor systems seen in the case of virtual robots. The ACT-R motor module, for example, was not used for motor output in the present case because it was originally designed to implement the sort of actions (e.g., keyboard presses and mouse movements) that a human subject might perform while seated in front of a desktop computer. Obviously, the kind of motor control capabilities required in the case of a non-humanoid virtual robot demand a set of specialized motor commands that are best handled by a custom module.

The custom module itself features two buffers: the 'robot-sensor' and 'robot-motor' buffers. The robot-sensor buffer contains information about the current sensory state of the robot within the virtual environment. The content of the buffer is populated automatically based on the messages received from Unity. In particular, when Unity posts a message containing updated sensory information, the message triggers the execution of a Lisp function that creates an ACT-R chunk containing all the sensor information contained in the message. This chunk is subsequently placed in the robot-sensor buffer.

The robot-motor buffer processes requests relating to the movement of the robot within the virtual environment. It accepts information (in the form of ACT-R chunks) about the desired state of a robot's motor system. When a chunk is placed into the buffer (as a result of rule execution), the robot module processes the chunk to extract the relevant motor information and then dispatches a JSON-formatted message containing the motor information to the JNI module for transmission to Unity. Once this message is received by Unity, it is handled by the ACT-R UI Framework: the relevant robot is identified based on the name of the ACT-R model that posted the message, and the message is then sent to the relevant ACTRAgent component for further processing. During the course of each update cycle of the game engine, the ACTRAgent will attempt to implement whatever motor instructions it receives from ACT-R. Typically, these instructions result in the robot moving forward at a particular velocity or turning to face a particular direction.

Phase I: Maze Exploration

In order to test the integrity of the integration solution and the performance of the cognitive model, a simple experiment was performed. The experiment consisted of two phases: an exploration (learning) phase and a navigation (testing) phase. In the exploration phase, the robot was placed at the start location (i.e., the base of the middle vertically-aligned arm), and it was then allowed to wander around the maze and learn about its structure. This resulted in the robot forming a cognitive map of the environment. Once the robot had explored all of the maze, the exploration phase was terminated and the cognitive map was saved to disk for later use.

In the subsequent navigation phase, the previously created cognitive map was loaded into declarative memory, and the robot was given a series of target locations to navigate towards. These target locations were situated at the ends of the vertically-aligned corridors that made up the long arms of the maze. As was the case for the exploration phase, the starting location of the robot for each trial of the navigation phase was the base of the smaller vertically-oriented arm.

The structure of the cognitive map formed by the robot during the exploration phase is shown in Figure 6. The white circles in this figure indicate the position of the place fields that were formed by the robot as it moved around the maze. The magenta trail represents the path of the robot. As can be seen from Figure 6, the robot explored the entire maze in a time of 3 minutes and 8 seconds. A total of 29 place fields were created in memory during the exploration phase. These served as the nodes in the graph that made up the cognitive map. The nodes were connected together into a single component network that consisted of 33 connections.

Phase II: Maze Navigation

The results of the maze navigation phase are shown in Figure 7. This figure shows the path taken by the robot as it navigated to each of the four target locations. It also shows the time taken by the robot to reach each of the target locations. For example,

Figure 6. The results of the exploration phase of the experiment. The white circles symbolize the location of place fields that are memorized by the robot as it explores the maze. The magenta trail shows the path taken by the robot.

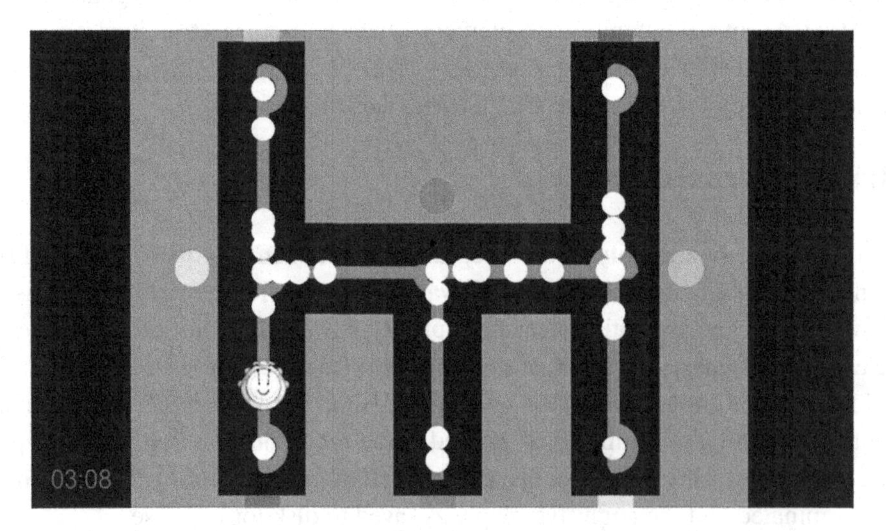

Figure 7. The results of the maze navigation phase of the experiment. The robot successfully navigated to each of the goal locations. The magenta trail indicates the path taken by the robot as it navigated towards the goal location.

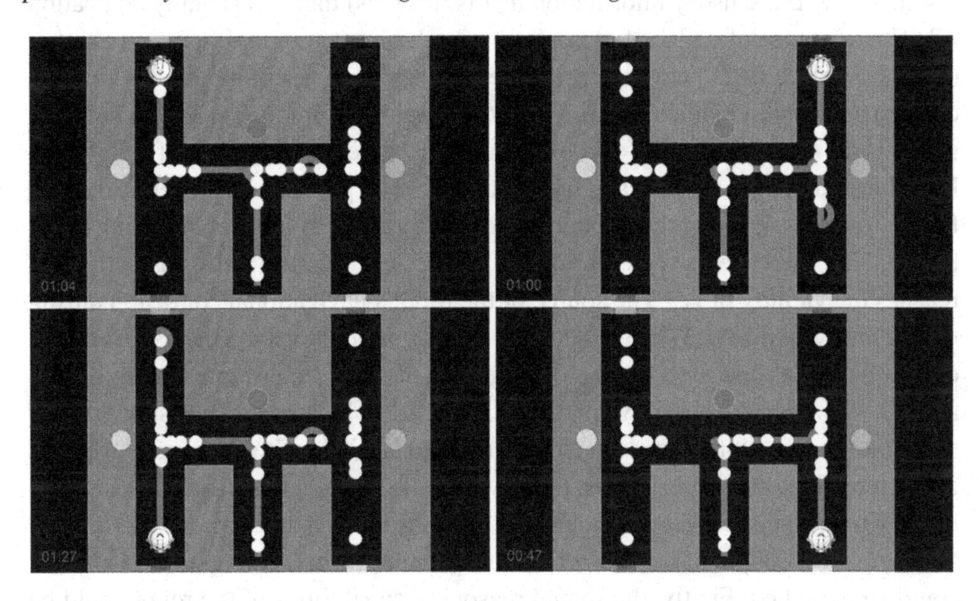

it took 1 minute and 4 seconds to reach the goal located at the top left of the maze. In all trials, the robot was able to successfully reach the target location – the robot identified its location at the start of the trial, plotted a route to the target based on the structure of the cognitive map, and then used this route to guide its movements towards the goal.

In spite of the ability of the robot to ultimately reach the target locations, Figure 7 shows some inefficiencies in the navigational decisions made by the robot. In particular, in three out of the four trials, the robot made an unexpected detour at some point along the route. For example, in the case of the goal at the top left of the maze (see top left image in Figure 7), the robot initially made a turn towards the right when it reached the horizontal corridor of the maze. In fact, the correct turn is towards the left, and the robot seemed to realize it's mistake when it subsequently made a 180 degree turn a few seconds later. The reason for this navigational anomaly is due to the fact that not all locations in the maze are uniquely identified by combinations of visual and tactile input. In Figure 8, for example, we can see that locations 2 and 4 (i.e., L2 and L4) are associated with exactly the same combinations of sensory input (a green cylinder to the east, a yellow cylinder to the west and walls to the north and south). Imagine that a robot has already encoded a link between L1 and L2 so that it knows that L2 is located to the east of L1. If the robot now heads west towards L4, it will pass through L1 *en route* to L4. As the robot encounters the stimuli associated with L1 it will recognize L1 as its current location. When it subsequently encounters L4, it will 'believe' (based on the combination of visual and tactile sensory information that is receives) that it is actually at location L2. The existence of a place field representing L2 in memory will prevent another (identical) place field being created to represent L4. In addition, the fact that the cognitive map used by the robot is a directed graph with unidirectional connections prevents the creation of a link encoding the otherwise confusing state-of-affairs that L2 is located both to the east and the west of L1. The problem, however, is that the failure to represent L4 with a separate place field results in the robot creating a link between the place fields representing L2 and L5 when it eventually passes L4 and reaches the vicinity of L5. The source of the robot's navigational confusion during testing is now revealed. When a route is planned to a goal location that is to the west of L4, the robot constructs a route that consists of place fields representing L1, L2 and L5. However, the cognitive map tells it that L2 is located to the east of L1, and so that is the direction it heads in when it reaches L1. When it subsequently arrives at L2, it realizes that L5 is located to the west of L2 and that it is therefore heading in the wrong direction. Hence the reason for the sudden about turn.

There are a couple of ways in which this peculiar form of navigational behavior could be remedied. Firstly, the spatial reasoning capabilities of the robot could be extended in order to support a sense of the relative positions of distinct locations. For example, with respect to Figure 8, the robot could reason that if it encountered

Figure 8. A series of locations in the 'H' Maze, only some of which are uniquely identified by combinations of tactile and visual input. Based on the sensory capabilities of the robot, the locations indicated by L2 and L4 are perceived as identical. This leads to the robot failing to create a separate place field corresponding to L4 (only locations indicated by white circles are associated with place fields). The result is that when a route is planned to L5, the robot will erroneously turn east and head towards L2 when it arrives at L1.

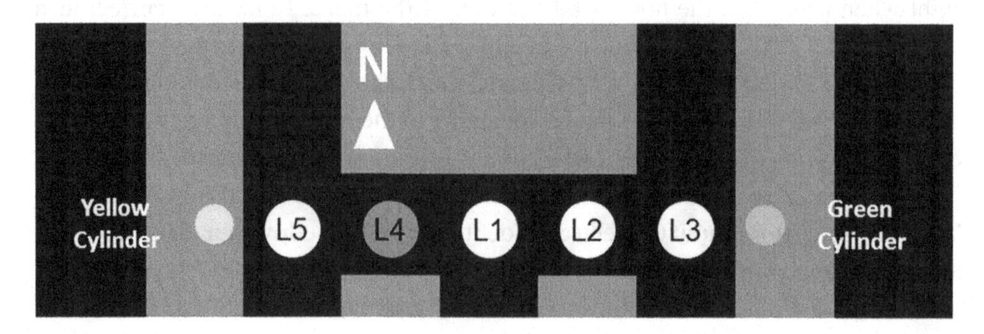

a location that was identical to L2 having already passed through L1 while heading in a westerly direction, and that L2 was to the east of L1, then the new location could not possibly be the same as L2. It could thus create a separate place field to represent L4, albeit with the same configuration of sensory information as that associated with the place field representing L2. An alternative strategy would be to improve the discriminative capabilities of the robot when it comes to an assessment of the stimuli associated with particular locations. The visual system of the robot could thus be enhanced to detect the relative size of visual landmarks within the visual field. This would enable a robot to discriminate between L2 and L4 in Figure 8 on the basis of the fact that the yellow cylinder would appear larger at L4 than it would at L2 (the reverse being true for the green cylinder).

CONCLUSION

ACT-R and Unity are the two most widely used systems in their respective domains of use. ACT-R is one of the most widely used cognitive architectures, with a long history of use within the human factors, cognitive psychology and (more recently) robotics communities. Similarly, Unity is a very popular game engine that has been used to develop a broad range of visualizations, applications, simulations and, of course, video games. Given the range of potential applications and research opportunities enabled by the ability to merge the capabilities of cognitive architectures with virtual environments, it is surprising that no discernible attempt has been made to develop an integration solution for ACT-R and Unity. As far as we are aware, the solution presented in this chapter is the only attempt thus far to integrate the two systems.

The integration solution described here draws on previous work by Hope et al (2014); however, it also extends that body of existing work by enabling ACT-R to inter-operate specifically with the Unity game engine. The solution is referred to as the ACT-R UI Framework: a set of components that can be used by Unity developers to link virtual game characters to distinct ACT-R models. The example implementation we have used to demonstrate the integration solution is a maze navigation problem in which a virtual robot is required to learn about the spatial structure of a maze and navigate to specific goal locations. This example is intended to showcase some of the features of the integration solution in a task context that exploits at least some of the cognitive features of the ACT-R architecture. Extensions of this work could obviously seek to improve the sophistication and complexity of the spatial learning and spatial navigation capabilities. For example, by integrating temporal information into the connections between place fields it should be possible to enable a virtual character to choose between multiple routes to the same target based on their relative time costs. In addition, by extending the range of movements of the virtual robot and the directional headings they are able to record, it should be possible to study navigational performance in a variety of other mazes; for example, the kind of mazes that are typically used by behavioral neuroscientists, as well as the more labyrinthine mazes encountered in many video games. Another form of extension relates to the perceptual processing capabilities of the virtual characters that are linked to ACT-R models. Clearly, the kind of low-level sensor processing seen in the case of the current example may not be appropriate for all situations where ACT-R is being used to control virtual characters, especially when the aim is to generate rapid behavioral responses in real-time. The advantage of a virtual environment, in this case, is that it is often possible to determine precisely what the outcome of sensor processing should be based on a knowledge of the properties of the sensor system and the character's location in the game. For example, in the case of visual input, the objects that are visible to a particular character can be determined based on the location of game objects relative to the character's view frustum. This obviates the need to engage in detailed processing of rendered images because high-level visual information regarding the location and properties of visual objects in the visual field can be computed independently of a render cycle. This strategy will no doubt be useful in situations where virtual characters must coordinate their behavioral output with respect to complex 3D scenes, although the use of low-level sensor processing routines may still be appropriate in cases where the primary purpose of the integration effort is to support scientific simulations.

Regardless of the kind of sensorimotor capabilities that are implemented for virtual characters, the ability to integrate ACT-R with Unity is likely to prove useful in a range of situations. The integration solution presented here is particularly useful in contexts where the aim is to use the capabilities of the Unity game engine to run simulations in which ACT-R models are effectively embedded in perceptually complex virtual environments. This is likely to be of particular relevance when it comes to computer simulations that seek to investigate the performance of virtual cognitive robots (as in the current case) or in situations where the aim is to advance our understanding of the role that agent-world interactions play in shaping human cognitive capabilities (Clark, 2008). It is also likely to be relevant to situations involving the interaction of multiple cognitive agents, as is the case in a number of recent scientific studies using the ACT-R architecture (Reitter & Lebiere, 2012; Smart et al., 2014).

Aside from the use of the integration solution to support serious scientific endeavors, there is also the case of cognitive architectures being used to create more human-like virtual characters. Such characters are particularly suited for applications that require synthetic agents to interact with their human counterparts, for pedagogic (e.g., Rickel & Johnson, 2000) or other purposes.

Finally, we encounter the more entertainment-oriented aspect of integration efforts: the use of cognitive architectures to enhance game play experiences by yielding characters with evermore intelligent capabilities (Arrabales et al., 2009). Given ACT-R's history of use by the scientific community specifically for the purposes of cognitive modeling, it is perhaps unsurprising that it is difficult to evaluate this particular use of the ACT-R/Unity integration solution. Further research needs to be done to determine whether ACT-R can produce the kind of intelligent responses that are currently difficult or impossible to produce using conventional game AI techniques. Needless to say, the first step in evaluating this possibility is the availability of a robust integration solution that enables ACT-R to be tested in conjunction with a state-of-the-art game engine. With such a solution now in place, and with Unity being used by an ever-greater number of professional game developers, there has never been a better time for us to explore the possibilities. Let's play!

ACKNOWLEDGMENT

This research was sponsored by the U.S. Army Research Laboratory and the U.K. Ministry of Defence and was accomplished under Agreement Number W911NF-06-3-0001. The views and conclusions contained in this document are those of the author(s) and should not be interpreted as representing the official policies, either expressed or implied, of the U.S. Army Research Laboratory, the U.S. Government, the U.K. Ministry of Defence or the U.K. Government. The U.S. and U.K. Governments are authorized to reproduce and distribute reprints for Government purposes notwithstanding any copyright notation hereon.

REFERENCES

Anderson, J. R. (2007). *How Can the Human Mind Occur in the Physical Universe?* Oxford, UK: Oxford University Press. doi:10.1093/acprof:oso/9780195324259.001.0001

Anderson, J. R., Albert, M. V., & Fincham, J. M. (2005). Tracing problem solving in real time: fMRI analysis of the subject-paced Tower of Hanoi. *Journal of Cognitive Neuroscience*, *17*(8), 1261–1274. doi:10.1162/0898929055002427 PMID:16197682

Anderson, J. R., Bothell, D., Byrne, M. D., Douglass, S., Lebiere, C., & Qin, Y. (2004). An integrated theory of the mind. *Psychological Review*, *111*(4), 1036–1060. doi:10.1037/0033-295X.111.4.1036 PMID:15482072

Anderson, J. R., Qin, Y., Jung, K.-J., & Carter, C. S. (2007). Information-processing modules and their relative modality specificity. *Cognitive Psychology*, *54*(3), 185–217. doi:10.1016/j.cogpsych.2006.06.003 PMID:16919255

Arrabales, R., Ledezma, A., & Sanchis, A. (2009). *Towards conscious-like behavior in computer game characters.IEEE Symposium on Computational Intelligence and Games*, Milan, Italy. doi:10.1109/CIG.2009.5286473

Barrera, A., & Weitzenfeld, A. (2007). *Bio-inspired model of robot spatial cognition: Topological place recognition and target learning.International Symposium on Computational Intelligence in Robotics and Automation*, Jacksonville, FL. doi:10.1109/CIRA.2007.382839

Best, B. J., & Lebiere, C. (2006). Cognitive Agents Interacting in Real and Virtual Worlds. In R. Sun (Ed.), *Cognition and Multi-Agent Interaction: From Cognitive Modeling to Social Interaction*. New York: Cambridge University Press.

Bringsjord, S., Khemlani, S., Arkoudas, K., McEvoy, C., Destefano, M., & Daigle, M. (2005) *Advanced synthetic characters, evil, and E.6th Annual European Game-On Conference*, Leicester, UK.

Chi, M. T. H., Glaser, R., & Farr, M. J. (Eds.). (1988). *The Nature of Expertise*. Hillsdale, NJ: Erlbaum.

Christel, M. G., Stevens, S. M., Maher, B. S., Brice, S., Champer, M., Jayapalan, L., & Bastida, N. et al. (2012) *RumbleBlocks: Teaching science concepts to young children through a Unity game.17th International Conference on Computer Games*, Louisville, KY. doi:10.1109/CGames.2012.6314570

Clark, A. (2008). *Supersizing the Mind: Embodiment, Action, and Cognitive Extension*. New York: Oxford University Press. doi:10.1093/acprof:oso/9780195333213.001.0001

Hope, R. M., Schoelles, M. J., & Gray, W. D. (2014). Simplifying the interaction between cognitive models and task environments with the JSON Network Interface. *Behavior Research Methods*, *46*(4), 1007–1012. doi:10.3758/s13428-013-0425-z PMID:24338626

Kurup, U., & Lebiere, C. (2012). What can cognitive architectures do for robotics? *Biologically Inspired Cognitive Architectures*, *2*, 88–99. doi:10.1016/j.bica.2012.07.004

Laird, J. E. (2012). *The SOAR Cognitive Architecture*. Cambridge, MA: MIT Press.

Laird, J. E., Newell, A., & Rosenbloom, P. S. (1987). SOAR: An architecture for general intelligence. *Artificial Intelligence*, *33*(1), 1–64. doi:10.1016/0004-3702(87)90050-6

Langley, P., Laird, J. E., & Rogers, S. (2009). Cognitive architectures: Research issues and challenges. *Cognitive Systems Research, 10*(2), 141–160. doi:10.1016/j.cogsys.2006.07.004

Lawson, G., & Burnett, G. (2015). Simulation and Digital Human Modelling. In J. R. Wilson & S. Sharples (Eds.), *Evaluation of Human Work* (4th ed.). Boca Raton, FL: CRC Press.

Mattingly, W. A., Chang, D.-j., Paris, R., Smith, N., Blevins, J., & Ouyang, M. (2012) *Robot design using Unity for computer games and robotic simulations.17th International Conference on Computer Games*, Louisville, KY. doi:10.1109/CGames.2012.6314552

Moon, J., & Anderson, J. R. (2012) *Modeling Millisecond Time Interval Estimation in Space Fortress Game.34th Annual Conference of the Cognitive Science Society*, Sapporo, Japan.

Morrison, J. E. (2003) A Review of Computer-Based Human Behavior Representations and Their Relation to Military Simulations. Institute for Defense Analyses. doi:10.1037/e427382005-001

Niehaus, J. (2013) *Mobile, Virtual Enhancements for Rabilitation.* Charles River Analytics.

Reitter, D., & Lebiere, C. (2012) *Social Cognition: Memory Decay and Adaptive Information Filtering for Robust Information Maintenance.26th AAAI Conference on Artificial Intelligence*, Toronto, Canada.

Rickel, J., & Johnson, L. W. (2000). Task-Oriented Collaboration with Embodied Agents in Virtual Worlds. In J. Cassell, J. Sullivan, & S. Prevost (Eds.), *Embodied Conversational Agents*. Cambridge, MA: MIT Press.

Rizzo, A., Hartholt, A., Grimani, M., Leeds, A., & Liewer, M. (2014). Virtual Reality Exposure Therapy for Combat-Related Posttraumatic Stress Disorder. *Computer, 47*(7), 31–37. doi:10.1109/MC.2014.199

Salvucci, D. D. (2006). Modeling driver behavior in a cognitive architecture. *Human Factors, 48*(2), 362–380. doi:10.1518/001872006777724417 PMID:16884055

Schreiber, G., Akkermans, H., Anjewierden, A., de Hoog, R., Shadbolt, N. R., Van de Velde, W., & Weilinga, B. (2000). *Knowledge Engineering and Management: The CommonKADS Methodology*. Cambridge, MA: MIT Press.

Shadbolt, N. R., & Smart, P. R. (2015). Knowledge Elicitation: Methods, Tools and Techniques. In J. R. Wilson & S. Sharples (Eds.), *Evaluation of Human Work* (4th ed.). Boca Raton, FL: CRC Press.

Smart, P. R., & Sycara, K. (2015a). *Place Recognition and Topological Map Learning in a Virtual Cognitive Robot.17th International Conference on Artificial Intelligence*, Las Vegas, NV

Smart, P. R., & Sycara, K. (2015b). *Situating Cognition in the Virtual World.6th International Conference on Applied Human Factors and Ergonomics*, Las Vegas, NV.

Smart, P. R., & Sycara, K. (2015c). *Using a Cognitive Architecture to Control the Behaviour of Virtual Robots*. EuroAsianPacific Joint Conference on Cognitive Science, Turin, Italy.

Smart, P. R., Sycara, K., & Tang, Y. (2014). Using Cognitive Architectures to Study Issues in Team Cognition in a Complex Task Environment. SPIE Defense, Security, and Sensing: Next Generation Analyst II, Baltimore, MD.

Smart, P. R., Sycara, K., Tang, Y., & Powell, G. (forthcoming). The ACT-R Unity Interface: Integrating ACT-R with the Unity Game Engine. Electronics and Computer Science, University of Southampton, Southampton, UK.

Sun, R. (2006a). The CLARION Cognitive Architecture: Extending Cognitive Modeling to Social Simulation. In R. Sun (Ed.), *Cognition and Multi-Agent Interaction: From Cognitive Modeling to Social Interaction* (pp. 79–99). New York: Cambridge University Press.

Sun, R. (Ed.). (2006b). *Cognition and Multi-Agent Interaction: From Cognitive Modeling to Social Interaction*. New York: Cambridge University Press.

Sun, R. (2007). Cognitive social simulation incorporating cognitive architectures. *Intelligent Systems*, *22*(5), 33–39. doi:10.1109/MIS.2007.4338492

Taatgen, N., & Anderson, J. R. (2010). The Past, Present, and Future of Cognitive Architectures. *Topics in Cognitive Science*, *2*(4), 693–704. doi:10.1111/j.1756-8765.2009.01063.x PMID:25164050

Thagard, P. (2012). Cognitive architectures. In K. Frankish & W. M. Ramsey (Eds.), *The Cambridge Handbook of Cognitive Science* (pp. 50–70). Cambridge, UK: Cambridge University Press. doi:10.1017/CBO9781139033916.005

Trafton, G., Hiatt, L., Harrison, A., Tamborello, F., Khemlani, S., & Schultz, A. (2013). ACT-R/E: An Embodied Cognitive Architecture for Human-Robot Interaction. *Journal of Human-Robot Interaction*, *2*(1), 30–54. doi:10.5898/JHRI.2.1.Trafton

ENDNOTES

[1] ACT-R is, in fact, somewhat unique in proposing a neuro-anatomical mapping for its structural elements. The goal module, for example, is deemed to map onto the anterior cingulate cortex, whereas the procedural module is deemed to map onto the caudate nucleus (see Anderson, 2007, pp. 74-86). This neurological mapping enables ACT-R models to be used in conjunction with brain imaging studies (Anderson et al., 2005). For example, the activity in specific modules (e.g., the goal module) can be used to predict activity in specific brain regions (e.g., the anterior cingulate cortex) at specific points in a cognitive task.

[2] Unlike the solution described here, Best and Lebiere (2006) used ACT-R in conjunction with the Unreal game engine.

[3] The notion of a process, here, refers to a computer process, i.e., an instance of a computer program that is being executed.

[4] ACT-R is, in fact, only one of a number of different cognitive architectures that can be used to support the design of intelligent virtual characters. The SOAR architecture is, of course, a popular alternative to ACT-R (e.g., Rickel & Johnson, 2000). In addition, a range of other, bespoke architectures, such as the RASCALS architecture (Bringsjord et al., 2005), have been the focus of efforts to combine cognitive architectures with virtual environments.

[5] A 'chunk' in ACT-R is the basic unit of information over which rule-based processes operate.

[6] Although the eight core modules (and their associated buffers) tend to form the basis of most ACT-R models, cognitive modelers are not restricted to the use of these modules. New custom modules can be added to implement additional functionality, as required. The JNI module described by Hope et al. (2014) is one example of such a module (see also below).

[7] The ACT-R website (http://act-r.psy.cmu.edu/) provides access to a broad range of academic publications covering areas such as problem-solving, learning, language processing, decision making, and perceptual processing.

[8] See http://unity3d.com/.

[9] http://www.monodevelop.com/

[10] Although we focus on the Unity game engine in the current chapter, the integration solution outlined here could be adapted to work with other game engines, such as Epic's Unreal game engine. All that is required, in this case, are components that replicate the functionality of the ACT-R Unity Interface Framework components (e.g., the ACTRAgent component).

[11] Note that this solution is preferable to an integration solution that attempts to re-implement ACT-R within the native programmatic environment of the

game engine. One issue to consider, here, is that ACT-R is subject to periodic updates and revisions. This makes the maintenance of ported code somewhat problematic. In addition, the scale and complexity of the ACT-R system complicates the attempt to duplicate ACT-R functionality within the native environment of a particular game engine.

[12] Although it might be appropriate for situations in which the ACT-R model is intended to model human player behavior and the display screen corresponds to the human's view of the game.

[13] Further information about the ACT-R UI Framework can be found in Smart et al. (forthcoming).

[14] The simulation described in the current chapter focuses on a single ACT-R cognitive model that controls a single virtual character; nevertheless, the extension to the multi-model case (i.e., where multiple ACT-R models control the behaviour of multiple virtual characters) is relatively straightforward. See Smart et al. (forthcoming) for practical steps on how to implement multi-model solutions using the ACT-R UI Framework.

[15] See http://youtu.be/zolWEO8PRQg

[16] To simplify the structure of the map, bidirectional connections between adjacent nodes were not permitted.

Chapter 3
A Graphical Tool for the Creation of Behaviors in Virtual Worlds

Andrea Corradini
Design School Kolding, Denmark

Manish Mehta
Accenture Technology Lab, USA

ABSTRACT

Typically, the creation of AI-based behaviors for Non-Playing Characters (NPC) in computer games is carried out by people with specialized skills in the broad areas of design and programming. This book chapter presents Second Mind (SM), a digital solution that makes it possible for novice users to successfully author behaviors. In SM, authors can use a graphical interface to define the behaviors of virtual characters in response to interactions with players. Authors can easily endow NPCs with behavior capabilities that make them act in different possible roles such as e.g. shopkeepers, museum hosts, etc. A series of user tests with human subjects to evaluate the behavior authoring process shows that Second Mind is easy to understand and simplifies the process of behavior production.

INTRODUCTION

In order to increase player experience, the virtual characters that populate video games and virtual worlds should exhibit believable behaviors (Umarov, Mozgovoy, & Rogers, 2012). This does not apply only to the player characters, which are the

DOI: 10.4018/978-1-5225-0454-2.ch003

characters that are controlled by the player, and as such are meant as more of a player's extension. This also applies to any other character that is controlled by the game, usually through artificial intelligence, such as townsfolk, vendors, enemies, allies, historical figures, bystanders, quest givers, etc. All these latter characters are referred to as non-player characters (NPCs) and are an integral part of the game. They exist to interact with the player characters, other NPCs, and the environment in a way that reflects their own distinct personalities and dialog capabilities.

In the context of synthetic characters, believability is all about making the users believe they are interacting with a living agent, whose existence is consistent with the virtual world the character is situated in. In essence, believability is the ability to suspend the users' disbelief, by providing an illusion of life (Bates 1994).

Creating behaviors for NPC usually requires specialized skills in both design and programming well beyond the skills of the average person (Magerko, Laird, Assanie, Kerfoot, & Stokes, 2004). Hence, the intrinsic nature of the behavior generation process is restricted to the individuals who have a certain expertise in this specific area. To date, there is little understanding of how individuals without programming and design experience (whom we will refer to as "novice users" throughout this paper) can carry out the behavior authoring process.

The behaviors of NPCs are usually scripted and are automatically triggered by certain actions, events, or dialogue. In that sense, player characters play the game, while NPCs display some facet of the game and help to further the storyline making the game what it is (Gillies & Ballin, 2004). In early digital games, NPCs display only basic non-interactive dialogue capabilities. Monologues realized with screens of text, floating text, voiceovers, cutscenes, text clouds and other non-branching dialogue techniques made it possible for NPCs to convey only an immediate impression of their personality in reaction to or to initiate interaction with the player. However, this lack of interactivity turns the player into a mere passive consumer of content. More recent, advanced games allow for interactive dialogue where the player can engage in a compelling conversation with NPCs that results in witty and dramatic storylines that fit the context at hand. In those games, a conversation is typically modelled using dialogue trees (Adams, 2012) where the player is presented with and can choose from a limited list of interaction options in response to an NPC's action. Each choice affects the course of the game in a different way since each of them may result in a different response from the NPC. Moreover, the possibility to choose between different options makes it necessary for the game designer to customize the NPCs' behaviors by modifying their default scripts or by creating entirely new ones.

Branching the overall storyline puts the player in charge of learning about the game by engaging him in conversation with NPCs. In this way, uncovering the storyline becomes part of an increased overall player experience (Li, Thakkar, Wang,

& Riedl, 2014). Besides engaging interactive narrative experiences (Yu & Riedl, 2014) with unfolding storylines, it is also necessary to have the tools that make it possible to design and create the appropriate behaviors. This is not just about making them funnier. It is also about creating more surprising and unpredictable behaviors within the game constraints so that two interactions with an NPC never turn out to be the same.

Digital games have much to gain from adopting artificial intelligence (AI) techniques to author complex behaviors. This is however, a difficult task for it requires a great deal of competence and resources. Usually, only the game creators or the people with specialized skills in both design and programming can carry out such a task. As a result, authoring behavior can even become a bottleneck in the process of creating the content displayed within digital worlds.

In this paper, we present Second Mind, a digital authoring tool solution that helps any players to populate virtual worlds with NPCs that can exhibit believable interactive behaviors. Our solution is based on a study that we carried out using paper prototypes to investigate how individuals without programming and design experience carry out the behavior authoring process. That study reveals some major limitations that we account for in the development of Second Mind. In Second Mind, novice users can easily design the behavior of graphical characters that impersonate different tradesmen such as shopkeepers, museum hosts, etc. in virtual worlds.

The rest of the paper is organized as follows. In the next section, we discuss related works and the challenges that behavior authoring poses to users, with particular focus on novice users. Thereafter, we present the results of the user study that we ran with 40 novice users to evaluate the feasibility of a paper-based prototype authoring system in a restricted context. We then detail the architecture of the digital authoring solution that we developed for novice users based on the findings discovered in running the user study. Eventually, we conclude with an outline of the work in process and our ideas for further improvement.

BACKGROUND AND RELATED WORK

Believable characters play an important role in many interactive stories and digital games (Rickel, Marsella, Gratch, Hill, Traum, & Swartout, 2002) because they can contribute to the overall narrative. Thus, the improvement and the control of their behaviors is propaedeutic to an increased game experience for players.

To ensure believability, virtual agents usually integrate multiple cognitive abilities of people such as language, gesture, emotion, and the problems associated with real-time navigation in and interaction with a simulated virtual world. Moreover, they also simulate many of the embodied and social aspects of human behavior as well as the need of autonomous behaviors, consistent with the character's personality, motivational state, and world context. Such an autonomy is usually implemented with an internal structure that allows the agent to have its own needs, motivation and interests. Based on these needs and interests the character can then dynamically select and generate goals that eventually result in autonomous behavior. An architecture to represent and manage internal driving factors in intelligent virtual character, using the concept of motivation, has been presented in (Avradinis, Panayiotopoulos, & Anastassakis, 2013). Because of the integration of human cognitive capabilities, virtual characters are gaining interest also as a methodological tool for studying and validating theories in human cognition (Gratch, Hartholt, Dehghani, & Marsella, 2013). At the same time several cognitive architectures have been put forward also for applications with virtual agents (Langley, Laird, & Rogers, 2009, Laird, 2008, Langley & Choi, 2006, Nuxoll & Laird, 2007, Sun 2007, Gluck 2010).

Common methods for authoring behaviors rely on heavy scripting, where a blend of dialogue scripts, hand-coded animation scripts, motion-captured characters, and hard-coded behaviors is used to exhibit a character's features. Usually, artists, designers and programmers must work together to detail a character-attitude and to portray the appropriate character's movements given a set of constraints. However, scripting is labor intensive, rigid because it does not adapt to all variations induced by interaction, and requires programming and design skills. An alternative approach is to use artificial intelligent (AI) techniques to adapt character behaviors in response to the different contexts that are induced by different interaction sessions. Despite its own inherent limitations, AI technology has been increasingly applied with success to digital game design and interactive narrative both to enhance interaction and to create novel gameplay scenarios (Riedl & Zook, 2013). AI techniques have been explored with diverse focus on character believability and expressivity in terms of e.g. synchronization of verbal and non-verbal modalities (Cassell, 2001, Corradini, Mehta, Bernsen, & Charfuelan, 2005), portrayal of emotion and personality (Allbeck & Badler, 2004, Aylett, Louchart, Dias, Paiva, & Vala, 2005), and so on.

The ease of the actual process of behavior creation has been given less relevance and explored by fewer researchers. The language ABL (Mateas & Stern, 2004) is an early effort in this direction. ABL is a Java-like programming language that can be used to manually script behaviors (Mateas & Stern, 2003, Gomes & Jhala, 2013). The production costs are high and the use of ABL requires advanced programming skills. Finite State Machines (FSMs) have also been used in digital games for many years as an alternative to scripting. However, FSMs can quickly become difficult to maintain and do not scale well with increasing game complexity (Schwab, 2009). Various FSMs modifications have been proposed to bypass part of these problems (Isla, 2005, Champandard, 2007). The use of AI techniques such as Layered Statechart-based AI (Dragert, Kienzle, & Verbrugge, 2012), Behavior Trees (Tomai & Flores,

2014), automated planning and scheduling strategies (Coman & Munoz-Avila, 2012, Kadlec, Toth, Cerny, Bartak, & Brom 2012, Kelly, Botea, & Koenig, 2008), has been explored as well. All these works require human intervention and a great deal of effort and skills to create and to debug the scripted behaviors. A recent attempt at eliminating manual scripting in favor of automatic script generation is presented in (Zhao & Szafron, 2014) where the content creation for cyclic behaviors is facilitated using a tiered behavior architecture model. This model features a scheduling algorithm to determine the objectives of virtual characters and to specify the roles that satisfy these objectives dynamically during game play. The results of a user study that compares the creation of manually scripted behaviors with the behaviors generated by an implementation of their model is also presented. Three metrics, and notably behavior completion time, behavior completeness, and behavior correctness, are used as a measure for assessing behavior reliability and designer efficiency.

Differently to most of the works presented previously, our solution provides a graphical user interface with different constraint types. While a few graphical interfaces have also been put forward to help people create scripts, these tools are aimed at story designers and at people who want to learn to program. Our implementation targets any person who wishes to create behaviors with no prerequisites in terms of skills and experience.

THE AUTHORING PROCESS AS A MENTAL ACTIVITY

Illustrative Example

Steven is an enthusiastic digital game player without any programming and design background. Steven is essentially an average end user and game player. He likes to buy interactive story games. Lately he has been playing an online game recently released that is based on the story "*And then there were none*" (Figure 1).

The story context is that the player has been invited to have dinner at a manor house. When the dinner is about to start, the player discovers that one of the guests has been murdered. After the introduction, the player is free to move around and interact with the other characters in the house. The goal of the game is to find out who committed the murder among the guests in the house.

After one of his many playing sessions, Steven wants to show off his creative skills by adding a set of simple scenes to the game in order to add more variability for other online players. Steven has a few scenes in mind that involve only characters with a particular personality, personal traits, emotions, and background. Authoring such experience is an example of the creative process that we wish to understand and enhance.

Figure 1. A snapshot of the commercial interactive story "And then there were none"

Using the traditional approach (i.e., hand-authoring scripts) would require a tremendous effort in authoring each one of the individual characters. Moreover, game companies usually provide toolkits to create only scenes/maps. The behavior creation is commonly carried out through scripts in a programming language, which restricts the process to people with a certain degree of programming skill. Steven does not have such a programming experience and thus he cannot add this kind of new content and experience to the game. If there was an environment that would allow more people like Steven to be involved in extending the game, it would provide them an opportunity to exercise their creative skills and make them available for other players. Consequently, the gameplay experience would improve given a larger base of users creating content and sharing it with each other.

AI Behavior Construction by Novice Users

Novice users bring little or no programming experience to the behavior construction task. However, they can employ a combination of real-world experiences, experience with other software, computers and electronic gadgets in general to understand the construction task. One of the fundamental capabilities of human cognition is the ability to relate something new to something similar that was already encountered and/or experienced.

Familiar interaction design approaches that allow users to create correspondences with a more familiar domain (Heckel, 1984) have been used in product design and have been shown to make a product or interface easier to understand. The key question, therefore boils down to identifying the right interaction design approach that can be employed as part of the presentation of the authoring process that can help novice users understand the construction process.

An AI behavior construction process requires an underlying representation formalism and vocabulary that is used to represent behaviors. Representation formalisms have been a keenly studied topic within the AI field for quite a long time. In fact, several representational formalisms have been put forward, ranging from predicate calculus through frames and scripts to semantic networks and more (Brachman & Levesque, 2004, Russell & Norvig, 2010). An AI scientist, who is familiar with the vocabulary of the underlying representational formalism, is adept at translation between the domain that is being represented and the formal language used to describe it. A novice user, however, would not have the necessary background to use the formal representational language. Several criteria can be utilized to classify different AI representation formalisms: expressivity is one of the most common among them. Clearly, the more expressive the formalism is, the easier it is to represent something with it. In our context though, the key requirements for the representation formalism and the corresponding vocabulary, are understandability and ease of use. Our idea is to forsake some level of expressivity for better understandability and ease of use. We contend that the representation vocabulary should speak the user's language and use words, phrases, and concepts familiar to the user, rather than terms more familiar to an AI scientist.

Assuming that a design solution that will enable novice authors to construct AI behaviors should provide the right interaction design approach and representation vocabulary to support them in the authoring activity, the main question that we have to address is what is a design solution that enables novice users to easily create AI behaviors for NPCs in a virtual world?

For the sake of simplicity, we focus on the construction of particular kinds of AI behaviors situated in a particular domain. Namely, we confine the scope of AI behavior creation as sequence of actions to NPCs impersonating tradesmen in a virtual world.

Representation Vocabulary

As a starting point in the process to identify the basic vocabulary and terminologies, we used an existing reactive architecture that we presented in (Corradini, Mehta, Bernsen, & Charfuelan, 2005, Zang, Mehta, Mateas, & Ram, 2007). The key property of such kind of architectures is their reactivity and responsiveness response to time-critical events, such as emergencies in a robot assistance domain, or dodging and avoiding unexpected obstacles in real-world navigation domains. This reactivity to time critical events makes them ideally suited for creating interactive avatars. Examples of reactive architectures include the Reactive Action Packages (Firby, 1987), the Gapps formalism (Kaelbling & Rosenschein, 1990), the Procedural Reasoning System (Georgeff & Lansky, 1987), the Soar architecture (Laird, 2012), and many more. A notable example of reactive architecture specifically developed for interactive believable characters is the A Behavior Language (ABL) (Mateas & Stern, 2004). ABL has been successfully used to author the central characters Trip and Grace for the interactive drama Facade (Knickmeyer & Mateas, 2005).

Based on our existing architecture and taking inspiration from the architectures previously mentioned, we identified the following terms:

- **Scene:** Provides a social context under which the behaviors of the virtual characters can be described. An example of a scene might be "a boy meeting a girl at the park for a date". Scenes may contain one or more characters, and can be of arbitrary length.
- **Behavior:** Represents the activities that the avatar can accomplish. Behaviors can be considered similar to plans in a traditional AI planning paradigm. An example of a behavior within the scene "a boy meeting a girl at the park for a date", the boy avatar could perform a "flirt with a girl" behavior. In general, behaviors consist of a sequence of other behaviors and actions so that "flirt with a girl" may contain e.g. the behavior "smile to a girl".
- **Action:** Represents the lowest level behaviors that can be performed by the avatar in the world and cannot be further decomposed. They represent the elementary units on which all behaviors are further built on.
- **Personality:** Represents a character's overall way of conducting himself. The personality is manifested in the way a specific avatar goes about performing the scenes through the kind of behaviors he can pursue, the way he carries out these behaviors and the way he conducts the basic actions.
- **Emotion:** Represents the avatar feelings in response to things happening in the environment.
- **Percept:** Represents basic events that the avatar detects from the environment.
- **Like/Dislike:** Represents things that the virtual characters likes or dislikes.

Behaviors are further associated with the following set of conditions:

- **Pre-Condition:** Detects when the behavior needs to be activated and carried out.

- **Success-Condition:** Detects when a behavior has been successful. This information is needed to identify when a behavior needs or not to be pursued again.
- **Non-Progress-Condition:** Detects when a behavior is not progressing well. If a behavior is not progressing well, the character would try to repair the situation by different means such as pursuing a different behavior to achieve the same or a similar behavior and/or have an emotional reaction in response.
- **Failure-Condition:** Detects if the behavior has failed to achieve its purpose.
- **Advancing-Condition:** Keep a check on whether the behavior is progressing well or not.

Success-condition and failure-condition are mutually exclusive. As such, they can be thought of as opposite to each other. They are used separately to provide more flexibility to the user to use either of them as needed. A similar reasoning went into providing the Non-progress-condition and the Advancing-condition as separate entities. As part of the authoring process, these conditions can be created by using basic sensory information through mathematical and logical operations that can be detected from the virtual world.

In our framework, the behaviors are defined bottom-up meaning that each behavior must resolve into a series of basic actions. Hence, during the authoring activity, a novice user must define each behavior in terms of its constituent actions. Likewise, in order for the framework to detect the conditions associated with a behavior, these conditions need to be defined in terms of the basic set of events that can be detected within the virtual world.

Next, we discuss the authoring process that we envisioned for the authors to carry out.

The Envisioned Authoring Process

The envisioned authoring activity is based on our previous works on the generation of interactive characters using reactive planning languages (Corradini, Mehta, Bernsen, & Charfuelan, 2005, Corradini, Mehta, & Charfuelan, 2005, Zang, Mehta, Mateas, & Ram, 2007). It can be carried out following three main steps.

The first step involves the creation of scenes with a hierarchical behavior structure. Scenes may contain one or more characters, and be of arbitrary length. As an example, we can consider a scene called "boy flirting with a girl in a bar". The given scene consists of two avatars: one representing a girl and the other representing a boy. The boy might be assigned a personality trait called "flirtatious" and "enthusiastic". The female character may be assigned a personality trait of being "shy". For such a scene, the behavior tree is shown in Figure 2 (top). The top-down structure linking the lowest level behaviors in terms of basic actions that can be carried out in the virtual world is also defined by the authors as shown in Figure 2 (bottom) for the behavior "Introduce". Such a behavior is made up of three actions in sequence, namely, a parameterized "Talk" action (the parameter being e.g. "I am Max"), followed by a "Smile" action, and concluded with another "Talk" action with parameter "how are you".

The second step in the authoring process consists in defining the various conditions for the behaviors as mentioned in the previous section. Various conditions for two example behaviors are shown in the Figure 3 (left). For example, the behavior "Go to a bar" is associated with an activation condition that the boy wants to chill out.

The third and final step involves linking higher level conditions with the basic level percepts available in the virtual world. For example, the condition "Girl is present in a bar" can be detected by performing an inside operation on location of bar and location of the girl as shown in Figure 3 (right). An exhaustive list of basic actions and percepts that are available to authors can be found in (Mehta, 2012).

Figure 2. (top) Graphical representation of the behavior tree structure created for a scene; (bottom) the behavior "Introduce" as linked sequence of basic actions available in the virtual world

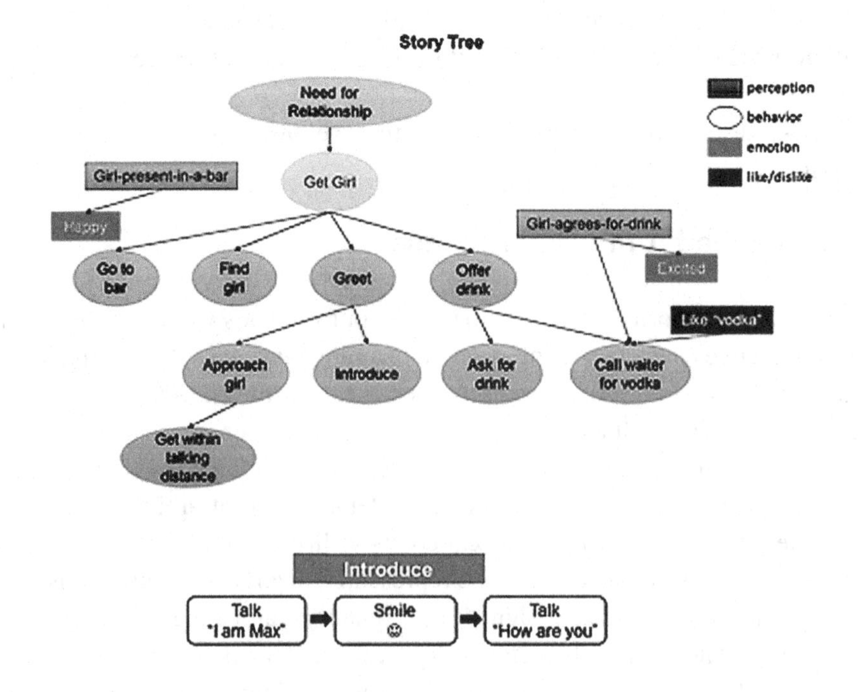

Figure 3. (left) Example of conditions for two behaviors; (right) percepts from the virtual world

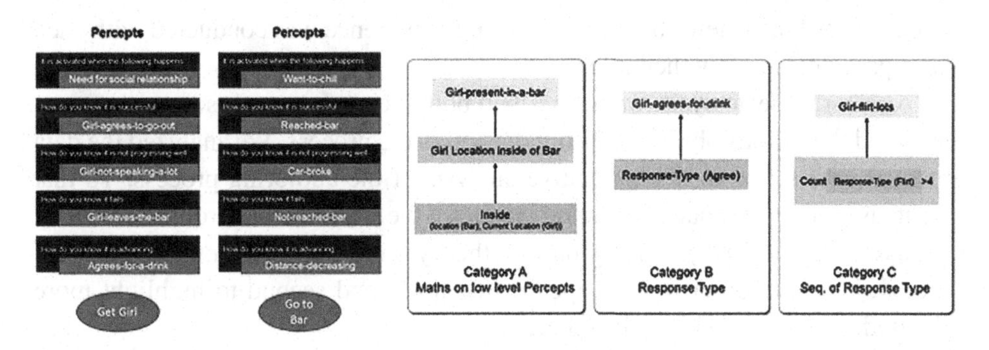

User Study: A Paper-Based Authoring Prototype

We conducted a user study in order to identify if novice users would be able to understand the representational vocabulary and carry out the authoring process following the steps listed above.

The user study was carried out using paper prototypes. We were particularly interested in figuring out which parts of the representation vocabulary are difficult for users to understand, what parts of the authoring process are difficult to conceptualize, and possibly what are the points at which the authoring process becomes tedious.

We recruited forty participants (referred to as P1 through P40): 10 females and 30 males. The participants were pre-screened to make sure they did not have any programming and design background. The participants ranged across multiple races, education levels, and ages (from 22 to 40 with an average age of 28). Each subject was paid 10 USD per hour rounded up to the nearest hour. Players signed a consent form at the beginning of the session. The authoring task was described to participants through example forms (as shown in Figure 2 and Figure 3).

Each participant was given a brief description of the scene setting. Two such scene settings were provided to the participants. The first scene was about a boy taking a girl to prom, while the second scene involved an armed robber who holds someone up. The list of basic actions and basic events that can be detected from the environment was also provided to the user. The participants were then asked to conduct the authoring activity using the paper-based prototype interface forms.

These forms were blanked out versions of the filled forms similar to those shown in Figure 2 and Figure 3.

During the authoring session, a researcher observed the player's interaction with the paper prototype and took notes. This included any unusual things such as "unable to fill a condition", or "having difficulty creating behaviors". On an average, a complete player interaction lasted approximately one hour. At the end of their interaction, subjects were asked to fill out a quantitative questionnaire. Moreover, an open-ended interview about the authoring experience was conducted with each participant immediately thereafter.

To perform the analysis, we transcribed both the user responses from the interviews and the actions observed during the authoring process. We analyzed the data obtained with focus on the qualitative analysis of the authoring process. To that extent, as qualitative analysis method we chose the well-known Grounded Theory (Strauss & Corbin, 1990). Using grounded theory principles, the transcriptions were processed, first in order to take open-ended notes and second to highlight more salient ideas expressed by each subject.

We generated codes, such as "User not able to fill precondition" or "User not able to complete scene creation process", and listed multiple user quotes for each category in an Excel spread-sheet. We then grouped related codes: for instance, all codes that referred to the problem of creating behaviors were initially grouped together into related concepts. We went through an iterative process of describing each concept and conceptually linking them to each other. We formed a single hierarchy of the concepts, and organized the same concept with a paragraph to describe each concept along with the top two to three quotes for it. This hierarchical organization of concepts served as an outline for the qualitative findings.

We also analyzed the results from the forms filled out by the users during the authoring process to identify any further issues during the authoring process and to get a better understanding of issues that emerged from the interview session.

The analysis of data collected during the study provided several key insights that can summarized into the following categories.

Conceptualization of Behaviors

The data analysis indicated that users had difficulty conceptualizing the behavior creation process as a hierarchical, top-down process. To that regards, P16 states:

..When you think of a scenario where X needs to happen, you can think of a few things that would happen, like A would happen and then B could happen and so on. Here (using the authoring interface) I am thinking of things in a different way. I am thinking like A would happen and then in order to accomplish A, I need to do these two other things and then I would have to break them down, that is kind of hard..

When users visualize the creation process, it seems very hard for them to think in a top down fashion and they seem more used to thinking in a sequential fashion. Subject P23 stated:

..When I do things, I do them in a very sequential fashion, it would be hard for me to plan out like I would do this and then it would be doing these other two things which would then involve other activities..

The difficulty of creating hierarchical structures is confirmed by high ratings extrapolated from the quantitative questionnaires. To the question of difficulty level in creating behaviors, our subjects provided an average rating of 4.03 (with 5 indicating "very difficult").

Feedback on Created Scene

Several users reported that they had difficulty seeing the results of their authoring process. Creating the behaviors on a paper prototype did not give them the opportunity to evaluate what they just created. Subjects mentioned that it would have been easier for them to create behaviors if the system could "play back a simulation of the behavior" (P13) so that they could "verify or correct their creations"(P15). Subject P2 mentioned:

..I was imagining how things would go when someone does these things, I could see that there could be some things which might not work exactly, I would say it would have been easier if things played out in a game, that would give good ideas on what works and what doesn't..

Users mentioned that they would want to see how "*the scene they had created would work in actual practice*" (P10). They felt that the scene "*might come out funny or some gesture might not work*" (P24). Looking at the avatar acting out the created behaviors would be very helpful in finding how well things would work in actual practice. Subject P19 mentioned:

..Seeing my creation being acted out would be great. That would tell me how well it works from the performance perspective..

During the authoring process, it is very hard for users to take into account all the possible situations that might occur in reality during gameplay. This task becomes even harder for novice users as they are new to carrying out the design task. Thus, it is very likely that while designing behaviors, the users would forget to take into account certain situations. It would be helpful if the users were able to observe a simulation of the created behaviors to identify immediately any issues with them.

Behavior Conditions

Some parts of the behavior condition creation process were left empty by the subjects. They had difficulty identifying things like the “progress-condition” (30 out of 40 users) and “failure-condition” (34 out of 40 users), the “advancing condition” (28 out of 40 users), and the “success-condition” (26 out of 40 users). Figure 4 (top) shows an example behavior condition form filled by one user who did not fill some of these conditions and provided only a very vague description for some of the others. Subject P24 mentioned:

..When my behavior needs to be activated was clear. I know when I want my character to perform something. Some of the other things were difficult like success

Figure 4. (top) A paper form related to behavior conditions as filled out by one user; (bottom) example of behavior condition that users had difficulty in connecting to sensory information from the environment

Behavior Template

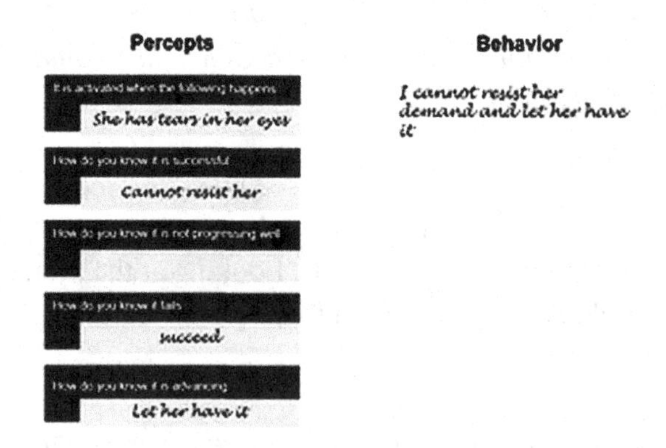

Perceptual State Detection
(relate it to the lowest level percepts)

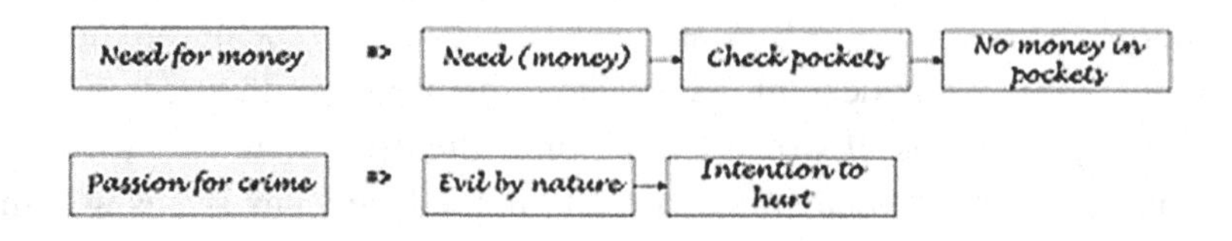

condition. I may not be able to know whether my behavior is successful or not by looking at things in the environment..

On the question of difficulty level in creating advancing condition, users provided an average rating of 4.1 (with 5 indicating "very difficult"). Similarly, for difficulty level in creating failure, success and progress condition users provided an average rating of 4.28, 4.35 and 4.55, respectively (with 5 always indicating "very difficult").

On the one hand, a simple way to address this problem would be to reduce the set of conditions. On the other hand, we are aware that this also takes away some of the expressivity of the system. Having novice users as target group, we decided to reduce the set of conditions in order to make the authoring process more accessible for them.

Additionally, subjects had difficulty linking behavior conditions with lowest level sensory information that was available in the environment. As P32 mentioned:

..Sometimes I could say when my behavior fails. But in general I wasn't sure if I could see these things in terms of what that was available in the environment..

The perceptions could become very abstract and users had difficulty relating it to sensory information that can be detected from the environment. Figure 4 (bottom) shows an example where "need for money" is broken down to "money in pocket" which a subject could not relate to the sensory information describing "no money in someone's pocket". Apart from that, in order to concretely connect the sensory information obtained from the world to higher level behavior conditions, mathematical operators needed to be defined in certain situations and contexts. Most of the times, subjects skipped this process. Subject P6 reported that "..creating mathematical formulas was the most difficult part..".

Similarly, subject P25 explicitly expressed that authors need technical skills in order to carry out this step:

..Defining it mathematically in terms of the lowest level things that are happening in the environment was unclear. I would think someone who has the technical skills would be able to understand it better than me..

One possibility would be to remove the notion of complex conditions defined using mathematical operators over the basic sensory information could be to provide a base set of parameterized conditions that can be used individually or combined together to result in more complex conditions.

SECOND MIND: A SYSTEM FOR BEHAVIOR AUTHORING

Rationale and Context

The previous user study explored a design solution that would enable novice users to successfully author AI behaviors. That study provided, among others, the following key insights and design directions for creating the authoring interface:

- The process to link behavior constraints to lower level percepts must be as simple as possible; the conditions themselves must be limited, clearly understandable and should be expressed with an easy-to-use formalism.
- The authoring task should be presented as a sequential process and not rely on deep hierarchical structures.
- Authors need continuous feedback on the authoring activity so that they can see the results of their ongoing process and thus promptly identify possible problems.

On one hand, reducing the set of conditions takes away some of the expressivity of the system, but on the other hand, it allows keeping the authoring process simple. Conditions help the avatar identify when, what, and how a certain behavior must be carried out.

We express conditions using what we refer to as triggers. We provide a set of triggers that can be used individually or combined together to define more complex constraints. An example of trigger is "User Clicks on (button label)", which can be used when the player wants the avatar to perform a behavior in response to a button click ("button label" is a placeholder that must be provided during the authoring process). Many other triggers are also provided such as "Avatar Close to (object name)", "Avatar Touches (object name)", "Player Close to (object name)", "Player Touches (object name)", "Behavior Ends (behavior name)", "Pause after Behavior Ends (behavior name, time)", "Run Event (event name)", and so on. Triggers can be composed using logical operators. For instance an "Trigger And(A, B)" can be used to specify a behavior for the avatar to perform when both a given trigger A and trigger B are specified.

In order to present the behavior creation task as a sequential process, we applied the notion of timelines. Timelines indicate a left-to-right sequential sequence, where each event occurs in chronological order. Various storyboard-authoring tools like Adobe Director and Apple's Garageband Timelines as well as AI-based interactive story authoring tools like Scribe (Medler & Magerko, 2006), U-Create (Sauer, Osswald, Wielemans, & Stifter, 2006) and Viper (Agamanolis & Bove, 2003) employ timelines as an interaction design approach. We use timelines to represent the ordering of story segments in a left-to-right sequence while also allowing story segments to be re-ordered on the timeline. In order to help the user conceptualize the task concretely, the authoring activity along the timeline is presented in a mind-like shape (Figure 5).

Eventually, in order to comply with the need to provide a continuous feedback to the user, we connect our authoring environment to an online virtual world named Second Life (Rymaszewski, 2009) where avatar's behaviors can be executed and displayed graphically in 3D. In Second Life, users can create, design and build their own clothes, buildings, artifacts, landscape, pets and much more. We chose Second Life as platform to display 3D-based user-generated content because it has a large user base, it allows for easy connection to external controllers through an API, and it provides a rich set of basic actions for the avatar in the form of emotions and physical actions. Using our authoring interface Second Mind, authors can create behaviors, execute them in Second Life, and immediately see the effects of their authored artifacts.

The Authoring Process in Second Mind

In order to use Second Mind, the user needs to create an account where to store, among other things, his/her Second Life login credentials. Once the account is created, the user is guided through a tutorial, which explains the terminology used in the authoring interface (like triggers, scenes, behaviors, actions, etc.). The user then can define one or more characters in Second Mind and connect each of them to a different avatar in Second Life.

At this stage, the user is ready to create a scene and its corresponding behaviors. A scene provides a social context such as, e.g. a furniture shop where the avatar acts

Figure 5. (left) Authoring of a greeting behavior as a left-to-right sequence of the basic actions "talk (text)", "salute" and "talk (text)" on the timeline; (right) Behaviors are shown in a sequence using the cover-flow approach

as a shopkeeper whose goal is to sell furniture items. The author creates a description of the scene by entering some information like the name, a description and by selecting one of the characters created in earlier steps (see Figure 6). The user also associate constraints a limited set of constraints. For instance, a constraint to apply to the scene in the shopkeeper context could be "likeable" or "salesman skills".

Beside the contextual constrains, all scenes are associated by default with two additional general constraints, namely "overall experience" and "avatar performance". The author uses a set of sliders to specify matching values on a 0 to 5 point scale for each of the constraints s/he wants to set. Moreover, the user can associate semantic tags to the scene. Now, avatar behaviors can be added within the scene.

Similarly, to the scene creation process, the author can create a complex behavior from a list of simple elementary behaviors. To define a new behavior, the author needs to enter a name that uniquely identifies it followed by the list of behaviors that make it up and by the specification of the constraints that characterize the behavior as a whole.

Using a set of sliders, the author can specify the values for the constraints. For example, in order to have an "enthusiastic" and "pushy" shopkeeper for a "greeting behavior", the user has to select first the corresponding values for those two constraints that give an idea of the avatar personality. These values influence the way (e.g. speed, spatial extension, etc.) a certain behavior is rendered. The user then can assign triggers to the behavior in order to specify the context in which it should be carried out by the avatar. The process is iterative, in the sense that each of the steps can be applied to define an even more complex behavior as well as to each single behaviors that are used to define the complex one. During the creation of a specific behavior, the author can also define a constraint for which it is the avatar that initiates the interaction. Once a behavior is created, the user can immediately play it out in Second Life and test its 3D graphical representation.

Figure 6. (left) Screenshot of the web version of Second Mind; (right) initial steps for the creation of a scene

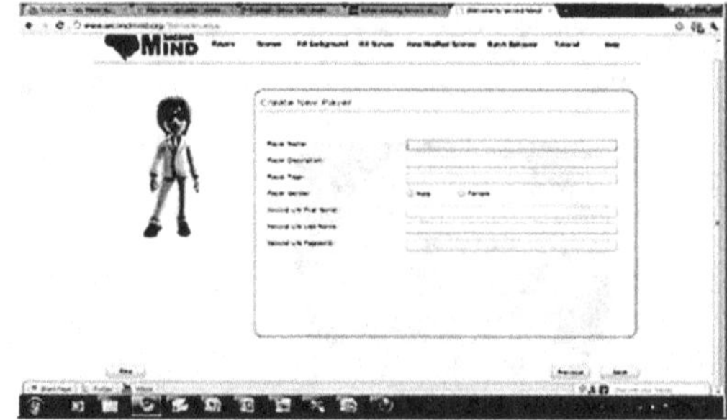

In order to interact with an avatar in Second Life, the player must click on an attachment on the avatar's left shoulder (see Figure 7). Upon clicking, the player is presented with a GUI containing buttons labeled with the names of the scenes that have been previously authored for the avatar (see Figure 7). Each of these buttons represents one "User Clicks on (button label)" trigger that is associated with the behaviors of a scene. To interact with the avatar, the player must simply click on a button.

System Architecture

With a basic understanding of the authoring activity, we can delve into the system architecture. As shown in Figure 8, Second Mind architecture has four core components and one application-specific component. The four core components are:

- **Second Mind Database:** A MS-SQL server based database that stores scenes, players, behavior elements such as triggers, basic action, etc.
- **Behavior Execution Engine:** An execution engine that loads the behavior elements and executes them in the external world. The behavior execution engine connects to the external world through the interface layer that is specific to the world that Second Mind is connecting to.

Figure 7. The player interaction with an avatar can be initiated by clicking on the device carried by the avatar on his left shoulder; clicking on the device causes a graphical interface to pop up with various scenes to choose from

Figure 8. Overview of Second Mind overall architecture

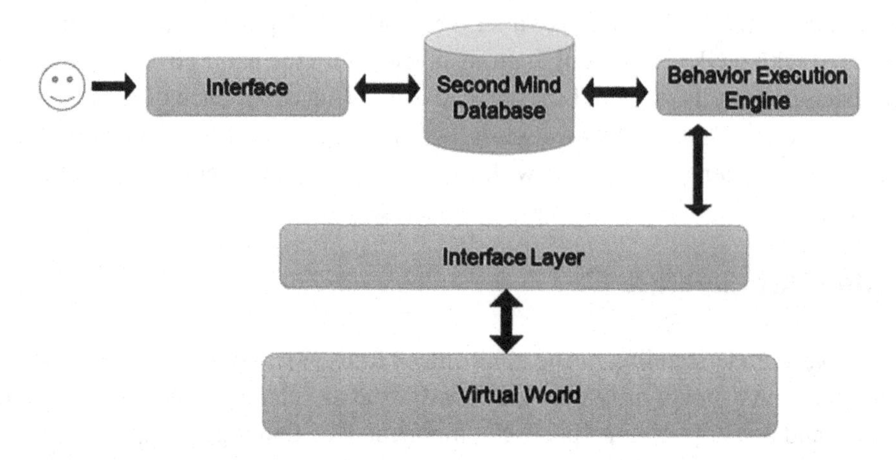

- **Graphical Interfaces:** Provides the interfaces for creating user account, scenes and behaviors as described in the previous section.
- **Behavior Recommendation Engine:** Provides suggestions during the behavior-authoring task.

In addition, in order to apply Second Mind to a given graphical world, one application-specific components is required: are:

- **Interface Layer:** The architecture of Second Mind is domain-independent and can be connected to different virtual world simulations via a custom interface layer. The interface layer provides the ability to connect Second Mind to the simulation platform, such as Second Life or other virtual world. The interface layer requires the definition of percepts, actions, and other application-specific elements along with the APIs to connect Second Mind to the simulation.

Behavior Execution Engine

The behavior execution engine (see Figure 9) loads the behaviors corresponding to the currently active scene, identifies the behavior(s) to carry out based on the current perceptual information from the world, and sends the information about these behaviors to the graphical world for execution through the interface layer.

The behavior execution engine can be correspondingly broken down into three main modules: interface:

Figure 9. Overview of behavior execution engine architecture

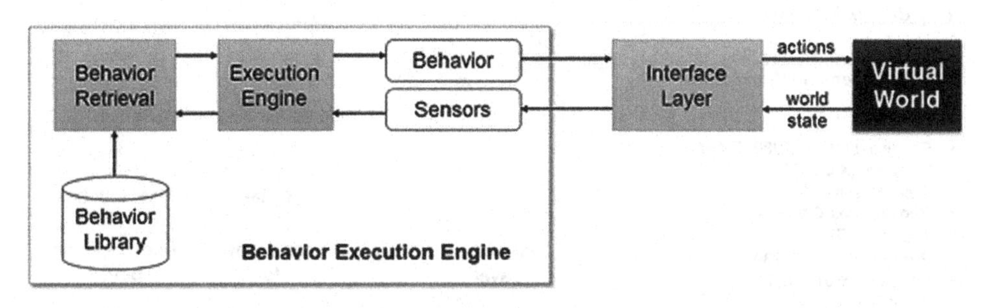

- **Behavior Library:** It is part of the Second Mind database that holds behavior elements for a specific application domain. For example, if the application domain were Second Mind in an educational setting, the library would contain personalities corresponding to teachers, students, etc. If instead the application domain were a household robot, the library would contain personality elements corresponding to cooking food, cleaning the house, etc.
- **Behavior Retrieval:** It is responsible for loading all the behaviors corresponding to the active scene from the behavior library. The active scene is the scene that the player selected when s/he was presented with all possible authored scenes associated with a certain avatar.
- **Execution Engine:** It receives the currently active domain independent sensors from the interface layer and checks the triggers for the behaviors corresponding to the active scene to see if any of the behaviors were activated. It also takes the current active behavior and sends it through the interface layer to the virtual world for the avatar to perform it.

Interface Layer

The Interface layer parses the domain dependent perceptual state into a set of domain independent sensors (see Figure 10). The domain dependent perceptual state from Second Life contains the following items:

- **Object Info:** Contains the information about various objects in the environment such as their current position, rotation, type (whether an avatar or a passive object), region to which the object belongs, etc.
- **User Button Click Info:** Contains information regarding button clicks made by the player while interacting with the avatar.
- **Event Info:** Contains the information regarding various events happening in Second Life such as the player moving around in the environment, the player

Figure 10. The interface layer parses the world state in second life into domain independent sensors

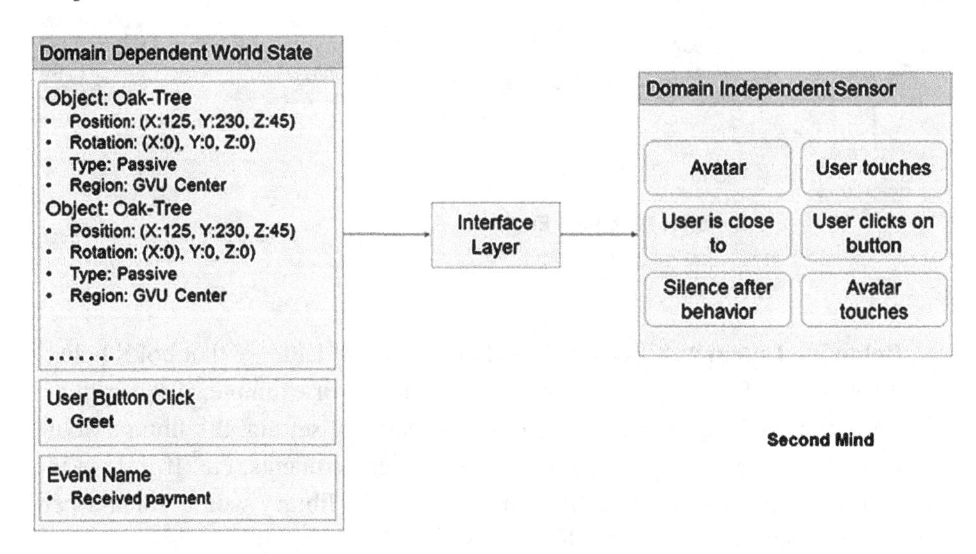

touching an object, etc. Objects in the environment can send various events to communicate with the avatar. If there are behaviors that have "Event happens (event name)" as triggers, avatar can react to events with name "event name".

The world specific information is parsed into a set of domain independent sensors that correspond to various triggers. This information is then sent to the execution engine, which checks the sensors against the triggers corresponding to the current set of behaviors to identify if any new behavior(s) has become activated. The interface layer module also receives the action corresponding to the active behavior and sends it to the virtual world. The interface layer provides an abstraction layer and maps the action name received from the execution engine to virtual world specific action name. For example, an action "*laugh*" in Second Mind is mapped to actions that express "*express_laugh*", which is the specific name used by Second Life virtual world. Table 1 shows the mapping of a few action names in Second Mind with the corresponding names in the virtual world Second Life.

Initially, in order to connect Second Mind to Second Life, we modified an open source version of Second Mind clients. We soon realized though that not all Second Mind users are willing to installed a modified version of Second Life as well. Therefore, we instead created scripts in Second Life's proprietary scripting language named Linden Scripting Language (LSL). Those scripts make it possible to send

Table 1. Action names in SM and corresponding virtual world action names in second life

Domain Independent Action Name	Virtual World Action Name
Laugh	express_laugh
Shrug	express_shrug_emote
Bow	avatar_hold_bow

Figure 11. Scripts organization for the connection of second mind and second life

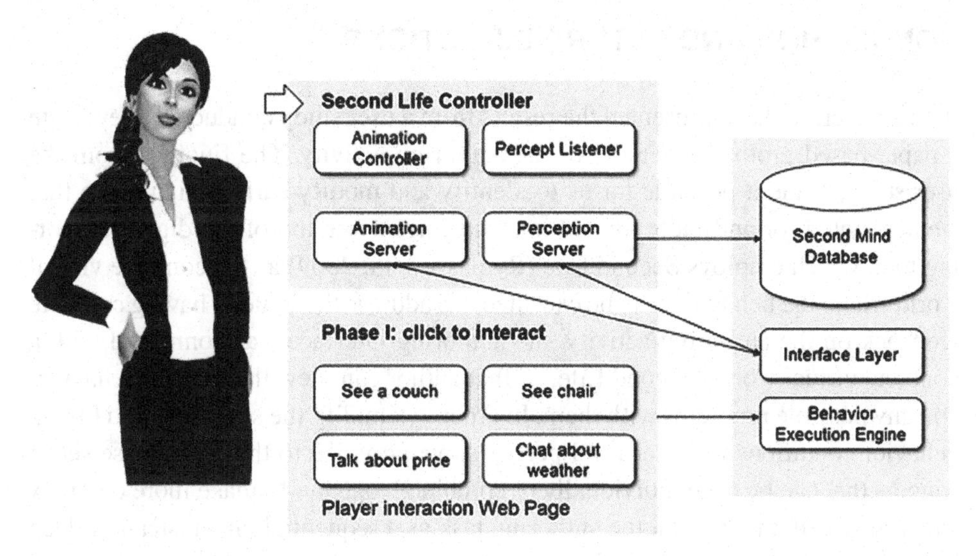

and to receive information through the Second Mind interface layer from/to Second Life for controlling the avatar with any genuine Second Life client (Figure 11).

The collection of scripts is packaged in a component called Second Mind Controller and distributed through the Second Life marketplace. Second Life marketplace is a social web portal from Second Life where all the items from various vendors are listed and available for purchase. Second Mind controller is listed on the Second Life marketplace and is available for free. In order to have their avatar controllable through Second Mind, players need to download it from the Second Life marketplace. Purchasing items from Second Life marketplace is a common activity performed by Second Life users. Once users have bought the items from Second Life marketplace. Second Mind controller is downloaded automatically onto their avatars, which they can then attach to their avatar in Second Life.

Behavior Recommendation

During the behavior authoring process, once the name of the behavior is entered, the user is presented with a list of recommended behaviors that he can use as a starting point. The behaviors are suggested based on similarity between the behavior name specified for the currently authored behavior and existing behaviors in the database. The semantic similarity is calculated using a WordNet-based algorithm (Simpson & Dao, 2014).

CONCLUSION AND FUTURE DIRECTIONS

In this paper, we have presented the results from a user study conducted to evaluate a paper-based prototype version of the authoring activity. The findings from the user study, made it possible for us to identify and modify early design flaws that we accounted for and incorporated into the current version of the digital authoring tool, which employs Second Life (Rymaszewski, 2009) as the concrete virtual world where the behaviors can be executed. To address the issue of having concrete feedback on the authoring activity, the authoring interface was connected with a concrete virtual world, Second Life, so that author can view the created behaviors and any possible problems with them. In order to simplify the steps required to link behavior conditions to lower level percepts, we provided to the user a base set of triggers that can be used individually or combined together to make more complex triggers. In order to present the authoring task as a sequential one, instead of deep hierarchies, we used a left to right sequential action creation approach represented on a timeline. Eventually, we outlined the authoring steps necessary to create and run a behavior along with the system architecture and its four core components.

We evaluated the Second Mind approach to check whether it can be understood and easily used by non-programmers. In order to evaluate our approach, we invited 65 subjects to create a certain set of behaviors for an avatar (e.g. act as a shopkeeper) on a Second Life island using the authoring system. We divided the subjects into three broad categories: User Interface designers, non-programmers, and programmers (with experience in Java). We transcribed data from the subjects' responses to interviews as well as from the notes taken while observing subject actions during the authoring process. To analyze this vast quantity of data collected from the user study, we adopted a combination of qualitative and quantitative analysis. Specifically, we employed again a Grounded Theory (Strauss & Corbin, 1990) approach to perform the qualitative analysis. A detailed report on the evaluation of the system is outside the scope of this paper. However, for the sake of completeness, the evaluation results show that non-programmers were able to easily create behaviors using the scaffolding provided by Second Mind. Moreover, no relevant statistical difference was noticed in the quantitative measures used to asses authoring skills among the different subject categories. This seems to indicate that designers, programmers, and non-programmers perform very similarly while authoring behaviors with Second Mind.

There is much work left to do in Second Mind, some of which is research, some of which is in development. There are still numerous open questions that form possible avenues of future directions for this research. Authoring support through demonstrating behaviors is something we are considering. Learning from demonstration is also a point we would like to investigate. Human learning is often accelerated by observing a task being performed or attempted by someone else. These capabilities of the human brain are also evident in computer games where players go through a process of training and imitating experienced players. These results have inspired researchers in artificial intelligence to study learning from imitation techniques. By observing an expert's actions, new behaviors can quickly be learnt and then successfully performed.

ACKNOWLEDGMENT

We would like to thank Ms. Maria Kikidou for the integral role she played in helping us with the editing process and the creation of the images in this book chapter.

REFERENCES

Adams, E. (2012). *Fundamentals of game design* (2nd ed.). New Riders.

Agamanolis, S., & Bove, V. M. (2003). Viper: A Framework for Responsive Television. *IEEE MultiMedia, 10*(1), 88–98. doi:10.1109/MMUL.2003.1218260

Allbeck, J., & Badler, N. (2004). Representing and parameterizing behaviors. In H. Prendinger & M. Ishizuka (Eds.), *Life-Like Characters: Tools, Affective Functions and Applications*. Springer. doi:10.1007/978-3-662-08373-4_2

Avradinis, N., Panayiotopoulos, T., & Anastassakis, G. (2013). Behavior believability in virtual worlds: Agents acting when they need to. *SpringerPlus, 2*(1), 1–11. doi:10.1186/2193-1801-2-246 PMID:23853745

Aylett, R. S., Louchart, S., Dias, J., Paiva, A., & Vala, M. (2005). FearNot! - An experiment in emergent narrative. *Proceeding of the International Conference on Intelligent Virtual Agents (IVA '05).*

Bates, J. (1994). The role of emotion in believable agents. *Communications of the ACM*, *37*(7), 122–125. doi:10.1145/176789.176803

Brachman, R., & Levesque, H. J. (Eds.). (2004). *Knowledge Representation and Reasoning*. Morgan Kaufmann.

Cassell, J. (2001). Embodied conversational agents: Representation and intelligence in user interfaces. *AI Magazine*, *22*(3), 67–83.

Champandard, A. (2007). Behavior trees for next-gen AI. *Proceedings of the Game Developers Conference*.

Coman, A., & Munoz-Avila, H. (2012). Plan-based character diversity. *Proceedings of the AAAI Conference on Artificial Intelligence and Interactive Digital Entertainment (AIIDE'12)*.

Corradini, A., Mehta, M., Bernsen, N.-O., & Charfuelan, M. (2005). Animating an interactive conversational character for an educational game system. *Proceedings of IUI*. ACM Press. doi:10.1145/1040830.1040872

Corradini, A., Mehta, M., & Charfuelan, M. (2005). Interacting with an animated conversational agent in a 3d graphical setting.*Proceedings of Workshop on Multimodal Interaction for the Visualization and Exploration of scientific data at ICMI*.

Dragert, C., Kienzle, J., & Verbrugge, C. (2012). Statechart-based game AI in practice. *Proceedings of the AAAI Conference on Artificial Intelligence and Interactive Digital Entertainment (AIIDE'12)*.

Firby, R. J. (1987). An investigation into reactive planning in complex domains. *Proceedings of the AAAI*.

Georgeff, M. P., & Lansky, A. L. (1987). Reactive reasoning and planning.*Proceedings of the AAAI*.

Gillies, M., & Ballin, D. (2004). Integrating Autonomous Behavior and User Control for Believable Agents. *Proceedings of the Third International Joint Conference on Autonomous Agents and Multiagent Systems*.

Gluck, K. A. (2010). Cognitive architectures for human factors in aviation. In E. Salas & D. Maurino (Eds.), *Human Factors in Aviation* (2nd ed.; pp. 375–400). New York, NY: Elsevier. doi:10.1016/B978-0-12-374518-7.00012-2

Gomes, P. F., & Jhala, A. (2013). AI authoring for virtual characters in conflict, *Proceedings of the AAAI Conference on Artificial Intelligence and Interactive Digital Entertainment (AIIDE'13)*.

Gratch, J., Hartholt, A., Dehghani, M., & Marsella, S. C. (2013). Virtual Humans: A new Toolkit for Cognitive Science Research. *Cognitive Science*.

Heckel, P. (1984). *The Elements of Friendly Software Design*. Alameda, CA: Sybex Inc.

Isla, D. (2005). Handling complexity in the Halo 2 AI. *Proceedings of the Game Developers Conference*.

Kadlec, R., Toth, C., Cerny, M., Bartak, R., & Brom, C. (2012). Planning is the game: action planning as a design tool and game mechanism. *Proceedings of the AAAI Conference on Artificial Intelligence and Interactive Digital Entertainment (AIIDE'12)*.

Kaelbling, L. P., & Rosenschein, S. J. (1990). Action and planning in embedded agents. *Robotics and Autonomous Systems*, *6*, 35–48.

Kelly, J. P., Botea, A., & Koenig, S. (2008). Off-line planning with hierarchical task networks in video games. *Proceedings of the AAAI Conference on Artificial Intelligence and Interactive Digital Entertainment (AIIDE'08)*.

Knickmeyer, R., & Mateas, M. (2005). Preliminary Evaluation of the Interactive Drama Façade. *Proceedings of CHI 2005*. ACM Press.

Laird, J. E. (2008). Extending the Soar cognitive architecture.*Proceedings of the Artificial General Intelligence Conference*.

Laird, J. E. (2012). *The Soar Cognitive Architecture*. MIT Press.

Langley, P., & Choi, D. (2006). A unified cognitive architecture for physical agents. *Proceedings of the twenty-first AAAI Conference on Artificial Intelligence*. Boston: AAAI Press.

Langley, P., Laird, J. E., & Rogers, S. (2009). Cognitive architectures: Research issues and challenges. *Journal of Cognitive Systems Research*, *10*(2), 141–160. doi:10.1016/j.cogsys.2006.07.004

Li, B., Thakkar, M., Wang, Y., & Riedl, M. O. (2014). Data-driven alibi storytelling for social believability. *Proceedings of the Foundations of Digital Games Workshop on Social Behavior in Games*.

Magerko, B., Laird, J. E., Assanie, M., Kerfoot, A., & Stokes, D. (2004). AI characters and directors for interactive computer games. *Proceedings of the sixteenth innovative applications of artificial intelligence conference*.

Mateas, M., & Stern, A. (2003). Facade: an experiment in building a fully-realized interactive drama. *Game Developer's Conference: Game Design Track.*

Mateas, M., & Stern, A. (2004). A Behavior Language: Joint Action and Behavioral Idioms. In H. Prendinger & M. Ishizuka (Eds.), Life-like Characters. Tools, Affective Functions and Applications. Springer.

Medler, B., & Magerko, B. (2006). Scribe: A general tool for authoring interactive drama. *Proceedings of the International Conference on Technologies for Interactive Digital Storytelling and Entertainment.* doi:10.1007/11944577_14

Mehta, M. (2012). *Construction and Adaptation of AI Behaviors for Computer Games.* (PhD thesis). Georgia Tech.

Nuxoll, A. M., & Laird, J. E. (2007). Extending cognitive architecture with episodic memory.*Proceedings of the twenty-second AAAI Conference on Artificial Intelligence.*

Rickel, J., Marsella, S., Gratch, J., Hill, R., Traum, D., & Swartout, W. (2002). Toward a new generation of virtual humans for interactive experiences. *IEEE Intelligent Systems, 17.*

Riedl, M. O., & Zook, A. (2013). AI for game production. *Proceedings of the IEEE Conference on Computational Intelligence in Games.*

Russell, S. J., & Norvig, P. (2010). *Artificial Intelligence: A Modern Approach* (3rd ed.). Prentice Hall.

Rymaszewski, M. (2009). *Second Life: The Official Guide* (2nd ed.). Sybex Inc.

Sauer, S., Osswald, K., Wielemans, X., & Stifter, M. (2006). U-create: Creative authoring tools for edutainment applications. *Proceedings of the International Conference on Technologies for Interactive Digital Storytelling and Entertainment.* doi:10.1007/11944577_16

Schwab, B. (2009). *AI Game Engine Programming* (2nd ed.). Course Technology.

Simpson, T., & Dao, T. (2014). *WordNet-based semantic similarity measurement.* Retrieved on March 03, 2015 from http://www.codeproject.com/KB/string/semanticsimilaritywordnet.aspx

Strauss, A., & Corbin, J. (1990). Basics of Qualitative Research: Grounded Theory Procedures and Techniques. *Sage (Atlanta, Ga.).*

Sun, R. (2007). The importance of cognitive architectures: An analysis based on CLARION. *Journal of Experimental & Theoretical Artificial Intelligence, 19*(2), 159–193. doi:10.1080/09528130701191560

Tomai, E., & Flores, R. (2014). Adapting in-game agent behavior by observation of players with learning behavior trees. *Proceedings of the International Conference on the Foundations of Digital Games.*

Umarov, I., Mozgovoy, M., & Rogers, P. C. (2012). Believable and effective AI agents in virtual worlds: Current state and future perspectives. *International Journal of Gaming and Computer-Mediated Simulations*, *4*(2), 37–59. doi:10.4018/jgcms.2012040103

Yu, H., & Riedl, M. O. (2014). *Personalized interactive narratives via sequential recommendation of plot points*. IEEE Transactions on Computational Intelligence and Artificial Intelligence in Games.

Zang, P., Mehta, M., Mateas, M., & Ram, A. (2007). Towards runtime behavior adaptation for embodied characters. *Proceedings of the IJCAI*, 1557-1562.

Zhao, R., & Szafron, D. (2014). Using cyclic scheduling to generate believable behavior in games. *Proceedings of the AAAI Conference on Artificial Intelligence and Interactive Digital Entertainment (AIIDE'14).*

Chapter 4
Learned Behavior:
Enabling Believable Virtual Characters through Reinforcement

Jacquelyne Forgette
Western University, Canada

Michael Katchabaw
Western University, Canada

ABSTRACT

A key challenge in programming virtual environments is to produce virtual characters that are autonomous and capable of action selections that appear believable. In this chapter, motivations are used as a basis for learning using reinforcements. With motives driving the decisions of characters, their actions will appear less structured and repetitious, and more human in nature. This will also allow developers to easily create virtual characters with specific motivations, based mostly on their narrative purposes or roles in the virtual world. With minimum and maximum desirable motive values, the characters use reinforcement learning to drive action selection to maximize their rewards across all motives. Experimental results show that a character can learn to satisfy as many as four motives, even with significantly delayed rewards, and motive changes that are caused by other characters in the world. While the actions tested are simple in nature, they show the potential of a more complicated motivation driven reinforcement learning system. The developer need only define a character's motivations, and the character will learn to act realistically over time in the virtual environment.

DOI: 10.4018/978-1-5225-0454-2.ch004

INTRODUCTION

In creating virtual environments, whether for entertainment or serious purposes, believable virtual characters have long been a challenge to developers and users. For large immersive virtual worlds, users have come to expect the presence of computer-generated characters as an essential element of the experience. These characters must act in a way that is reasonable and consistent with their personae, the world, and the context in which they are operating; in other words, they must act believably.

The requirements for achieving a believable virtual character are extensive and non-trivial, but the rewards in terms of immersion and satisfaction make this an important problem to solve (Bailey & Katchabaw, 2008). As written in (Rizzo et al, 1997), characters "are considered believable when they are viewed by an audience as endowed with thoughts, desires, and emotions, typical of different personalities". This definition of believability is not necessarily a definition of character but an illusion of life, permitting the audience's suspension of disbelief and acceptance of the virtual as real, at least for a time. While this idea of believability has long been studied in other disciplines, its difficulty in relation to virtual environments invariably comes back to the required interactivity of a virtual character (Loyall, 1997). This level of interactivity requires autonomous and flexible behavior that is not defined a priori. Traditionally, character behavior has been driven by static scripting, pre-coded expert knowledge, or search algorithms. These approaches, however, are simply not sufficient to enable believability; only with narrow problem domains are they sufficient as a solution (Rabin, 2010). Consequently, a more robust cognitive solution is clearly necessary.

In this chapter, we present an approach to character believability enabled by reinforcement learning (Mitchell, 1997) in which behavior is guided by a model of a character's own unique motivations. In this way, developers are able to develop virtual characters by directly referencing character traits as defined by the underlying narrative or the needs of the virtual environment. The characters will, in turn, learn what actions they must take to benefit their respective motives, in real-time. Not only does this result in interesting and realistic unscripted behavior tuned to the particulars of each individual character, but it does so without the complexity and expense of an extensive and exhaustive programming exercise, as appropriate behaviors can be learned without explicit coding.

The theory behind this approach is based on work done by Reiss (Reiss, 2004; Reiss & Havercamp, 1998), in which motives are the reason that causes a person to initiate and perform a voluntary behavior, and unusually strong or weak desires are used to characterize an individual. In Reiss's model, motives capture the basic desires of an individual, including power, curiosity, independence, status, social contact, vengeance, honor, idealism, physical exercise, romance, family, order, eating, acceptance, tranquility, and saving. Studies have shown that this theory can be used to describe diverse higher-order aspects of behavior such as religious beliefs (Reiss, 2000), athleticism (Reiss et al, 2001), and lack of scholastic achievement (Reiss, 2004). In general, motives are said to affect perception, cognition, emotion, and ultimately the resultant behaviors. A person must have a motive to perform any particular action, even if that person is not consciously aware of the motive, and so this model is a reasonable basis for believable virtual characters (Bailey et al, 2012). In this approach, when a particular motive is far from satisfied, a character will take appropriate actions that address this shortfall. Similarly, when a motive is fulfilled more than desired, the character will take actions that compensate and avoid actions that contribute to its further advancement.

To enable this approach, actions to be selected by virtual characters are fully defined in terms of how they affect other objects, characters, and places in the game world, including how they influence a character's set of motivations. In this case, the developer plays the role of teacher by making complicated actions available to the characters, enabling a wide variety of rich and compelling scenarios to be engaged by the characters. Through exploration, characters will learn the effects of actions in various contexts, including whether they support, work against, or have no impact on their particular motivations. Over time, characters will become better at exploiting their acquired knowledge and balancing their motivations, keeping them satisfied, but not over-stimulated. Doing so will have the appearance to users that the characters are acting independently, according to their unique motives, thus creating an illusion of believability.

This chapter delves into both theoretical and practical aspects of using reinforcement learning to create believable virtual characters for modern virtual environments and applications, such as computer and video games, for example. This chapter is sectioned as follows:

- **Background:** Introducing the reader to the study of personality, desire, motivation, and their applications to creating believable virtual characters.
- **Designing for Believability:** An examination of how to design and structure virtual characters to combine models of motivation with machine learning to create believable behavior.
- **A Prototype System:** Practical aspects of implementing this design, using our own software system as an example.
- **Case Studies:** With implementation in hand, we run our believable virtual characters through a number of increasingly complex and realistic scenarios, assessing their behavior both quantitatively and qualitatively. These studies

leverage a variety of reinforcement learning algorithms (Sarsa, Q-learning, and Dyna-Q), with different action selection methods (ε-greedy and Softmax), and a large number of configurable parameters to assess behavior and performance.

- **Conclusion:** A summary of the chapter, and discussion of future directions for work in this area.

BACKGROUND

Many approaches have been taken in the direction of believable virtual characters. This is only natural as the requirements for character believability are both diverse and steep (Loyall, 1997). Such requirements include elements such as personality, emotion, self-motivation, social relationships, consistency, the ability to change, and the ability to maintain an "illusion of life", through having goals, reacting and responding to external stimuli, and so on (Loyall, 1997).

As a result, some techniques specifically target key aspects of believability, including social behavior (Bailey et al, 2012; Guye-Vuilleme & Thalmann, 2000; Prada, 2012; Rizzo et al, 1997), personality traits (Bevacqua, 2010; Doce, 2010; Read et al, 2010) and even emotional responses (Marinier & Laird, 2008). Other works explore perception; focusing on how a character can detect human user intentions (Doirado & Martinho, 2010), and how to project emotion and intention to the user (Ruijten et al, 2013). Much of this work focuses on systems, modeling, and giving appropriate knowledge, rules, and other supports to characters to enable believability. The approach described in this chapter, in contrast, provides a framework by which a character can learn or acquire this on its own.

This section provides a brief overview of personality psychology, and desire theory, along with related work using motivation and learning in this domain.

Theories of Personality in Psychology

The mission behind personality psychology theory is to use mechanisms to explain a person's thought patterns, emotions, and behaviors. The most notable problem in personality psychology is the fragmented nature of its many solutions. Integrating the specialized research topics, like developmental, social, cognitive, and biological, into a unifying understanding of a person, is a particularly difficult challenge (Funder, 2001).

The historical basis of personality psychology includes many different approaches, resulting in different paradigms like psychoanalytic, trait, behaviorist, and humanistic. Through developments in these main fields, other new paradigms were created. The trait and behaviorist paradigms evolved into social-cognitive and biological approaches. There is also evolutionary psychology, a relatively new paradigm (Funder, 2001).

Regardless of the paradigm followed, in order to empirically study personality, three elements are key: person, situation, and behavior (Funder, 2001). The personality triad is imbalanced, because proper attention has not been given to the situation variable. While the person variable has been explored by all the various personality paradigms, the situation variable has not received the same attention (Funder, 2001). While this is a problem in real-life situations, it should not be a problem in virtual environments. In virtual environments, the situation consists of sensory representation of the virtual world as it relates to the character in question, and this information is generally readily available from the software supporting the virtual world. The problem then becomes how to define the person, or in this case, character.

All the research into personality, while diverse, can be grouped roughly into two main categories. The first studies the statistical structure of personality across large groups of people, and is also called inter-individual personality structure. The second group, intra-individual personality structure, studies the dynamics of personality within a person, to explain the processing systems responsible for their personality (Read et al, 2010).

With this in mind, numerous psychological models have been leveraged to define personality for use in believable virtual characters. Such models include the Myers-Briggs Type Indicator (Myers et al, 1998) and the Five Factor/OCEAN model (Tupes & Cristal, 1961), as well as meta-approaches to blend elements of various models together, as discussed in (Bailey et al, 2012), for example. Directly driving behavior and action selection using these models can be challenging, however, as it is not always clear how different elements of personality contribute to the process (Acton, 2009).

As motivations are more easily linked to behavior, they are perhaps a more natural approach for virtual characters (Acton, 2009). While they do not directly model a character's personality, at least some personality traits can be approximated or inferred by their motivations and resultant actions. For example, traits of introversion and extroversion are common in personality models, including those listed above; such a notion can be reasonably captured through a socialization motivation, as found in models such as the one proposed by Reiss (Reiss, 2004), as discussed further in the next section.

Desire Theory and Motivation

Voluntary behavior is influenced most heavily by a person's motives. The motives themselves affect a person's perception, cognition, emotion, and behavior. Motives are said to be either intrinsic, or extrinsic depending on their purpose. Psychologists have defined intrinsic motivation as being moved to do a particular behavior, because it is inherently enjoyable. Extrinsic motivation is defined as wanting to do a particular action, because of a specifically desired outcome.

Most people have heard of the concept "ends versus the means", which divides a person's motives based on the purpose of performing the behavior (Aristotle et al, 1976). An end motive is when an individual will perform behavior for no apparent reason, other than it is what they desire to do. A motive is considered a means if it is only needed to fulfill another end motive. For instance, a student may be motivated to get high grades in order to please their parents. In this example, the motive to get high grades is a means, and the end motive is to please their parents. A behavior chain is a series of means that ultimately, by definition, finishes with an end (Reiss, 2004).

There are numerous ways in which a person can seek to accomplish an end. The number of possible means is also large, constrained only by the imagination. In contrast to means, ends are finite by human nature. The classification of ends is important, given that they reveal an individual's ultimate goals.

The classification of end goals is split into two different perspectives. Multifaceted theory postulates that end goals are unrelated to each other, even genetically distinct with different evolutionary paths. Unitary theory, in contrast, proposes that end goals can be roughly grouped into a manageable number of categories, based on common characteristics (Reiss, 2004).

The separation of end goals into drives and intrinsic motives is currently popular with psychologists. Drive theory defines reward as reducing a state of deprivation. When a character's hunger drive is in a state of deprivation, food becomes a powerful reward, increasing motivation to learn actions that produce food. In (Hull, 1943), Hull identifies four types of drives: hunger, thirst, sex, and escape from pain. A large disadvantage to drive theory's validity is the fact that it does not explain exploration (curiosity), autonomy, and play. Unitary intrinsic motivation theory is an alternative to drive theory that explains the motives that drive theory could not. For example, it theorizes that competence is the origin of curiosity and autonomy. That said, while there may be a correlation, it is unrealistic to say that if a person has an above average amount of curiosity, then they must also have an above average amount of autonomy (Reiss, 2004).

The multifaceted theory of end goals has been explored by many different angles throughout the years. Very early on, Aristotle defined 12 end motives (Aristotle et al, 1976): confidence, pleasure, saving, magnificence, honour, ambition, patience, sincerity, conversation, social contact, modesty, and righteousness. Evolution may also play an important role in multifaceted theory. Different intrinsic motives may originate from distinct survival needs, embedded in different genes. The need to build nests for survival from weather or predators, suggests the evolutionary motivation efficacy. Autonomy, being the desire for freedom, motivates an animal to leave the nest and search for food in a large area (Reiss, 2004).

Reiss's theory of 16 basic desires is an important multifaceted model of intrinsic motivation, outlined in Tables 1 and 2. As mentioned earlier, studies have shown that the theory can be used to describe religious beliefs (Reiss, 2000), athleticism (Reiss et al, 2001), and lack of scholastic achievement (Reiss, 2004). For example, religious motivation was found to be motivated by above average honour and family, and below average vengeance and independence.

Emotions, Motivations, and Learning in Cognitive Architectures

Emotions can influence human behavior by altering our perception of people and events. Beyond altering perception, the concept of feelings can be used to drive reinforcement learning towards quicker learning and better policies. In emotion-driven reinforcement learning (Marinier & Laird, 2008), virtual characters learn behaviors that make them feel good, and avoid behaviors that make them feel bad. In this context, specific goals can lead to a good feeling, to make the characters want to reach the goal through intrinsically motivated reward. This application of reinforcement learning has been shown to produce improved learning in limited test cases in a maze type environment (Marinier & Laird, 2008) using the SOAR cognitive architecture (Laird, 2012).

Table 1. Hypotheses of Reiss's theory of 16 basic desires

Hypothesis 1	Each of the basic desires is a trait motive.
Hypothesis 2	The 16 basic desires motivate animals as well as people (except maybe idealism and acceptance).
Hypothesis 3	The 16 basic desires are considered genetically distinct with different evolutionary histories.
Hypothesis 4	Satisfying a basic desire produces an intrinsically valued feeling of joy. Each desire produces a different feeling of joy.
Hypothesis 5	Individuals prioritize each desire differently.
Hypothesis 6	Each basic desire is a continuous range between opposite values.
Hypothesis 7	What is motivating are the discrepancies between the amount of an intrinsic satisfier that is desired and the amount that was recently experienced.
Hypothesis 8	Basic desires influence: attention, cognition, feelings and behavior into a coherent action.

(Reiss, 2004)

Table 2. Motives in Reiss's theory of 16 basic desires

Name	Motive	Intrinsic Feeling
Power	Desire to influence	Efficacy
Curiosity	Desire for knowledge	Wonder
Independence	Desire to be autonomous	Freedom
Status	Desire for social standing including desire for attention	Self-importance
Social contact	Desire for peer companionship (desire to play)	Fun
Vengeance	Desire to get even (including desire to compete to win)	Vindication
Honour	Desire to obey a traditional moral code	Loyalty
Idealism	Desire to improve society (including altruism and justice)	Compassion
Physical exercise	Desire to exercise muscles	Vitality
Romance	Desire for sex (including courting)	Lust
Family	Desire to raise own children	Love
Order	Desire to organize (including desire for ritual)	Stability
Eating	Desire to eat	Satiation
Acceptance	Desire for approval	Self-confidence
Tranquility	Desire to avoid anxiety/fear	Safe/relaxed
Saving	Desire to collect/value of frugality	Ownership

(Reiss, 2004)

The CLARION cognitive architecture (Sun, 2006) leverages motivation to provide goals for action-centered learning and decision-making processes. This architecture allows for both low and high-level primary drives that are built-in and relatively unalterable, as well as secondary or derived drives that can be acquired, learned, or received through instruction. Reinforcement signals may be used in various contexts for learning actions to perform, tuning drive strength, and adjusting goal setting (Sun, 2009).

The ACT-R cognitive architecture (Anderson, 2007) is a robust cognitive architecture, capable of supporting a number of learning activities and methodologies. This includes reinforcement learning and reward-related learning, and supports motivations in these contexts, as in the architectures discussed above.

Traditional reinforcement learning has also been extended with intrinsic motivations to promote autonomous development of skill hierarchies, called motivated reinforcement learning (Barto et al, 2004; Chentanez et al, 2004; Merrick, 2008; Merrick, 2011; Merrick & Maher, 2006; Merrick & Maher, 2009; Merrick & Shah, 2011). This approach uses a motivation function based on interest to calculate the reinforcement function. Characters calculate the differences between past and present states to compute the reward signal that is responsible for directing learning (Merrick & Maher, 2006). The motivation function is not dependent on the domain; instead it is based on the concept of interest to calculate the intrinsic motivation signal. The skills the character develops depend on the environment and its experiences, allowing different characters to learn different skills and adapt to changes in the environment (Merrick & Maher, 2006). Motivational profiles discussed in (Merrick & Shah, 2011) show how learning can be guided by broad tendencies for achievement, motivation, and power. During a risk-taking test, power motivated characters would select goals with high incentives; whereas achievement motivated characters would select goals with regards to higher difficulty.

In contrast to emotional and motivated reinforcement learning, the method discussed in this chapter uses intrinsic motivations to directly guide the reinforcement learning and action selection of the virtual character with the express purpose of believability. The focus is not on the efficiency of the reinforcement learning method, but on how the character's motivations should be incorporated into the situation, guiding the character's action selection with believability in mind. In effect, characters learn how to act according to their motivations, a key trait of believability, as opposed to using motivations to guide or adjust other learning activities.

This approach is quite uncommon in the literature, unfortunately. The work in (Ehrenfeld et al, 2015) appears to follow this strategy to create an interesting and interactive Mario agent, but details are scant, and it is not quite clear how motivations drive their Mario agent or the role they play in learning. The work nonetheless is quite promising.

DESIGNING FOR BELIEVABILITY

In this section, we present our approach to believable virtual characters, using reinforcement learning so that the characters can learn to act according to their own unique motivations. In this approach, one defines the reward function, commences the learning algorithm, and an action-selection policy is refined that will maximize the rewards received over time. While this is the best possible case, the process is rarely so simple in reality. If the rewards are not clear, the state is not properly represented, or the learning parameters are wrong for the given task, the learned policy will be sub-optimal.

General Architecture

Figure 1 shows the basics of character-environment interaction, where the character chooses an action based upon the current state of the environment. This action produces a change in the environment, either to characters, objects, or places. The reinforcement function evaluates the changes caused by the action, and assigns a reward based upon the character's goals. The goals of a character are defined by this reinforcement function; by carefully setting the necessary positive or negative reinforcements, the goal is outlined to the character in a way that is understandable and achievable. The reinforcement function defines a mapping from state-action pairs to rewards, thereby determining what action is good in the short-term (Sutton & Barto, 1998).

The action-selection policy defines how the character behaves at any given time in the game. It maps the character's perceived state of the environment to actions. The policy must be learned if it is to be effective. The reinforcement learning process seeks to discover the optimal policy, given the current goals and environment (Harmon & Harmon, 1996; Sutton & Barto, 1998). At each step, there is one action that leads to the greatest estimated future cumulative reward, as determined by the current learned policy. This action is the greedy action, and it is selected if the learning algorithm chooses to exploit the knowledge it currently possesses. If an action that is non-greedy is selected, the algorithm is choosing to explore, potentially obtaining greater knowledge that can ultimately be used to achieve greater overall reward. It is not possible to both explore and exploit with one single action selection; proper balance is needed between the two extremes, as determined by an appropriate action selection algorithm.

Traditionally, reinforcement learning has been solely considered a technique for learning, but more recently, it is being used in effective planning. When planning, the character must keep a model of the environment, to predict how the environment will react to any given action. Given a state and an action, the model must predict the next state and the next reward. Planning and learning are very similar; learning uses real experience generated by the environment, while planning uses simulated experience generated by a model. Online planning introduces interesting interactions between planning and learning. New interactions with the environment are used to alter the model, thereby affecting the planning. Planning and learning can be computationally expensive and, as a result, computational resources may need to be divided between them. The most intuitive approach is for planning and learning to occur in parallel, utilizing shared resources between both processes (Sutton & Barto, 1998).

This general approach is flexible and open, allowing us to use, configure, and experiment with many different reinforcement learning algorithms and parameters in our work. Not every algorithm used in this paper performs planning as discussed above, and in such cases, the corresponding character-environment interactions do not occur. (In this chapter, Dyna-Q employs planning whereas the other approaches do not, for example.)

Figure 1. Character-environment interaction

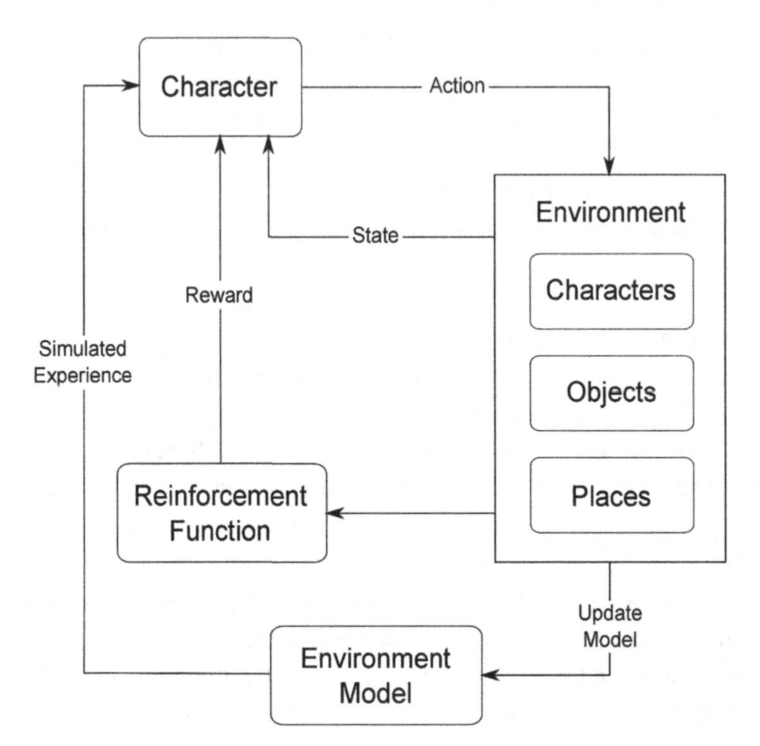

Inside the Character

In this work, we are taking a simplified approach to virtual character modeling, using a subset of our previous cognitive architectures from (Bailey et al, 2012; Kope et al, 2013). This allows us to focus on the key issues of this work, namely learning behavior through a model of character motivation, while maintaining compatibility and easy integration with more complex and robust models, including not only our own work, but other architectures such as CLARION, SOAR, and ACT-R, discussed above. Such integration is currently being explored.

Figure 2 illustrates the main elements of a character required to support the current work. Two main data collections are necessary for a character: a model of the environment and the state of the character itself. The environment model, as discussed above, is a representation of knowledge that the character has about other characters, objects, and places in the world. Character state, on the other hand, collects a number of attributes describing the current condition of the character, including psychological, sociological, and physiological elements.

Each character also possesses four pipelined processing subsystems and their associated data. The sensing subsystem retrieves information about the environment

Figure 2. Main character elements

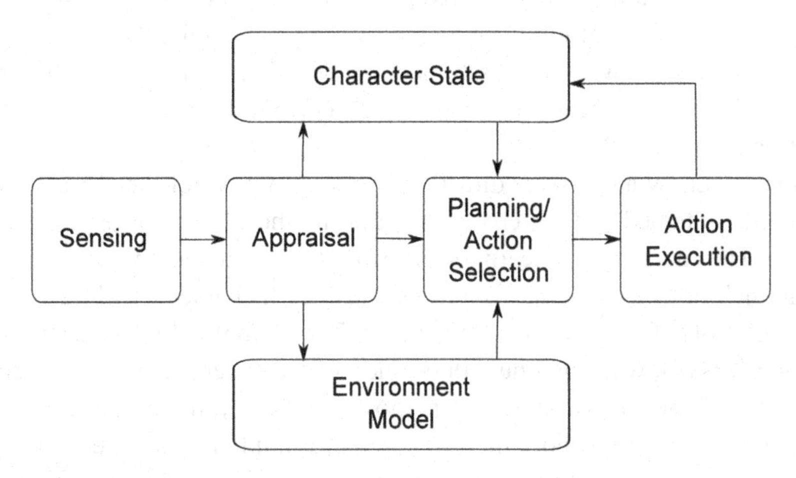

and activity occurring around the character. Generally, this is limited to what the character can perceive, but more global information can be retrieved as necessary. The appraisal subsystem filters the information stream from the sensing subsystem and determines the relevance of this information to the character; relevant items are integrated into the environment model as necessary or trigger changes to the character's own state. The planning/action selection subsystem determines what the character will do based on the state of the world and the character itself; this is driven by the reinforcement learning mechanisms discussed in the previous section. Lastly, the action execution subsystem will carry out the actions as determined by the planning/action selection subsystem. These actions may affect the environment, as shown in Figure 1, or they may impact the character's own state. Resulting changes to the character's state are noted immediately, while changes to the environment must first be perceived by the sensing subsystem.

Integrating Motivations

Rather than having motivation drive skill learning as in motivated reinforcement learning, as discussed above, our approach tracks the state of a character's motivations, a key aspect of character state from Figure 2 in this work, to learn actions that will benefit the character's self-interest. This approach, while drawing inspiration from Reiss's theory of 16 basic desires (Reiss, 2004), is not constrained by any particular motivational model. In this way, any motivation, including both means and ends (Aristotle et al, 1976), can be created and used to suit the developer's needs. According to Reiss, as discussed above, basic motives include power, curiosity, independence, status, social contact, vengeance, honor, idealism, physical exercise, romance, family, order, hunger, acceptance, tranquility, and saving; this set is more than sufficient for our initial work here. (This said, there is nothing in particular that ties this work to Reiss's model; it would be fairly simply to substitute a new model from the research community or a custom-defined one tailored to the application in question.)

Actions taken by a character directly affect any number of the character motivations, as maintained by their character state, and must be specified in the action definition. The character will learn to perform actions that benefit itself, according to its internal motivation values and thresholds. Each character will have its own distinct thresholds for individual motivations. For example, if a character's social interaction falls below its minimum threshold, the character should learn to choose actions that will increase its social interaction. Conversely, if the character's social interaction value is raised above the maximum threshold, the character should seek the opposite, isolation, to return within acceptable limits. Setting different thresholds for different motivations, and different characters, allows for a unique representation of a person, which is essential to believability. (This is also consistent with Hypothesis 5 of Reiss's model as presented in Table 1.)

A motive, then, consists of a current value, $m_{current}$, a maximum desired value $m_{max} < 100$, and a minimum desired value $m_{min} > 0$, where $m_{min} < m_{max}$. (See Figure 3 for a graphical representation of a motive.) The minimum and maximum desired values are also called thresholds. A motive value will naturally decay at a linear rate over time, consistent with the theories in (Reiss & Havercamp, 1998). When this value falls below m_{min}, the desire reasserts itself and the character is motivated to take an action to increase the value. Likewise, if an action increases $m_{current}$ above m_{max}, an action is taken to decrease the value or actions that increase the value are avoided until the value decays naturally. The minimum and maximum desirable motive values remain fixed, representing the particular tendencies and needs of the character. The current desire values are constantly changing, and should therefore be a part of the situation variable of the character, used in learning and action selection.

For example, consider the hunger motivation from Reiss's model. Suppose a character's hunger level falls below its m_{min} threshold and the character is now starving. At this point, the character is motivated to eat to raise $m_{current}$ for its hunger motivation above m_{min}. This may be done in a single action of eating, or may require

Figure 3. Representing motivations in characters

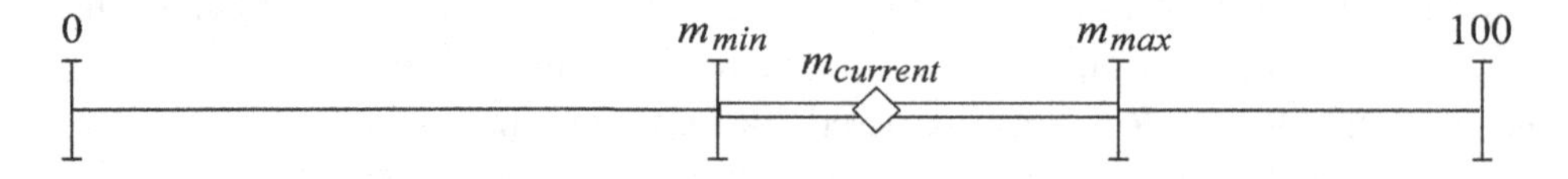

more than one action to do so. When this is done, the character is rewarded for satisfying its needs accordingly, and over time effectively learns the action or chain of actions that produces the desired result. If the character overeats to the point where $m_{current}$ exceeds m_{max}, the character will stop eating and allow its hunger to decay naturally to an acceptable level. (If actions existed to relieve the overeating, the character would perform them, but for most simulations, we likely would not want to provide such an option to the character for the hunger motivation.) Note that the character need not wait until it is starving to eat; it could be advantageous to eat earlier, because it has the opportunity to do so, or perhaps because it has nothing better to do. Likewise, the character need not eat until it has overeaten; it is free to stop eating once it is no longer starving. As a character learns, it will know to eat before it starts starving and stop before it has overeaten to maintain its hunger at an acceptable level between its thresholds.

To handle cases where motivations have no upper limit, we set $m_{max} = 100$, and to handle where motivations have no lower limit, we set $m_{min} = 0$. For example, suppose a character is extremely social. The m_{max} threshold for their social motivation could be set to 100 to indicate that they could never have excessive socialization; it could never go high enough for them to remove themselves from a social situation or to seek seclusion. (Their m_{min} threshold would also be set fairly high so that in the absence of social contact, it does not take long for them to desire to seek it out again.) Likewise, suppose a character is very introverted and strongly prefers being on their own. The m_{min} threshold for their social motivation could be set to 0 to indicate that they would never be dissatisfied by the lack of social contact in their life; they would never take an action to seek out social contact. (Their m_{max} threshold would also be set low so that it would not take much inadvertent or accidental socialization for them to actively seek out seclusion once again.)

A PROTOTYPE SYSTEM

As proof of concept, a prototype system has been developed using our approach to motivated believable characters. The system itself was written in C++ using Microsoft Visual Studio and was developed for the Microsoft Windows platform. Our prototype system allows for the execution of multiple characters in a virtual world with no graphical interface, and no user input. This enables the collection of quantitative results through observing and recording interactions between characters and their environment. This approach facilitates large-scale experimentation, allowing the exploration of multiple algorithms and parameters with ease.

The ultimate goal of characters in our system is to stay as close as possible to motivational equilibrium, with all motivations within maximum and minimal thresholds. During a simulation, a number of statistics are gathered by the system in regards to each individual character. The metric for direct evaluation is the sum of the rewards received throughout the simulation. As reward is the parameter with which reinforcement learning is enforced, this metric allows the most accurate evaluation of the learning capabilities of this approach. Though maximizing the total reward is highly desirable, it is also good to maximize the number of simulation steps with all motives within bounds. Both metrics are recorded as percentages (percent of total reward and percent of steps within bounds) allowing for direct comparison between two runs with different numbers of simulated steps.

Virtual worlds in our prototype system are defined through XML files that are parsed by our system at the beginning of the game. Using XML allows us to have a standardized format for representing worlds that can be readily deployed in simulations and games across multiple platforms and engines. Each virtual world is a collection of characters, objects, and places. Places contain both characters and objects, while objects can only contain other objects (for example, a refrigerator contains food). Characters can be defined with any number of motives. All actions are defined globally, but an action instance is created for each character at the beginning of the simulation. Actions include the information needed to affect changes to objects (position, ownership, erased, etc.), places (add a character or object), or characters (position, gaining a possession, change in motive values, etc.). If an action includes changes to a motive not present in the current character, the motive changes are ignored. While loading the game world, if the character's actions have already been created in a previous simulation, with no changes to its motives, the learned action policies are simply loaded from memory.

Built into our prototype system is the means to use different reinforcement learning algorithms (Sarsa, Q-learning, and Dyna-Q), with different action selection methods (ε-greedy and Softmax). This allows a collection of simulation scenarios to be tested with different combinations of methods and various parameter values easily, as detailed in the next section. Where necessary, artificial neural networks were used for function generalization and approximation, based on the Fast Artificial Neural Network library (FANN).

CASE STUDIES

With our prototype system, there are a large number of parameters and algorithms from which to choose in studying the creation of believable virtual characters. To that end, a set of variables must be found that will perform well in most situations, and so extensive testing is needed to understand the impact of each variable on a character's learning and performance. Given the large number of variables involved in the testing process, used for fine-tuning rewards, learning rates, and so on, it is not feasible to exhaustively test every possible variable combination. Instead, an incremental approach was used to determine what variables produced the best overall results. This manner of testing was repeated for each simulation scenario, with the understanding that each variation introduces a more complex learning task, potentially requiring certain parameters to be different. Understanding the parameters needed for each increasingly complex situation enables the possible discovery of more optimal values.

While this incremental approach required time and planning, it produced superior results at a considerable savings over more exhaustive testing. Test harnesses developed and built into our prototype system successfully automated much of this process as well, making this approach easier to use and deploy by other developers.

In the course of this work, many thousands of experimental trials were conducted in a scientifically rigorous fashion. This chapter summarizes and highlights key findings of these experiments; for complete results, the reader is urged to consult (Forgette, 2013).

Given the sheer number of experiments required, experimentation was conducted using a private cluster of computers without human input. This provided a more stable testing environment, but is somewhat artificial, as randomness and unpredictability added by human players is absent. While extended training of characters should improve robustness and believability even in the presence of human players, further study with human players is clearly necessary in the future.

Simulation Scenarios

Experimentation with our prototype system was conducted across five game-like simulation scenarios of increasing complexity. Successive scenarios added more characters, objects, and places to the environment, as well as more characters motives and a richer and more complex set of possible actions to execute, including some with delayed effects.

Game One

The first simulation scenario is the simplest. In this case, the virtual world consists of a one-room house that is occupied by a character named Tarzan, and an object named food. Every simulation step, Tarzan has the option to eat food, or do nothing, with the goal of ultimately satisfying his only motive, hunger. The eat food action increases Tarzan's current hunger motive value, and takes a single simulation step to complete. If Tarzan decides to do nothing, all motives decay according to a linear decay function. The learning task for Game One is to develop an action selection policy that knows when it is best to eat, and when it is best to do nothing (for example, to avoid over-eating) based on rewards given out during the scenario.

Game Two

The second virtual world adds complexity through multiple objects, three motives, and actions with delayed reward. The character Tarzan resides in his one-room house, with a television, a treadmill, and a refrigerator. The refrigerator contains one healthy food and one greasy food, and must be opened before the food becomes accessible. Tarzan's motives include hunger, health, and entertainment. (Note that health and entertainment are a fusion of elements from Reiss's original model.) Watching the television provides entertainment at the expense of healthiness. Using the treadmill increases healthiness, but makes Tarzan hungry at the same time. The refrigerator has no impact on motives, but the food it contains satisfies hunger and either increases or decreases healthiness, depending on whether the healthy or greasy food is selected. The purpose of this scenario is to learn combination actions, such as opening the refrigerator to eat food, while trying to satisfy multiple motives using actions that can impact more than one motive at the same time.

This is a significant increase in complexity over the first scenario, and resulted in some rather interesting learned behaviors. For example, Tarzan would binge on unhealthy and greasy food, and then hit the treadmill to compensate by working it off. Other times, Tarzan would reward himself after extended exercise on the treadmill by eating some unhealthy greasy food. In such cases, Tarzan would ignore the healthy food almost entirely. The result is some rather interesting and typical human-like behavior. This said, Tarzan learned different behaviors and performed at different levels, depending on his exploration of the action space and the tuning of various parameter settings.

Under the best performing policy, for example, Tarzan would spend most of the game watching television. Since his entertainment motive had no maximum threshold, Tarzan could watch a considerable amount of television without exceeding his maximum desired entertainment value. Watching this much television, however, decreases his healthiness. When his health motive value moves close to the desired minimum value, Tarzan will stop watching television and instead use the treadmill in an effort to be healthier. When Tarzan's hunger motive is close to the minimum desirable amount, he will open the refrigerator and take and eat healthy food from it. In this case, Tarzan does not eat any greasy food during the entire game, even though it is also available from his refrigerator, as he has learned that greasy food satisfies his hunger motive less than healthy food, and also decreases his healthy motive value. (He would have to use the treadmill more and watch television less, which apparently is undesirable in this instance.)

The worst performing policy under these parameter settings, as another example, had the highest discount rate among the optimal parameters selected for study here, and so Tarzan will consider long-term rewards to be much more important than short-term rewards. The only action with delayed reward in this game scenario is opening the refrigerator, and this reward is only delayed by two game steps when food is eaten. In fact, in this case, Tarzan continuously opens the refrigerator door, ignoring the other motive values that need to be satisfied. This type of behavior is interesting, but not surprising. There are no negative consequences associated with opening the refrigerator (for example, shortening the life span of the food or using electricity), and so this behavior is deemed to be perfectly acceptable. Furthermore, this approach optimizes access to long-term rewards, as Tarzan will be able to eat food faster when he gets hungry; after all, chances are the refrigerator door is already open, and an additional action is not necessary to satisfy this motive in this case.

Game Three

This simulation scenario introduces the possibility of interactions and socialization between characters. There are now three characters: Tarzan, Jane, and Bob. Tarzan and Jane each have their own private one-room houses. Bob is the bar tender at the local bar, and does not have the ability to leave. Both Tarzan and Jane have a social motivation, and a hunger motivation. Bob's only motivation, however, is social. Jane and Tarzan must go to the bar for social interaction, and return to their respective houses to eat. In order to eat food, the refrigerator needs to be opened first.

Managing social interactions is the most difficult aspect of this simulation scenario. When any of the characters talk to another character, they receive an increase to their current social motivation value, and also increase the social motivation value of the other character. In this scenario, a character must learn that doing nothing is not enough to decrease their social interaction level. Since other characters contribute to their social interaction, characters must learn to remove themselves from the social situation, and return home to where no social interaction can take place.

All characters have different minimum and maximum thresholds for their social motivation. This allows for a direct comparison of how different motivational profiles would learn using the same reinforcement learning algorithm. The purpose of this learning task is to learn inter-character interactions, force travel between different places to satisfy motivations, and to compare qualitative results from two different motivation profiles (Tarzan vs. Jane).

For example, as Bob is quite social yet confined to his establishment, he socializes constantly with either Tarzan or Jane when they enter his bar. He does this to the point where he drives Tarzan and Jane away, so they can avoid excessive social exposure. Because Bob monopolizes the socialization in this case, Tarzan and Jane sadly never learn to talk to each other while at the bar, as that only exacerbates their socialization problems. As Jane is more social than Tarzan, she will go to the bar earlier and linger longer.

Game Four

The fourth game scenario adds complexity by introducing an action with extremely delayed reward. A new locale is introduced—the grocery store. In this scenario, the character must go to the grocery store, buy food (healthy or greasy), return home, access the refrigerator, and finally, eat the food. All five steps must be completed to increase the current hunger motive value. In this case, Tarzan can only move between his house and the grocery store; Jane and Bob and their locales do not appear in this scenario. While in some ways this scenario is simpler than the previous scenario, the multi-step process of eating is much more complex than previous actions, and so it was analyzed separately.

Additional rules were added to the food system for this scenario. Only one food item can be owned at any time, and a food item must belong to Tarzan (ownership) before he can eat it. Furthermore, food is delivered to the refrigerator on being bought from the grocery store. Without these constraints in place, Tarzan would learn some rather terrible habits and manners. For example, with no financial system in place and therefore no spending limits, Tarzan would purchase food in extremely large quantities to avoid having to go to the grocery store frequently. (This had the side effect of making it difficult to determine if he was actually learning the multi-step food consumption process.) Without the concept of ownership, Tarzan would simply steal food and eat it rather than going through the purchase process. (After all, he got the food either way, so why take the extra step of purchasing it?) Lastly, if food was not delivered for him, Tarzan would simply eat his purchased food at the grocery store as it was faster and more convenient. While we experimented with adding negative social consequences to this behavior, Tarzan would simply compensate by trying harder to socialize with other characters.

Tarzan was able to learn the correct food eating process, and could avoid mistakes that would otherwise make his actions unrealistic. For example, if Tarzan had opened the fridge on a previous occasion and eaten his food, Tarzan learned that there was no point in going back to the fridge for food without making a trip to the grocery store. (There would be no food to eat and hence no reward for his actions in such a case.) Adding delayed reward mechanisms could give the characters the illusion of memory in certain circumstances, such as this. Note that this does not take the place of a more robust memory system, as in (Kope et al, 2013), and we are exploring the integration of motivations, learning, and memory in parallel work (Patrick et al, 2015).

While we can create interesting scenarios and character behaviors, the above issues highlight that a great deal of thought and planning is necessary to design a good balance of motivation, action, and consequence to maintain believability, especially in complex situations.

Game Five

The fifth, and final, game scenario, is the culmination of all previous game scenarios. This scenario, naturally, is the most complex of all, with characters having to successfully manage all of the elements introduced in the previous scenarios simultaneously. It includes inter-character actions, delayed reward, and multiple motives. In this case, the virtual world consists of four places: Tarzan's house, Jane's House, the bar, and the grocery store. Both Jane and Tarzan have the option of going to the bar, or the grocery store, but not to each other's houses. Bob is meant to be the bartender, but he now also has the option of going to the grocery store. When a character talks to another character, they receive increases to social and entertainment motive values, and give similar increases to the other character. Since Jane has no entertainment motive, any motive change for entertainment has no effect on her. Jane only has social and hunger motives, while Bob only has a social motive. Tarzan, on the other hand, has social, hunger, entertainment, and health motives. Objects function similarly as in previous scenarios.

Results

For brevity, since Game Five encompasses all of the features of the previous simulation scenarios, we will focus our discussion here on the results from experimentation in that scenario, specifically those using Sarsa and ε-greedy. Again, for detailed results across all scenarios, the reader should consult (Forgette, 2013).

Our prototype system allows for both training and testing simulation runs. Training runs permit exploratory actions that are key to learning, while testing runs do not and instead rely upon a previously created action selection policy. The number of steps in each run is configurable, and defaults to 1000 simulation steps in our experimentation. Successive runs are allowed to build upon each other. The simulation run *trainN* is the *Nth* training simulation of a particular testing configuration. The simulation run *testN* follows the policy obtained through the *Nth* training simulation run. For example, a training run, *train10*, begins with 9000 simulation steps of training from previous simulations, building upon the training from *train9*. Real characters should maintain a low-level exploration rate to promote the discovery of better policies, given sudden shifts, or changes in the virtual world.

Initial experimentation with Game Five was conducted to determine reasonable parameter values for longer and more rigorous testing in this scenario. To assess the prototype in this scenario, 10 repeated tests were run with the same discovered parameter values to assess consistency of performance. In this case, 200 training simulations were used in succession, with the percent of reward of the last 5 training games (train196, train197, train198, train199, and train200) for every test number given in Table 3. While the system did well in achieving decent levels of reward in general, it was only able to satisfy all motives simultaneously in just under 10% of simulation steps on average, as can be seen from Table 4.

Taking a closer look at actual action selection sequences shows how well the character Tarzan learned how to balance all four of his motives. Figure 4 shows the normalized motive values obtained by the action selection policy that resulted in achieving the highest percent of reward in the last five training simulations (*train198* of repeated test number 7). The entertainment motive was easily learned given there were no bounds on the maximum desirable value. Much of the character's time was spent moving between the grocery store and the bar in search for social interaction from either Bob or Jane. Other motives were satisfied as it was convenient to do so. In general, Tarzan fared well in this run.

Other simulation runs were less optimal, such as *train200* of repeated test number 1, as can be seen in both Table 3 and in Figure 5, which shows the normalized motive values obtained in this run. During that game run, Tarzan went to the grocery

Table 3. Percent of award statistics from game five

	Percent of Reward					
Test	**train196**	**train197**	**train198**	**train199**	**train200**	**Mean**
1	95.64	97.65	97.81	89.75	60.51	88.27
2	95.62	97.26	96.84	87.41	96.17	94.66
3	97.19	97.37	97.48	96.46	98.45	97.39
4	94.99	83.72	84.33	91.62	93.76	89.69
5	96.57	96.11	96.37	78.11	87.57	90.95
6	98.09	85.41	89.77	83.53	85.16	88.39
7	97.79	95.04	98.50	96.66	96.45	96.89
8	78.35	80.20	84.39	95.90	98.04	87.38
9	87.33	76.77	81.22	88.72	86.00	84.01
10	94.42	96.62	95.95	96.98	96.21	96.03
Std	6.17	8.16	6.68	6.34	11.46	4.61

Table 4. Percent of steps in bound statistics from game five

	Percent of Steps in Bound					
Test	**train196**	**train197**	**train198**	**train199**	**train200**	**Mean**
1	9.56	9.77	9.78	8.97	6.05	8.83
2	9.56	9.73	9.68	8.74	9.62	9.47
3	9.72	9.74	9.75	9.65	9.84	9.74
4	9.50	8.37	8.43	9.16	9.38	8.97
5	9.66	9.61	9.64	7.81	8.76	9.09
6	9.81	8.54	8.98	8.35	8.52	8.84
7	9.78	9.50	9.85	9.67	9.65	9.69
8	7.83	8.02	8.44	9.59	9.80	8.74
9	8.73	7.68	8.12	8.87	8.60	8.40
10	9.44	9.66	9.59	9.70	9.62	9.60
Std	0.62	0.82	0.67	0.63	1.15	0.46

store to buy food, but was continuously talked to by the other characters, resulting in significantly higher social interaction than desired and issues with addressing his other motives as well as he attempted to compensate.

Nevertheless, despite this, performance was reasonable across all tests shown in Table 3, with an average of 91.37% of rewards achieved.

CONCLUSION

The approach to believable virtual characters proposed in this chapter introduces reinforcement learning guided by a character's motives, with the purpose of creating believable character behavior. In doing so, we could create a character that learns to take actions that satisfy its motives and study how various parameter values influenced the character's learning. After extensive testing, results showed that a reinforcement learning character is able to select actions to satisfy up to four motives, even when the environment has actions with delayed reward, and interactions between characters.

Programming believable virtual characters is a non-trivial problem that involves creating a character that acts realistically in the eyes of the user. This work demonstrates significant promise in addressing this challenge, with the potential to both ease the development of believable characters and enable new types of characters and gameplay as a result.

Figure 4. Motive values from best performance of game five

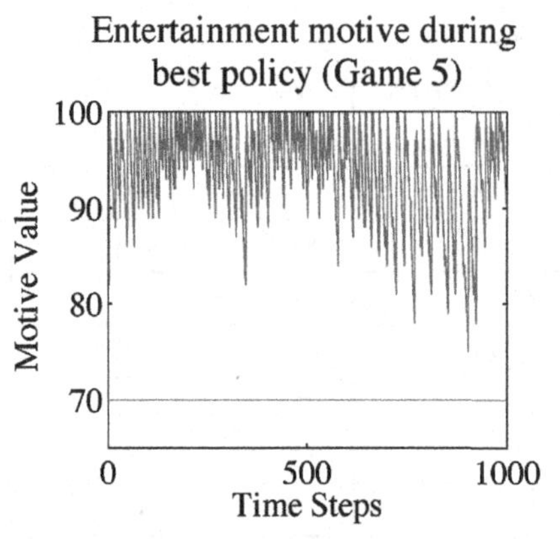

(a) The *entertainment motive* values for Tarzan in *train198*, repeated game number 7.

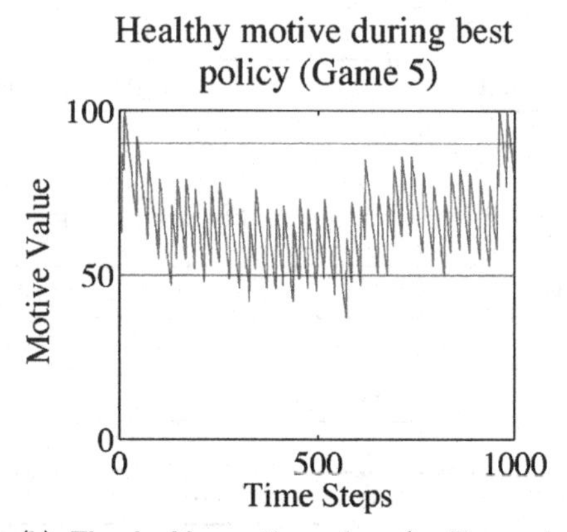

(b) The *healthy motive* values for Tarzan in *train198*, repeated game number 7.

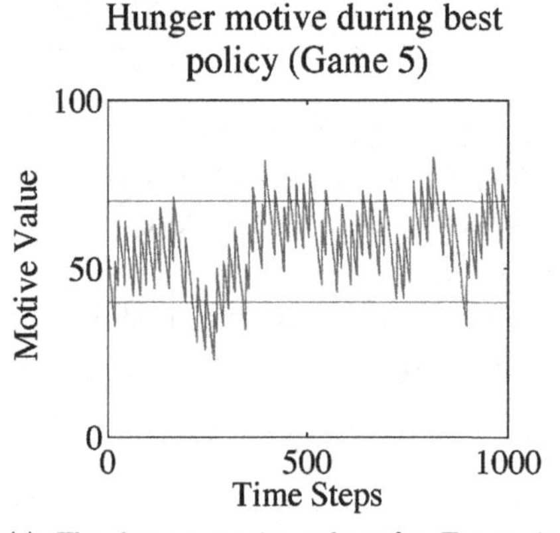

(c) The *hunger motive* values for Tarzan in *train198*, repeated game number 7.

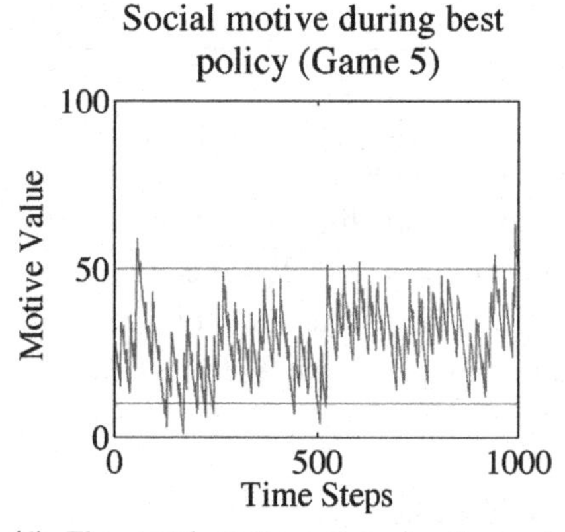

(d) The *social motive* values for Tarzan in *train198*, repeated game number 7.

This said, there are still many opportunities to improve on the proposed method, including the addition of more advanced reinforcement learning techniques, and the addition of other motive elements. Currently, all motives decay at the same rate according to a linear decay function; if different decay functions were available, it would add another way of making motives unique across multiple characters. So far, the actions available to characters are fairly basic; more complex actions or longer sequences of actions that could evolve into skills are other avenues worthy of exploration. Other reinforcement functions can also be examined; for example, one that assigns more importance to motives that are further from their bounds. This change might decrease the possibility of one motive being completely ignored over satisfying others. The addition of more motive elements and more advanced

Figure 5. Motive values from worst performance of game five

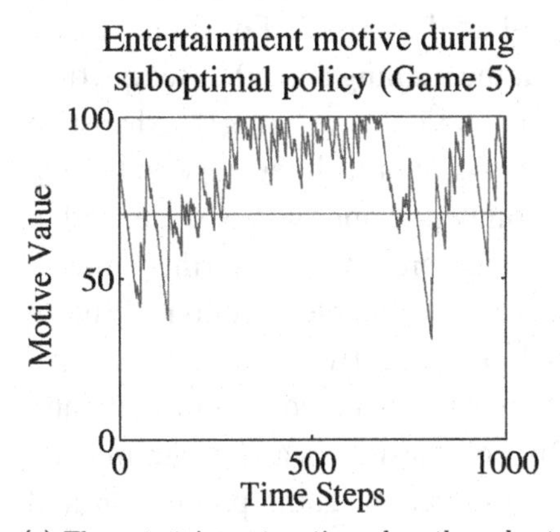

(a) The *entertainment motive* values throughout the final training game with the least amount of reward for Tarzan.

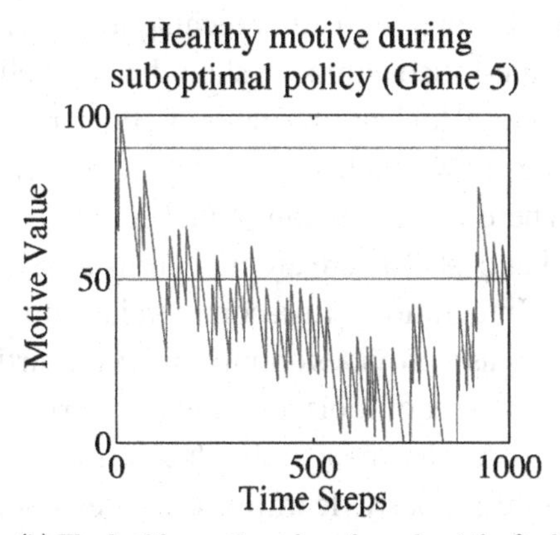

(b) The *healthy motive* values throughout the final training game with the least the amount of reward for Tarzan.

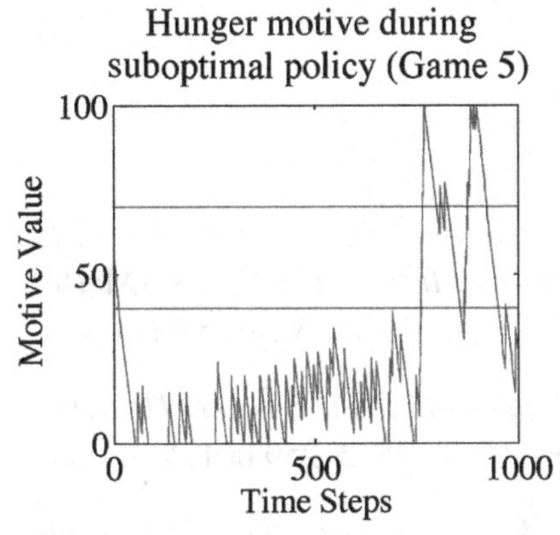

(c) The *hunger motive* values throughout the final training game with the least amount of reward for Tarzan.

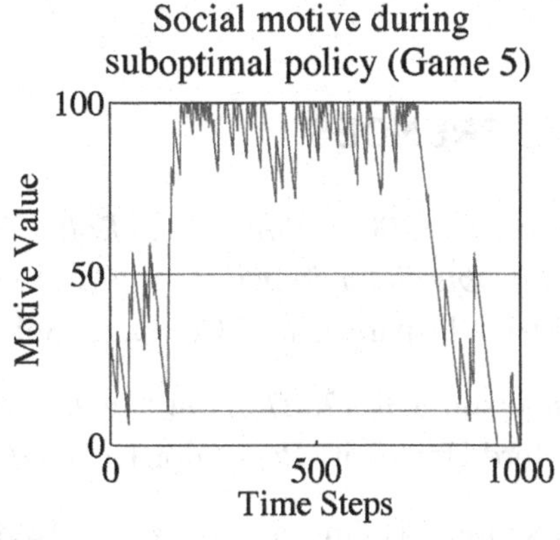

(d) The *social motive* values throughout the final training game with the least amount of reward for Tarzan.

reinforcement learning techniques would enable more diverse character specifications, and more realistic action sequences.

Ultimately, this method also requires a comprehensive user analysis, aimed at determining the perceived believability of a motivated virtual character, particularly in the presence of potentially unpredictable human input as discussed earlier. Believability can be best, and perhaps only, assessed through human evaluation of our virtual characters and their behaviors. While this work is promising, further study is clearly necessary, with human subjects observing and interacting with the virtual characters in an appropriate virtual environment.

It would also be interesting to compare the reinforcement-learning approach of this chapter with other techniques, such as a rule-based approach. For simple cases, a rule-based approach should work well, without the training and other setup effort required for learning methods. For larger, more complex cases, creating an exhaustive rule set would become intractable, giving a learning approach the edge in theory. Where is this tipping point? How well do the competing approaches scale, and at what costs? There are numerous questions to be answered through further study.

Integration with a commercial game engine is also a valuable direction for future work as it provides opportunities for further development and wider testing in games. A solid set of tools, prefabs, and scripts for such a platform could also simplify and ease development and reduce effort required by both designers and programmers. Done properly, learning activities could occur throughout game production and testing, giving characters ample opportunity to grow and mature in time for release. This kind of integration exercise is currently being explored with the popular Unity engine through parallel work in this area (Patrick et al, 2015).

REFERENCES

Acton, G. (2009). *Playing the Role: Towards an Action Selection Architecture for Believable Behaviour in Non Player Characters and Interactive Agents*. (Masters Thesis). Department of Computer Science, The University of Western Ontario.

Anderson, J. R. (2007). *How Can the Human Mind Occur in the Physical Universe?* Oxford University Press. doi:10.1093/acprof:oso/9780195324259.001.0001

Aristotle, , Thomson, J. A. K., & Tredennick, H. (1976). *The Ethics of Aristotle: The Nicomachean Ethics*. Penguin.

Bailey, C., & Katchabaw, M. (2008, November). An emergent framework for realistic psychosocial behaviour in non player characters. In *Proceedings of the 2008 Conference on Future Play: Research, Play, Share*. doi:10.1145/1496984.1496988

Bailey, C., You, J., Acton, G., Rankin, A., & Katchabaw, M. (2012, December). Believability through psychosocial behaviour: Creating bots that are more engaging and entertaining. In P. Hingston (Ed.), *Believable Bots: Can Computers Play Like People?* Springer.

Barto, A., Singh, S., & Chentanez, N. (2004, October). Intrinsically motivated learning of hierarchical collections of skills. In *Proceedings of the 3rd International Conference on Developmental Learning*.

Bevacqua, E., de Sevin, E., Pelachaud, C., McRorie, M., & Sneddon, I. (2010, March). Building credible agents: Behaviour influenced by personality and emotional traits. In *Proceedings of International Conference on Kansei Engineering and Emotion Research.*

Chentanez, N., Barto, A., & Singh, S. (2004, December). Intrinsically motivated reinforcement learning. In *Proceedings of the Eighteenth Annual Conference on Neural Information Processing Systems.*

Doce, T., Dias, J., Prada, R., & Paiva, A. (2010, September). Creating individual agents through personality traits. In *Proceedings of the 10th International Conference on Intelligent Virtual Agents*. doi:10.1007/978-3-642-15892-6_27

Doirado, E., & Martinho, C. (2010, May). I mean it! Detecting user intentions to create believable behaviour for virtual agents in games. In *Proceedings of the 9th International Conference on Autonomous Agents and Multiagent Systems.*

Ehrenfeld, S., Schrodt, F., & Butz, M. (2015, January). Mario lives! An adaptive learning AI approach for generating a living and conversing Mario agent. In *Video Proceedings of the 29th Conference of the Association for the Advancement of Artificial Intelligence (AAAI 2015).*

Forgette, J. (2013). *Reinforcement Learning With Motivations For Realistic Agents.* (Masters Thesis). Department of Computer Science, The University of Western Ontario.

Funder, D. C. (2001). Personality. *Annual Review of Psychology*, *52*(1), 197–221. doi:10.1146/annurev.psych.52.1.197 PMID:11148304

Guye-Vuilleme, A., & Thalmann, D. (2000). A high-level architecture for believable social agents. *Virtual Reality (Waltham Cross)*, *5*(2), 95–106. doi:10.1007/BF01424340

Harmon, M., & Harmon, S. (1996). *Reinforcement Learning: A Tutorial.* Retrieved March 16, 2015, from http://citeseerx.ist.psu.edu/viewdoc/summary?doi=10.1.1.33.2480

Hull, C. L. (1943). *Principles of Behavior.* New York: Appleton-Century-Crofts.

Kope, A., Rose, C., & Katchabaw, M. (2013, October). Modeling autobiographical memory for believable agents. *Proceedings of the 9th AAAI Conference on Artificial Intelligence and Interactive Digital Entertainment (AIIDE'13).*

Laird, J. (2012). *The SOAR Cognitive Architecture*. MIT Press.

Loyall, A. B. (1997). *Believable Agents: Building Interactive Personalities*. (PhD Thesis). Stanford University.

Marinier, R. III, & Laird, J. (2008, July). Emotion-driven reinforcement learning. *Proceedings of the 30th Annual Meeting of the Cognitive Science Society*.

Merrick, K. (2008). Modeling motivation for adaptive nonplayer characters in dynamic computer game worlds. Computers in Entertainment, 5(4), 5:1-5:32. doi:10.1145/1324198.1324203

Merrick, K. (2011). A computational model of achievement motivation for artificial agents. In *Proceedings of the 10th International Conference on Autonomous Agents and Multiagent Systems*.

Merrick, K., & Maher, M. (2006, May). Motivated reinforcement learning for non-player characters in persistent computer game worlds. In *Proc. of the 2006 ACM SIGCHI International Conference on Advances in Computer Entertainment Technology*. doi:10.1145/1178823.1178828

Merrick, K., & Maher, M. (2009). *Motivated Reinforcement Learning: Curious Characters for Multiuser Games* (1st ed.). Springer Publishing Company, Incorporated. doi:10.1007/978-3-540-89187-1

Merrick, K. & Shah, K. (2011, February). Achievement, affiliation, and power: Motive profiles for artificial agents. *Adaptive Behavior: Animals, Animats, Software Agents, Robots, Adaptive Systems, 19*(1).

Mitchell, T. M. (1997). *Machine Learning*. McGraw-Hill.

Myers, I., Briggs, McCaulley, M., Quenk, N., & Hammer, A. (1998). MBTI Manual: A Guide To The Development And Use Of The Myers Briggs Type Indicator. Consulting Psychologists Press.

Patrick, A., Gittens, C., & Katchabaw, M. (2015, October). The virtual little Albert experiment: Creating conditioned emotion response in virtual agents. In *Proceedings of 2015 IEEE Games, Entertainment, and Media Conference*. doi:10.1109/GEM.2015.7377228

Prada, R., Raimundo, G., Dimas, J., Martinho, C., Pena, J. F., Baptista, M., & Ribeiro, L. L. et al. (2012, June). The role of social identity, rationality and anticipation in believable agents. In *Proceedings of the 11th International Conference on Autonomous Agents and Multiagent Systems*.

Rabin, S. (2010). *Introduction to Game Development*. Game Development Series. Course Technology.

Read, S., Monroe, B., Brownstein, A., Yang, Y., Chopra, G., & Miller, L. (2010). A neural network model of the structure and dynamics of human personality. *Psychological Review*, *117*(1), 61–92. doi:10.1037/a0018131 PMID:20063964

Reiss, S. (2000). Why people turn to religion: A motivational analysis. *Journal for the Scientific Study of Religion*, *39*(1), 47–52. doi:10.1111/0021-8294.00004

Reiss, S. (2004). Multifaceted nature of intrinsic motivation: The theory of 16 basic desires. *Review of General Psychology*, *8*(3), 179–193. doi:10.1037/1089-2680.8.3.179

Reiss, S., & Havercamp, S. (1998). Toward a comprehensive assessment of fundamental motivation: Factor structure of the Reiss Profiles. *Psychological Assessment*, *10*(2), 97–106. doi:10.1037/1040-3590.10.2.97

Reiss, S., Wiltz, J., & Sherman, M. (2001). Trait motivational correlates of athleticism. *Personality and Individual Differences*, *30*(7), 1139–1145. doi:10.1016/S0191-8869(00)00098-2

Rizzo, P., Veloso, M., Miceli, M., & Cesta, A. (1997). Personality-driven social behaviors in believable agents. In *Proc. of the AAAI Fall Symposium on Socially Intelligent Agents*.

Ruijten, P., Midden, C., & Ham, J. (2013, April). I didn't know that virtual agent was angry at me: Investigating effects of gaze direction on emotion recognition and evaluation. In *Proceedings of the 8th International Conference on Persuasive Technology*. doi:10.1007/978-3-642-37157-8_23

Sun, R. (2006). The CLARION cognitive architecture: Extending cognitive modeling to social simulation. In Cognition and Multi-Agent Interaction. Cambridge University Press.

Sun, R. (2009). Motivational representations within a computational cognitive architecture. *Cognitive Computation*, *1*(1), 91–103. doi:10.1007/s12559-009-9005-z

Sutton, R., & Barto, A. (1998). *Reinforcement Learning: An Introduction*. MIT Press.

Tupes, E., & Cristal, R. (1961). *Recurrent Personality Factors Based on Trait Ratings*. Technical Report ASD-TR-61-97. Lackland Air Force Base, TX: Personnel Laboratory, Air Force Systems Command.

KEY TERMS AND DEFINITIONS

Action-Selection Policy: The action-selection policy, or simply policy, defines how a character behaves at any given time in its virtual environment. It maps the character's perceived state of the environment and itself to actions. The policy must be learned if it is to be effective. The reinforcement learning process seeks to discover the optimal policy, given the current goals and environment.

Believable Character: A character in an interactive simulation can be considered to be believable if they portray an illusion of life; that is, they have thoughts, feelings, personality, needs, and wants, in a way that is distinctive, individual, and realistic in the given context.

Goals: The goals of a reinforcement learning character are defined by a reinforcement function. By carefully setting the necessary positive or negative reinforcements, or rewards, the goal is outlined to the system in a way that is understandable and achievable. The reinforcement function defines a mapping from state-action pairs to rewards, thereby determining what action is good in the short-term. Reinforcement learning reward functions must remain unchanged by the reinforcement learning system, and serve as a basis to change the policy. The reinforcement learning system designer is responsible for defining the reinforcement function.

Motives: Motives are the reason that causes a person to initiate and perform a voluntary behavior. As such, they make a solid foundation for reinforcement learning to generate realistic actions for a believable character.

Reinforcement Learning: Reinforcement learning is a sub-field of machine learning, focused on goal directed learning through trial and error interactions between a character and its environment. Though many types of reinforcement learning algorithms exist, their goals revolve around developing an action-selection policy that maximizes the reward received across multiple steps. The character learns from its environment by taking actions, receiving rewards, and adapting its action-selection policy based on past, present and future rewards. Reinforcement learning is a very popular learning technique because of its unsupervised nature. One must simply define the reward function, commence the learning algorithm, and an action-selection policy is refined that will maximize the rewards received over time.

Virtual Character: A virtual character, or simply character, is an active entity inside a virtual environment. For this work, we are primarily concerned with characters that are being controlled through some form of artificial intelligence as opposed to direct user control and characters that are human, humanoid, or human-like in behavior. (This work can readily be applied to other types of characters, though, through the use of different models of motivation.)

Virtual Environment: For the purposes of this work, a virtual environment, also sometimes referred to as a virtual world, is an interactive simulated environment executing on a computing device of some kind. (A video game is a common example of this for entertainment purposes.)

Chapter 5
Personality-Based Cognitive Design of Characters in Virtual Environments

Maryam Saberi
Simon Fraser University, Canada

ABSTRACT

Personality-based cognitive architectures should yield consistent patterns of behaviour through personality traits that have a modulatory influence at different levels: These factors affect, on the one hand, high-level components such as 'emotional reactions' and 'coping behaviour', and on the other hand, low-level parameters such as the 'speed of movements and repetition of gestures. In our hybrid cognitive architecture, a deliberative reasoning about the world (e.g. strategies and goals of the 3D character) is combined with dynamic real-time response to the environment's changes and sensors' input (e.g. emotional changes). Hybrid system copes dynamically with changes in the environment, and is complicated enough to have reasoning abilities. Designing a cognitive architecture that gives the impression of personality to 3D agents can be a tremendous help making 3D characters more engaging and successful in interactions with humans.

DOI: 10.4018/978-1-5225-0454-2.ch005

1 INTRODUCTION

During daily human to human interactions, people evaluate the personality of others, to predict their behavior, to understand them, to help or to motivate them (Campbell & Rushton, 1978; Funder & Sneed, 1993). One of the important sources people refer to when attributing personality to others is nonverbal behavior such as gestures, body stance, facial expressions and gaze behavior. Psychological data show a correlation between the perceived personality and nonverbal behaviour. For instance, the speed at which someone moves their hands or the direction of their gaze reveals information about their personality (Campbell & Rushton, 1978; Funder & Sneed, 1993; Borkenau & Liebler, 1992). Likewise, people tend to assess the personality of 3D virtual agents (McRorie et al., 2012; Carney, Colvin & Hall, 2007). Our goal is to design and develop a cognitive architecture for generating nonverbal behaviour to express personality for 3D virtual agents. We are focusing on the nonverbal behaviour of the agent and there is no speech involved.

The hybrid computational architecture which models the agent's personality, consists of two components: an 'Event-based' component and an 'Emotionally-Continuous' component (Saberi, Bernardet & DiPaola, 2014). The Event-based component will generate the agent's communicative gestures based on different states of the interaction and the user's behaviour. The behavioural scope is limited to strategic turn-taking interaction between the agent and the user. The agent's facial expressions will be controlled by the Emotionally-Continuous component. Opposite to communicative gestures, which are triggered when an environmental event is triggered, emotions are continuously updated based on internal and external status. The emotional weights of the gestures are also specified by the Emotionally-Continuous component. Using MATLAB/Simulink and Stateflow structure, we simulate the real-time continuous behavior in addition to event-based behavior which is responsive to changes of states of the interaction. MATLAB's Model-Based Design, facilitates model level access to components of the system which makes it very easy for future researchers to tune parameters of the system and see how that affects the results of the simulation. Matlab/ Stateflow's graphical charts are developed as modular and encapsulated libraries that can be reused across multiple charts and models.

The intended contribution of this architecture is to preserve the believability of the 3D agent over time by generating consistent behaviour while being responsive to the user. Believability is defined as providing the illusion of life and provokes the audience's suspension of disbelief (Mateas, 1999). Our proposed architecture aims at maintaining the consistency in behaviour, emotions and thoughts of the 3D virtual agents while interacting with humans. Hierarchical structuring of our proposed architecture will address behaviour, emotions and thoughts of the 3D virtual agent while modulating and encapsulating these three parts. In addition, most of the architectures designed thus far either have not been tested in real-time with high quality 3D characters. Our generic architecture is capable of a wide range of functionalities such as sensing the environment, processing the data and react dynamically and in real-time to the sensor inputs. Designing an architecture that gives the impression of personality to 3D agents through nonverbal behaviour can be a tremendous help making 3D characters more engaging and successful in their interactions with humans. It can lead to a the wider use of embodied interface agents in different areas such as story-telling systems, interactive dramas, training systems, therapy systems, museum guides, and web based receptionists.

2 THEORETICAL BACKGROUND ON PERSONALITY

Personality is the pattern of characteristic thoughts, feelings, and behaviour that distinguish one person from another and persists over time and situations (Pervin, Cervone & John, 2005). During daily human to human interactions, people evaluate the personality of others e.g. to predict their behaviour, to understand them, to help or to motivate them. One of the important sources of information people rely on when attributing personality to others is nonverbal behaviour such as gestures, body stance, facial expressions, and gaze behaviour. For instance, the speed of body movement or duration of direct gaze affects how people perceive personality of a person (Campbell & Rushton, 1978; Borkenau & Liebler, 1992). Likewise, during human and virtual humanoid character interactions, people attribute personality to virtual characters by using clues from their nonverbal behaviour (McRorie et al., 2012). Yet, in many architectures designed for virtual humanoid characters, nonverbal behaviour is generated for communication purposes and decision making processes while ignoring the importance of personality in the generation of behaviour. This leads to virtual humanoid characters with behaviours that are not consistent through time and do not follow human behavioural patterns.

2.1 Models of Personality

The science of personality is a controversial domain in the psychology discipline. Different theories on personality models have emerged by considering the effects of variables such as individual differences, the environment, varying situations, mental skills, and intelligence levels. Following is a brief review of three widely used personality models.

BIS/BAS Model of Personality

Extraversion and neuroticism traits have been associated with Gray's two-dimensional model of impulsivity and anxiety. Gray proposed that people differ in the sensitivity of their Behavioural Approach System (BAS, responsible for impulsivity) or Behavioural Inhibition System (BIS, responsible for anxiety) (Gray, 1987). People with BAS are sensitive to signals of reward and desired events, while those with BIS tend to be more sensitive to moving away from unpleasant events and punishments.

Circumplex Model of Personality

A simplified version of the Five Factor Model (FFM) is Wiggins' Circumplex model of personality, which presents the two dimensions of Affiliation, and Dominance for personality (Wiggins, 2003). This model represents the FFM's extraversion and agreeableness factors in a circumference map. By using a combination of two factors, resulting personalities can be distributed into a plane, and each point represents a specific personality (PhD, 1996). Zammitto, DiPaola & Arya (2008) used a multidimensional hierarchical approach to model Circumplex personality systems into a parameterized facial character system.

Big Five Model of Personality

The Five Factor Model (FFM) or Big Five is a comprehensive model that is widely used and validated in several studies (McCrae & John, 1992). In FFM, personality is categorized according to the following traits: Openness to experience (inventive/curious vs. consistent/cautious), Conscientiousness (efficient/organized vs. easy-going/careless), Extraversion (outgoing/energetic vs. solitary/reserved), Agreeableness (friendly/compassionate vs. cold/unkind), and Neuroticism (sensitive/nervous vs. secure/confident). These values range from 0 to 100. There are 6 "facets" of personality associated with each mentioned factor (Costa & McCrae, 1992). For this study, we narrow the work to two particularly important traits: extroversion and neuroticism, since we found more psychological data on how these traits affect the behavior and how they affect the emotion generation and expression. However, we tend to explore the effect the other three traits in future. Extraversion shows how outgoing and social a person is. Extraverts are referred to as people who enjoy being with others and participating in social activities; they have more energy, and like to engage in physical activity (Eysenck, 1967); they also like to express themselves and join conversations. By comparison, introverts are less outgoing, participate less in conversations, and show less engagement in social activities; they seem to need less stimulation and more time alone (Costa & McCrae, 1992). Neuroticism is referred to emotional instability and the tendency to experience negative emotions, such as stress and depression. As a result, same situations can be interpreted as more threatening for neurotics. Neurotics' negative emotional reactions also tend to last for unusually longer periods of time. On the other hand, people with a lower score in neuroticism are more emotionally stable. However, stability does not necessary equate to positive feelings. Research suggests that extraversion and neuroticism are negatively correlated (Norris, Larsen & Cacioppo, 2007).

2.2 Big Five Expressed through Movements

Psychological studies show a significant correlation between the impression of personality and movements. Here, we are reviewing some of the discovered links between behaviour and traits of the five factor personality model: extraversion, neuroticism, agreeableness, openness, and conscientiousness. To limit the scale of the model, we concentrate on extraversion and neuroticism traits. In future, other traits can be implemented and added to the system. Extraverts have extensive smiling (Borkenau & Liebler, 1992). They show more body movements and facial activity (La France, Heisel & Beatty, 2004), and have more frequent hand and head movements (Borkenau & Liebler, 1992). They show more gesturing, more head nods, and faster general speed of movement (Borkenau & Liebler, 1992; Borkenau, Mauer, Riemann, Spinath, & Angleitner, 2004). Based on Gill & Oberlander's empirical study, extraverts show more direct facial postures and eye contacts (Gill & Oberlander, 2002). Extraverts are sensitive to signals of rewarding and they show heightened emotional reactivity to positive-mood induction (Carver, & White, 1994; Larsen & Ketelaar, 1991). In coping situations, they show positive thinking and rational actions (McCrae & Costa, 1986). Tankard's study showed that people who looked straight seemed more active than people who looked downwards (Tankard, 1970). Based on the work of Larsen et al., gaze-avoidant women were viewed by others as not extraverted (Larsen & Shackelford, 1996). Extraverts have shorter dwelling time and a higher number of fixations (Rauthmann, Seubert, Sachse, & Furtner, 2012). They also tend to position themselves closer to others in conversation and have direct eye contact (Farabee, Nelson & Spence, 1993).

Based on Campbell and Rushton's study, neurotics were associated with touching oneself such as touching the face or the head (Campbell & Rushton, 1978). Neurotics show signs of tension or anxiety and express insecurity or sensitivity, hostility, self-pity, and guilt. They seek reassurance, behave in a fearful or timid manner, are irritable, and try to sabotage or obstruct interactions (Funder & Sneed, 1993). Additionally, highly anxious patients generated significantly more stroking, twitches, and tremors. They also maintained eye contact for significantly less time on each gaze. Low-anxiety patients smile more frequently and engaged in more manual signaling (Waxer, 1977). Neuroticism is correlated with sensitivity, nervousness, and low confidence. In Tankard's study, people who looked straight seemed more secure than people who looked downwards (Tankard, 1970). Neurotics are sensitive to signals of non-reward and punishment (Carver & White, 1994). They experience more negative emotions (e.g. anxiety and, guilt) (Funder & Sneed, 1993). In coping situations, they show the use of withdrawal, indecisiveness, and passivity (McCrae & Costa, 1986). Cook et al. showed that those who maintain lower gaze duration are considered nervous and less confident (Cook & Smith, 1975). Less eye contact is also assigned to anxiety in communication oriented research (Northey, 2008; O'Hair, Friedrich & Dixon, 2011). Multiple psychological works also revealed that as the amount of eye contact increases, people will be perceived as more self-confident (Droney & Brooks, 1993) and less anxious (Napieralski, Brooks & Droney, 1995). People with higher self-esteem would maintain eye contact for longer periods of time, and break eye contact less frequently, as compared to people with lower self-esteem (Vandromme, Hermans & Spruyt, 2011).

Borkenau & Liebler's empirical study showed that agreeable people have extensive smiling (Borkenau & Liebler, 1992). As shown in on Funder & Sneed's research, they have friendly and self-assured expressions (Funder & Sneed, 1993). Disagreeable people, on the other hand, show less visual attention but more visual dominance. They also display less back channeling (Smith, Brown, Strong & Rencher, 1975). Agreeableness is a personality trait correlated with friendliness and, being compassionate vs. cold and unkind. Tankard (1970) demonstrated that people who looked straight seemed more receptive than people who looked downwards. With a normal gaze amount of 50%, the eyes are perceived as friendly (Fukayama, Ohno, Mukawa, Sawaki, Hagita, 2002).

Conscientious people have a predominance of upward looks (Exline, 1965), and high eye contact (Funder & Sneed, 1993). They tend to avoid negations, try to control interactions, have high enthusiasm and energy levels, and engage in constant eye contact. They express warmth and show genuine interest in intellectual matters. They appear relaxed and comfortable, and offer advice to conversational partners (Funder & Sneed, 1993). Borkenau & Liebler showed that conscientious people do not show fast movements (Borkenau & Liebler, 1992), do not have frequent hand movements (Borkenau & Liebler, 1992), and do not touch themselves frequently (Borkenau & Liebler, 1992). Conscientiousness is correlated with efficiency and, being organized vs. being easy-going and careless. Gaze avoidant women were viewed by others as less conscientious (Larsen & Shackelford, 1996). Openness is associated with being relaxed and comfortable. Open people have high enthusiasm and energy levels and seem to enjoy the interactions. They engage in constant eye contact and do not behave in a fearful, timid, reserved, or unexpressive manner (Funder & Sneed, 1993). Openness is linked to inventive and curious behaviour vs. consistent and cautious manners.

3 COMPUTATIONAL BACKGROUND ON COGNITION AND PERSONALITY-EXPRESSIVE BEHAVIOUR

In the first section, we review a few influential cognitive architectures designed for modeling affective behavior. In the second section, we mainly focus on computational models that specifically address expression of the personality.

3.1 Cognitive Architectures and Affective Models

A cognitive architecture is a broad and generic computational cognitive model of the structure and process of the human mind which can be used to analyze the behavior in various domains on different levels (from microscopic physiological level to macroscopic sociological level) (Newell 1990, Sun 2002). Our proposed personality-based cognitive architecture yields consistent patterns of behaviour through personality traits that have a modulatory influence at different levels: These factors affect at one hand high-level components such as 'emotional reactions' and 'coping behaviour', and on the other hand low-level parameters such as the 'speed of movements and repetition of gestures. The architecture provides a concrete framework for modeling of personality, through two main components: an 'Event-based' component and an 'Emotionally-Continuous' component (Saberi, Bernardet & DiPaola, 2014). The Event-based component will generate the agent's communicative gestures based on different states of the interaction and the user's behaviour. The behavioural scope is limited to strategic turn-taking interaction between the agent and the user. The agent's facial expressions will be controlled by the Emotionally-Continuous component. Opposite to communicative gestures, which are triggered when an environmental event is triggered, emotions are continuously updated based on internal and external state. For details of implementation please see section 4.

Following is a review of a few influential cognitive architectures designed for modeling affective behavior. In next section, we mainly focus on computational models that specifically address expression of the personality. Fum and Stocco's developed an extension of ACT-R to reproduce Gambling Task's results (Fum & Stocco, 2004). ACT-R stands for Adaptive Control of Thought-Rational (Anderson, 2007; Anderson et al., 2004). In this model two knowledge representations are used:

declarative (consist of facts) and procedural (consist of productions which are knowledge about how humans do things). Productions are matched on perceptions which lead to an action in response to the environment or to change declarative memory. In their model emotional weight is considered as a risk probability. Emotional strength is added as a parameter to ACT-R memory activation formula. Memories associated with risk have higher probability of being recalled. Cochran et al.'s ACT-R extension (Cochran, Lee & Chown, 2006) supports arousal and valence model of affect (Kleinsmith & Kaplan, 1963). In this model, arousal parameter is added to base activation formula. Base activation decreases gradually if tagged with low arousal and increases if it is tagged with a high arousal. WASABI (WASABI Affect Simulation for Agents with Believable Interactivity) is based on BDI (Believe-Desire-Intention) cognitive theory (Becker-Asano & Wachsmuth, 2009). BDI hierarchical structures include beliefs about the environment, desires the agent wants to achieve, and intentions the agent plans to perform. On each cycle, agent decides whether to continue executing its current intention or to select a new intention. In WASABI model, BDI planning processes is affected by emotional states triggered by PAD-based emotional system. The agent's emotional state limits the set of possible next actions and goals. Thus, emotion is the crucial component of this model. EMA (Marsella & Gratch, 2009) is also designed based on BDI theory. It consists of a series of cognitive operators that explain the emotional changes during a sequence of events. Plan steps are informed by appraisal frames which lead to either emotion derivation or coping (change of strategies believes, desired and intentions). In FAtiMA (Fearnot AffecTIve Mind Architecture) information received from environment update the memory and trigger appraisal process (Dias, Mascarenhas, & Paiva, 2011). The result of the process is saved as affective state and influences the action selections. Lim, Dias, Aylett, & Paiva (2012) used FAtiMa as a base for ORIENT (Overcoming Refugee Integration with Empathic Novel Technology) architecture as an interface for interaction between users and 3D virtual agents. In this model, personality is mainly revealed to modulate an emotional framework and the appraisal process. In Soar-Emote (Marinier, Laird & Lewis, 2009), emotion is effective on three levels: biological, cognitive and social. On the biological level, it addresses physiology and body emotion system. On the cognitive level, it includes appraisal rules (Scherer, 1999), cognitive emotion system and emotion-focused coping. Finally, on the social level problem-focused coping and perception of external physiology of others are addressed. In this model, knowledge influences but not determines the emotions and feelings. FLAME which stands for "Fuzzy Logic Adaptive Model of Emotions" is based on Ortony et al. (Ortony et al.1988) and Roseman et al.'s (Roseman et al. 1990) event-appraisal models of emotion. The model uses a fuzzy-logic method to map emotional states to remembered events. The model uses learning algorithms for learning patterns of events. A computer simulation of a pet is used to evaluate the system (El-Nasr et al. 2000).

Our proposed architecture is not developed as a part of any previous cognitive architecture. However, we found some similarities between how it is structured and ORIENT architecture (Lim, Dias, Aylett, & Paiva, 2012). Similar to our work, they considered using a hybrid structure as a combination of reactive versus deliberative systems. However, we are mainly focused on realistic behavior of 3D humanoid characters. Additionally, in our design, personality not only directly affects behavior; but also it affects the generation and expression of emotion and the copying behavior of the agent. Few works exists that specifically investigate the role of personality in the cognitive model and how to generate various personality impressions for humanoid 3D characters. Also, many of the mentioned architectures are not used in an actual real-time scenario in which a virtual agent is interacting with humans while is reactive to the changes in the environment. In a few of the architectures that are empirically tested, the focus is not to create a believable and human-like behavior for the virtual agent. Thus, it is still a necessity to increase the ability of 3D character agents to behave displaying consistent bodily and facial behaviour while interacting with users in real-time. Our work addresses this gap by performing an imperative study on our designed personality model. A combination of high resolution body and facial expressions are used to enhance the expressiveness of the 3D virtual agent.

3.2 State of the Art on Computational Models for Expressing Personality

Currently, several computational models have been designed that consider personality traits as weighting parameters in different frameworks, e.g., emotional frameworks, conversational agents, and non-verbal behaviour such as gestures during the speech. Personality dimensions are also used to modulate behaviour at a decision-making level. The following is a review of related background literature on various computational models for generating the impression of the personality. André, Klesen, Gebhard, Allen & Rist (2000) developed computational models of emotions and personality for children's virtual puppet theatres, virtual sales presentations, and virtual guides for internet websites to make the interaction more enjoyable and closer to communication styles in human-human conversations. Zammitto et al. (2008) proposed a multidimensional hierarchical approach to model a parameterized facial character system, which only focused on facial features to express personality.

McRorie et al.'s (2012) work is part of a European project (SEMAINE) with the aim of developing a system that facilitates human interaction with conversational and Sensitive Artificial Listeners (SAL) agents. They designed an architecture in which personality affects the agent's non-signaling gestures during speech and appearance. The main focus in this work is the content of the conversation and behaviour during the conversation. The study empirically examines how users rate videos and images of 3D virtual agents' expressive behaviour, but no real-time interaction between humans and the agent is tested. Read et al. (2007) proposed a neural network model of structure and dynamics of personality based on research about the structure and neurobiology of human personality. Differences in the sensitivities of motivational systems, the baseline activation of specific motives, and inhibitory strength are used to model the given personality traits. The model is designed for portions of behaviour such as "Tease and Make Fun of", "Gossip and Talk about Others" and "Ask for Date" as well as for situational parameters such as "At Home" and "In Conference Room". Neff, Toothman, Bowmani, Tree & Walker (2011, January), limited their study to investigate the coloration between FFM's neuroticism trait and changes in conversations and nonverbal behaviour. They found that the presence of self-adaptors (movements that often involve self-touch, such as scratching) made agents look more neurotic.

In another study, ALMA (A Layered Model of Affect) (Gebhard, 2005) was designed to provide a personality profile with real-time emotions and moods for 3D virtual agents. ALMA is part of the 'VirtualHuman' project, which creates interactive 3D virtual agents with conversational skills. Appraisal rules and personality profiles are specified in an XML based affect modeling language. The emotions and moods are computed based on the appraisal of relevant inputs. The concentration in this study is on modulating the appraisal process, but there is no mapping between nonverbal behaviour and personality traits. Alternatively, Kshirsagar (2002) devised a personality model of emotional 3D virtual agents. They used Bayesian Belief Networks and a layered approach for modeling personality, moods and emotions. They also integrated the system into a chat application in which the developer is able to design and implement personalities, and in which users can interact with the agent. The focus in this work was also only on emotional personality. Similarly, Su, Pham, & Wardhani (2007) designed a model to control affective story characters with parameters for personality and emotion. They developed a hierarchical fuzzy rule-based system to control the body language of a story character with personality and affect. In this system, story designers specify a story context with personality and emotion values with which to drive the movements of the story characters. Likewise, Poznanski and Thagard (2005) developed a neural network model of personality and personality change: SPOT (Simulating Personality over Time). Personality-based predispositions for behaviour, moods/emotions, and environmental situations specify the output behaviour. In their model, personality develops over time, which is in turn based on the situations encountered. The focus of the study is on modeling personality change, with nine behaviour mapped to personality via output tags, e.g., "talk" or "avoid help".

Reviewing the state of the art shows that there is still a necessity to increase the ability of the 3D character agents to behave in consistent bodily and facial behaviour while interacting with users in real-time. Our work addresses this gap by performing an imperative study on our designed personality model. For the details please see section 5.1. High resolution facial models are used to display expressive movements properly. In addition, a combination of body and facial expressions are used to enhance the expressiveness of the 3D virtual agent. Interacting dynamically with a 3D virtual agent, capable of expressing unique personality, creates a rich and effective experience for users and adds to the agent's believability.

4 THE PROPOSED PERSONALITY-BASED COGNITIVE ARCHITECTURE

4.1 Architecture of the System

The hybrid computational architecture proposed to model the agent's personality consists of two components: an 'Event-based' component and an 'Emotionally-Continuous' component. The Event-based component will generate the agent's communicative gestures based on different states of the interaction and the user's behaviour. The behavioural scope is limited to strategic turn-taking interaction between the agent and the user. The agent's facial expressions will be controlled by the Emotionally-Continuous component. Opposite to communicative gestures, which are triggered when an environmental event is triggered, emotions are continuously updated based on internal and external status. The emotional weights of the gestures are also specified by the Emotionally-Continuous component. The architecture consists of inputs, parameters, outputs, and two main components being an Emotionally-Continuous System and an Event-based system (Figure 1). In conjunction, the Emotionally-Continuous System and the Event-based system generate the gestures, postures and facial expressions that are expressing personality. The environmental inputs will be fed to the Event-based system through the sensors. The Event-based-system is responsible for the logic and flow of the interaction. Based on rules and configuration parameters defined at the beginning or during the interaction and goals and strategies specified for the agent, the Event-based system determines the generated gestures. Emotionally-Continuous system is responsible for generating facial expressions and gestures which continuously calculates the emotion by comparing the agent's goals with inputs such as the user's actions. In addition to facial expressions and gestures, Emotionally-Continuous system also affects the

Figure 1. Architecture of the proposed system (Personality traits affect mainly four parts of the architecture: 1- Gestures e.g. extraverts show faster hand and head movements; 2- Facial expressions e.g. extroverts filter less and show more facial emotions; 3- Coping mechanism e.g. neurotics coping strategy is to withheld; 4- Emotional reactions e.g. neurotics experience more negative emotions in general.)

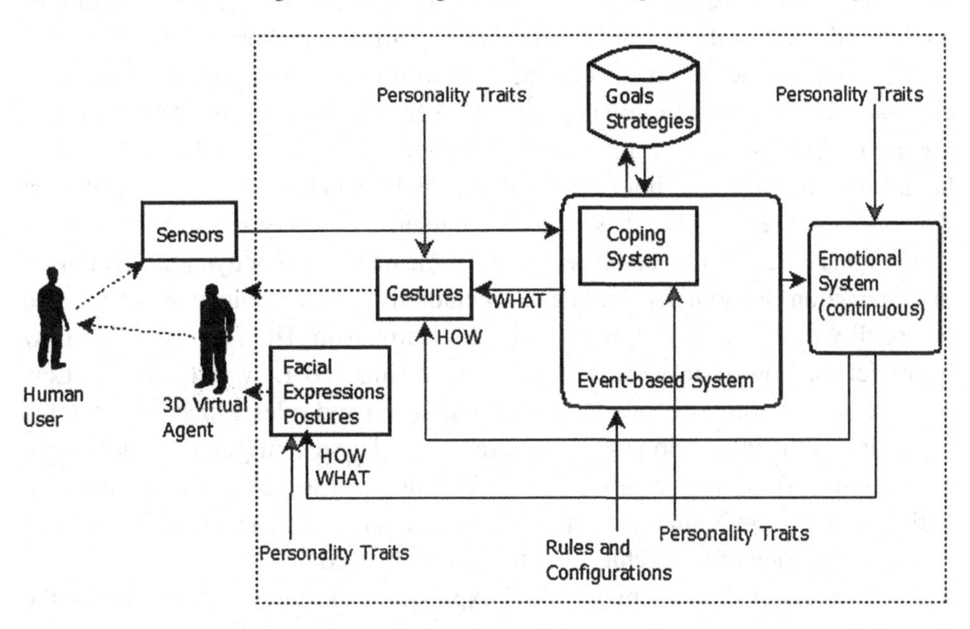

quality of the gestures and how they are expressed. For example, an angry head nod is faster than a sad head nod. The combination of gestures, facial expressions and postures are fed to the animation toolkit continuously and are presented to the user.

Personality in the System

We are using the FFM model, which is a widely used method in affective computing. All five factors have convergent and discriminant validity across instruments and observers (McCrae & Costa, 1986). The FFM also has a biological basis (Fritz & Angleitner, 1994). For instance, there is a large body of research on genetic influences on neuroticism and extraversion (Eaves, Eysenck & Martin, 1989). In addition, there is empirical evidence regarding the link between the five traits and nonverbal behaviour (Borkenau & Liebler, 1992; Campbell & Rushton, 1978). Personality traits affect mainly four parts of the system. In the lowest level personality affects the gestures, facial expressions and postures of the agent. As mentioned before, types and parameters of gestures are different in extraverts and neurotics. For instance, extraverts show faster hand and head movements. Personality also affects the expression of facial emotions. People with different personalities tend to filter their emotions differently. For example, extroverts tend to express their feelings more easily as they do less filtering (Borkenau & Liebler, 1992). Another aspect through which personality is revealed is the coping mechanisms and emotional reactions of the agents to users' behaviour. Next section will give a detailed description of how coping and emotional reactions are different in different personalities.

The inputs of the system are continuously received from sensors installed in the environment. Various sensors, such as the Kinect 3D camera and overhead cameras, are used to send streams of inputs such as users' coordination in the space, users' heights or environmental information such as noise and light. The agent will react dynamically in real-time to these inputs. Parameters of the system can be tuned by the designer of the system at the beginning of the interaction or dynamically during the interaction. Personality parameters include the four combinations of the two personality traits (High Extroversion-High Neuroticism, High Extroversion-Low Neuroticism, Low Extroversion-High Neuroticism, and Low Extroversion-Low Neuroticism. In addition, interaction configurations and rules which are set at the beginning of the interaction are input parameters. These configurations determine the rules specific to the scenario of the interaction, such as the desired coordination of the user in the environment. Goals and strategies of the interaction are hard-coded in the first version of the architecture to narrow the study.

The outputs of the system are facial expressions, postures and gestures of the 3D virtual character, which are dynamically generated and fed into the animation toolkit before being displayed to the user. Gestures of the agent, such as pointing or nodding, are dynamically controlled based on the events that occur during the interaction. For example, if the user performs a certain task, the agent nods. On the other hand, facial expressions and postures of the agent are controlled continuously and based on the continuous feedback from the environment. In addition, the emotional value of gestures and how they are performed are controlled with the Emotionally-Continuous component. For instance, if the agent is angry, he shakes his head faster.

Event Based System

We limit the behavioural scope of the architecture to strategic turn-taking interaction between the agent and the user. Because of the discrete and turn-based nature of Event-based system, this component of the system is implemented using the finite state machine (FSM). A state machine is a set of input events, output events and states. FSM is in one state at a time. The state the FSM is in at any given time is called the current state. The machine can change from one state to another (which is called transition) when an event or condition is triggered. A FSM is defined by a list of its states, and the triggering condition for each transition (Selic, Gullekson, & Ward, 1994). The state machine in our architecture determines what the agent's gestures and gaze behaviours are in correspondence to the events, conditions, goals and strategies. In addition, the FSM controls the turn-taking behaviours of the interaction. Based on the scenario the turn-taking behaviour can be synchronized using users' actions or environment inputs. For example, turn-taking between the agent and the user can be synced using the coordination of the user in the space. It is the agent's turn to act only if the user is standing still (his coordination is not changing). In FSM, specific events or information triggers corresponding gestures from the agent. For instance, if the agent wants to guide the user to move to a place, he points to the location and gazes at the user to encourage him. Based on the scenario of the interaction, predefined goals and strategies, and inputs the agent gets from the environment, he decides what state he should transfer to and what gestures he should express in response to the user. For example, if the goal is to lead the user to a specific coordination (and if, based on the sensor value, the user is not yet in that coordination), the agent's next action is to convince the user to move there by pointing. If the user does not comply by moving to the desired place, the agent may try different strategies or give up, based on the coping strategy he selects

Coping

Based on Lazarus and Folkman (1984), coping is a mechanism of dealing with problems while trying to minimize conflict. The effectiveness of a coping mechanism depends on the circumstances of the events and the individual's personality. Therefore, when the environment's feedback is not compatible with the agent's goals, the agent strategizes and acts accordingly. If the feedback does not change, regardless of the agent's effort, the agent becomes distressed and will deal with the situation based on his personality. In our architecture, coping is a part of the Event-based system. Based on the situation and personality of the agent, the agent picks a specific coping mechanism. As explained in the previous sections, coping mechanisms are different for different personalities. Extraverts tend to think positively and react rationally, while neurotics tend to withhold and become passive in coping situations.

Emotionally-Continuous System

The Emotionally-Continuous component is for controlling the emotional reactions to the environment. Emotions are updated continuously and are not necessarily based on specific events. Event-based gestures are only triggered based on specific states of the interaction so they do not care about the period of the time passed. Emotional valence and arousal on the other hand is changing based on how much time is passed being in a specific state. For instance, if the agent is continuously receiving negative feedbacks from the environment, his disappointment can continue to increase. If he cannot find a way to avoid the negative feedback, after a certain amount of time, he may start finding a way to cope with this situation. Event-based states, on the other hand, are timeless. Emotionally-Continuous module is mainly responsible for generating facial expressions but it also affects the gestures and poses of virtual character. In the next two sections we go over the details of how personality affects the emotion generation in Emotionally-Continuous module. Then we go over the details of how these generated emotions will be expressed by the 3D virtual character.

Emotional Responses which Express Personality

The emotional reactions of extravert and neurotic agents are different in several aspects, including coping mechanism and reaction to positive and negative feedbacks. Based on psychological data, extraverts show a stronger emotional response to compliance than denial (Carver & White, 1994). Thus, in our architecture, when the user's compliance is positive (meaning the user is acting based on the agent's desire), the valence curve has a high exponential raise. However, extraverts are not sensitive to negative feedbacks so when compliance is negative, the decrease of valence is not as fast. In addition, based on psychological studies, extraverts experience more positive emotions in general (Larsen & Ketelaar, 1991), and their coping mechanism involves acting rationally and thinking positively (McCrae & Costa, 1986). Neurotics, on the other hand, show a stronger response to denial than compliance (Carver & White, 1994). Therefore, in the model, the valence value is decreased with a higher rate in response to negative feedbacks. Neurotics experience more negative emotions in general (Funder & Sneed, 1993) and their coping mechanism is to withhold and be passive (McCrae & Costa (1986).

Generating Facial Emotions for 3D Characters

Emotion can be activated by external or internal environmental effects and lasts for a short amount of time (Ortony, Norman, & Revelle, 2005; Moffat, 1997). One of the most widely used dimensional models is Russell's circular configuration of Affect (Russell, 1980). Russell proposed a two-dimensional model that includes two polars of valence (displeasure-pleasure) and arousal (relaxation-excitation) (Figure 2). The Circumplex model of affect includes all possible combinations of arousal and valence. For example, happiness is expressed as a pleasure-high arousal.

In the Circumplex model, (Figure 2), each emotion can be understood as a linear combination of valence (pleasure) and arousal (Russell, 1980). Based on the personality of the agent, some of the emotions will be filtered and considered as inter-

Figure 2. Circumplex model of emotion (Russell, 1980)

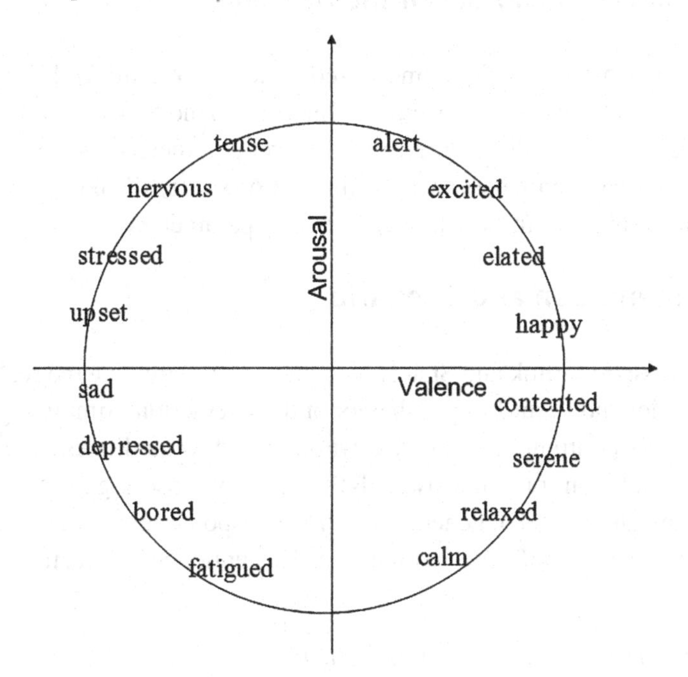

nal emotional states, and some will be demonstrated to the users. For the emotions that should be revealed externally to the user, generated values of valence and arousal in the Emotionally-Continuous System will be mapped to Ekman's Action Coding System to generate facial expressions for the agent (Ekman, 1977). The Facial Action Coding System (FACS) was the earliest approach to systematically describe facial actions in terms of small Action Units (AUs) such as left-eye-lid-close and jaw-open. Paul Ekman and W.V. Friesen proposed the original FACS in the 1970s by verifying how the contraction of each facial muscle (individually and in combination with other muscles) changes the appearance of the face. The goal of proposing FACS was to produce a proper and reliable way to categorize facial actions. They used videos of facial actions to recognize the differences caused by muscles' interactions, and how to detect them. Studying anatomy, reproducing the appearances, and palpating their faces led Ekman and Friesen to speculate on relationships between appearance modifications and the effects of muscles. FACS measurements are described in terms of Action Units and, not in the activations of the muscles themselves, since in some cases, each AU is a composition of several muscles activations. Otherwise, some individual muscle movements may not result in recognizable facial changes. Additionally, sometimes one muscle produces an appearance modification that decomposes to two or more AUs in order to show the independent actions of different parts of the muscles.

4.2 Hierarchical Structure of the System

The model has a hierarchical parameterized structure (Figure 3). It includes low-level parameters such as “face movements” and “torso movements”. A combination of the low-level data results in intermediate-level parameters such as “smile” and “nod”. Higher-level parameters such as “interaction strategy” and “personality” are formed from combining these intermediate-level parameters.

4.3 Implementation and Scenario

Using MATLAB/Simulink and Stateflow structure, we simulate the real-time continuous behavior and event-based behavior of the interaction. Simulink provides an infrastructure for continuous simulation while Stateflow facilitates the event-based design. The simulation continuously reads the values of sensors, goes though states of interaction, updates the GUI, generate emotional response and sends behavior commands to the behavior realizer (see Figure 4). As mentioned before, the components

Figure 3. Hierarchical structure of the system

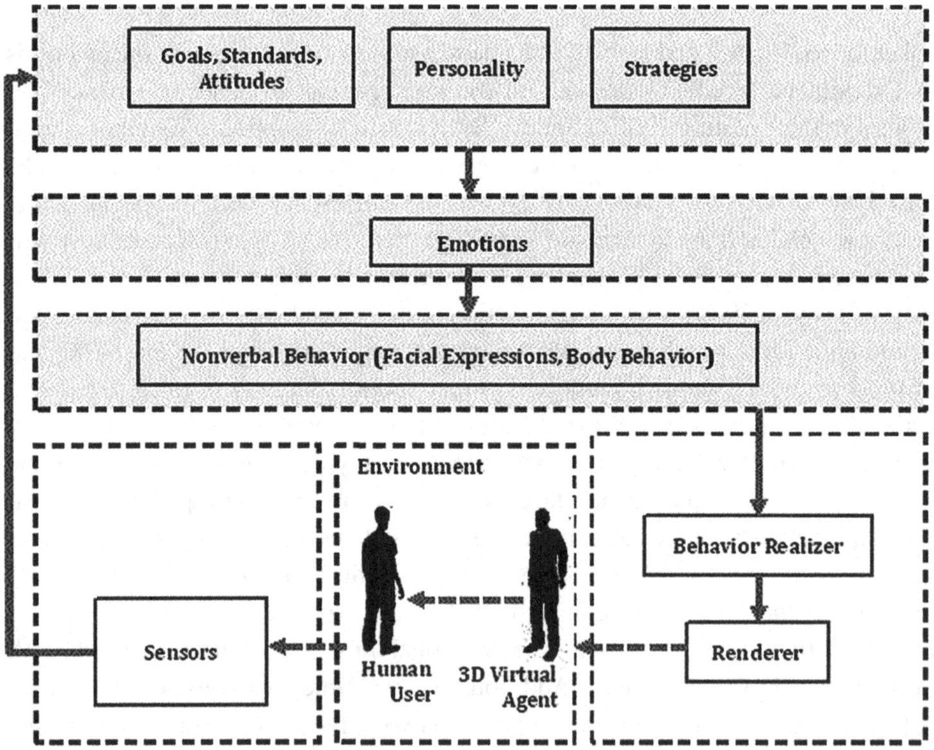

Figure 4. Implementation of the system

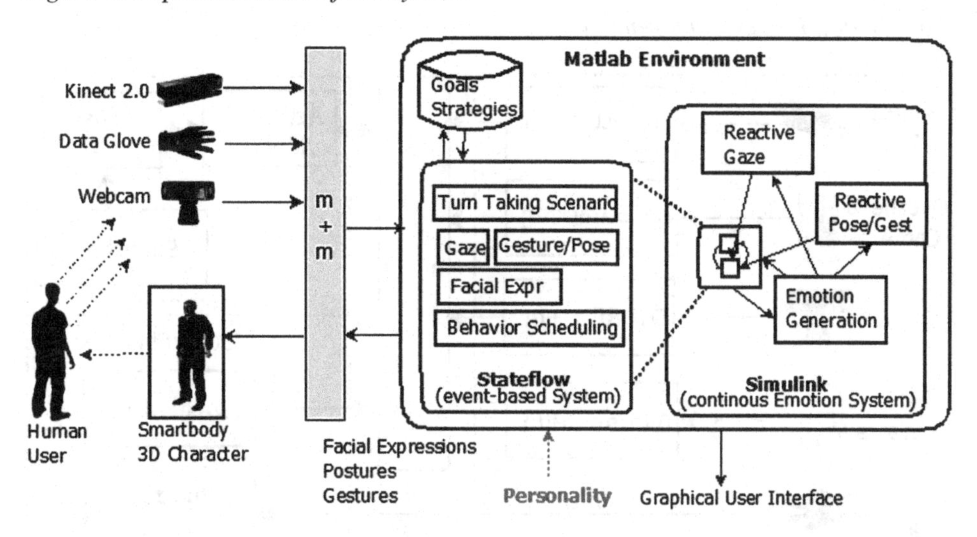

of the system which have event-based nature such as scenario of the interaction, gaze, gesture, pose and facial expression of emotion are coded using Stateflow. On the other hand, the component which has continuous nature such as generation of emotion and reactive gaze and pose to the environment changes are implemented using Simulink blocks. All event-based behavior are queued and prioritized in the scheduler module which uses a fixed three priority level. Behavior which are reactive to the environment such as "agent smiles if user smiles" are in the highest priority; behavior responsive to the scenario of the interaction such as playing a rock paper or scissor hand is in the second priority; Idle behavior are in the lowest priority. Generated behavior commands are sent to the animation realizer. Movement + Meaning (m+m) is used as a platform to facilitate acquiring data, managing and rendering movement data (see Figure 5). Key features of the m+m middleware are portability between platforms, low latency and high bandwidth (http://mplusm.ca).

5 FUTURE WORK AND CONCLUSION

5.1 Empirical Testing in Real Time Interaction

To truly evaluate how users perceive the personality of virtual humanoid character based on their nonverbal behavior, a real-time interaction between the user and the virtual character with focus on nonverbal behavior is required. Here we propose a

Figure 5. Movement and Meaning (m+m) platform is used to facilitate acquisition, process, and render of movement data

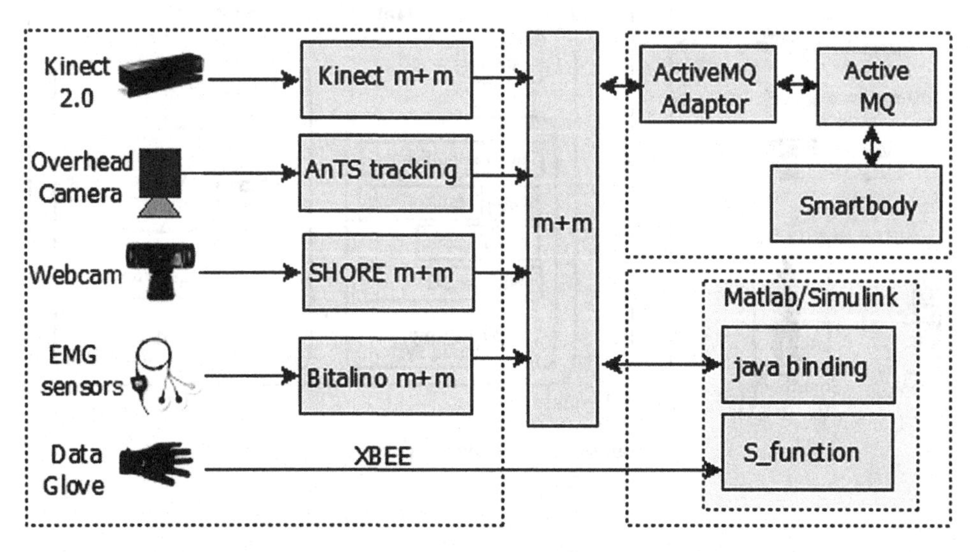

test case scenario for evaluating the personality model, with the goal of creating an easy-to-learn and engaging scenario that provides an interactive environment with minimum conversation. Our testing paradigm is a scenario of an interactive game between a user and a virtual humanoid character. Rock-Paper-Scissors game is proposed as a demo scenario with the goal of creating an easy-to-learn and engaging scenario that provides an interactive demonstrator with minimum conversation (see Figure 6). The demo user stands in front of a monitor demonstrating a virtual character. Smartbody an academic 3D character animation toolkit is used as our animation rendering system which provides locomotion, gazing and nonverbal behavior in real time under our scripted control via the Behavior Markup Language (Thiebaux et al, 2008). In each game session, 10 rounds of Rock-Paper-Scissors are played between the user and the character. The character's hand gestures (rock, paper or scissors) are generated randomly. A data glove sensor is designed to capture the hand gesture of the user. A webcam feeds the facial information of the demo user to SHORE application to detect user's smile (Brinkman, 2015). Microsoft Kinect 2.0 feeds the joint information of the user to the system. Using this data, the system locates the user and extracts the user's gestures. At the end of each session user evaluates the personality of the character. After or during the interaction users will rank the personality of the virtual character using the TIFI personality test. Results will be compared with parameters specified before the experiment.

Figure 6. A user plays with a virtual character which behaves based on commands in receives from a Matlab application

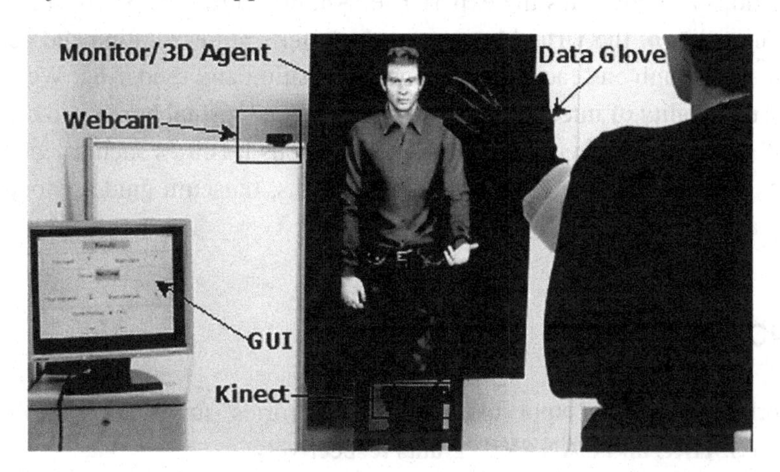

5.2 Summary and Conclusion

Consistency and coherence in behaviour are key concepts for creating virtual characters which follow the behaviour patterns of humans. Explicitly endowing virtual characters with a personality individualizes them by ensuring they exhibit consistent behaviour (Wang & Mckenzie, 1998). We propose a computational architecture to generate perception of personality through nonverbal behaviour of a virtual character during dynamic interactions with human users. A hierarchical hybrid model is designed which combines logical behaviour and continuous updates of the emotional expressions of the virtual character while encapsulate different layers of the system. Two main component of the system Emotionally-Continuous system and the Event-based system in conjunction generate the gestures and facial expressions that express personality. The Event-based system is responsible for the logic and flow of the interaction and generates the virtual character's communicative gestures based on different states of the interaction and the user's behaviour. Emotionally-Continuous system is continuously updated and controls the virtual character's facial expressions. The system continuously calculates the emotion by comparing the virtual character's goals with inputs such as the user's actions. In proposed architecture personality traits have an influence at four levels: 1) At generation of non-verbal behaviour 2) At the level of filtering of emotions. 3) At the coping behaviour of the virtual character. 4) At the emotional reactions of the virtual character.

Different aspects of our architecture can be extended and enriched in future. Right now the inputs from the environment are limited to user's coordination and height. Our system has the possibility to feed more channels of inputs from the environment such as noises and light changes. Being responsive to users' gestures and emotions also enriches the experience. Another future step is to increase the graphical quality of the virtual humanoid character's representation and generated animations via high end facial expressions, animation and rendering. We hope increasing the quality of interaction between human and virtual humanoid characters lead to the wider use of virtual characters in various faculties such as educational systems, therapy systems, games, training systems, museum guides, story-telling and interactive dramas.

ACKNOWLEDGMENT

This work was partially supported by "Moving Stories" and "Moving+Meaning" Canadian SSHRC and CANARIE grants respectively.

REFERENCES

Anderson, J. (2007). *How can the human mind occur in the physical universe?* Oxford University Press. doi:10.1093/acprof:oso/9780195324259.001.0001

Anderson, J. R., Bothell, D., Byrne, M. D., Douglass, S., Lebiere, C., & Qin, Y. (2004). An integrated theory of the mind. *Psychological Review*, *111*(4), 1036–1060. doi:10.1037/0033-295X.111.4.1036 PMID:15482072

André, E., Klesen, M., Gebhard, P., Allen, S., & Rist, T. (2000). Integrating models of personality and emotions into lifelike characters. In *Affective interactions* (pp. 150–165). Springer Berlin Heidelberg. doi:10.1007/10720296_11

Becker-Asano, C., & Wachsmuth, I. (2009). Affective computing with primary and secondary emotions in a virtual human. *Autonomous Agents and Multi-Agent Systems*, *20*(1), 32–49. doi:10.1007/s10458-009-9094-9

Borkenau, P., & Liebler, A. (1992). Trait inferences: Sources of validity at zero acquaintance. *Journal of Personality and Social Psychology*, *62*(4), 645–657. doi:10.1037/0022-3514.62.4.645

Borkenau, P., Mauer, N., Riemann, R., Spinath, F. M., & Angleitner, A. (2004). Thin Slices of Behaviour as Cues of Personality and Intelligence. *Journal of Personality and Social Psychology*, *86*(4), 599–614. doi:10.1037/0022-3514.86.4.599 PMID:15053708

W. P. Brinkman, J. Broekens, & D. Heylen (Eds.). (2015). Intelligent Virtual Agents. In *15th International Conference, IVA 2015*, (Vol. 9238). Springer.

Brooks, R. A. (1986). A robust layered control system for a mobile robot. *Robotics and Automation. IEEE Journal of, 2*(1), 14–23.

Campbell, A., Rushton, J. (1978). Bodily communication and personality. *The British Journal of Social and Clinical Psychology, 17*(1), 31-36.

Carney, D. R., Colvin, C. R., & Hall, J. A. (2007). A thin slice perspective on the accuracy of first impressions. *Journal of Research in Personality, 41*(5), 1054–1072. doi:10.1016/j.jrp.2007.01.004

Carver, C. S., & White, T. L. (1994). Behavioural inhibition, behavioural acti-vation, and affective responses to impending reward and punishment: The BIS/BAS scales. *Journal of Personality and Social Psychology, 67*(2), 319–333. doi:10.1037/0022-3514.67.2.319

Cochran, R., Lee, F., & Chown, E. (2006). Modeling emotion: Arousal's impact on memory.*Proceedings of the 28th Annual Conference of the Cognitive Science Society* (pp. 1133–1138).

Cook, M., & Smith, J. M. (1975). The role of gaze in impression formation. *The British Journal of Social and Clinical Psychology, 14*(1), 19–25. doi:10.1111/j.2044-8260.1975.tb00144.x PMID:1122344

Costa, P. T. Jr, & McCrae, R. R. (1992). Four ways why five factors are basic. *Personality and Individual Differences, 13*(6), 653–665. doi:10.1016/0191-8869(92)90236-I

Damasio, A. (1994). *Descartes' error: Emotion, reason, and the human brain*. New York: Putnam.

Dias, J., Mascarenhas, S., & Paiva, A. (2011). FAtiMA Modular: Towards an agent architecture with a generic appraisal framework.*Proceedings of the International Workshop on Standards for Emotion Modeling*.

Droney, J. M., & Brooks, C. I. (1993). Attributions of Self-Esteem as a Function of Duration of Eye Contact. *The Journal of Social Psychology, 133*(5), 715–722. doi:10.1080/00224545.1993.9713927 PMID:8283864

Eaves, L. J., Eysenck, H. J., & Martin, N. G. (1989). *Genes, culture, and personality: An empirical approach*. New York: Academic Press.

Ekman, P. (1977). *Biological and cultural contributions to body and facial movement*. Academic Press.

El-Nasr, M. S., Yen, J., & Ioerger, T. R. (2000). Flame—fuzzy logic adaptive model of emotions. *Autonomous Agents and Multi-Agent Systems*, *3*(3), 219–257. doi:10.1023/A:1010030809960

Exline, W. (1965). Affect Relations and Mutual Gaze in Dyads. In S. Tomkins & C. Izard (Eds.), *Affect, Cognition and Personality*. Springer.

Eysenck, H.J., (1967). *The Biological Basis of Personality*. Thomas.

Farabee, D., Nelson, R., & Spence, R. (1993). Psychosocial Profiles of Crim-inal Justice- and Noncriminal Justice-Referred Substance Abusers in Treatment. *Criminal Justice and Behavior*, *20*(4), 336–346. doi:10.1177/0093854893020004002

Fukayama, A., Ohno, T., Mukawa, N., Sawaki, M., & Hagita, N. (2002). Messages embedded in gaze of interface agents — impression management with agent's gaze. *Proceedings of the SIGCHI Conference on Human Factors in Compu-ting Systems, CHI '02* (pp. 41–48). New York: ACM. doi:10.1145/503376.503385

Fum, D., & Stocco, A. (2004). Memory, emotion, and rationality: An ACT-R interpretation for gambling task results.*Proceedings of the Sixth International Conference on Cognitive Modeling* (pp. 106–111).

Funder, D. C., & Sneed, C. D. (1993). Behavioural manifestations of person-ality: An ecological approach to judgmental accuracy. *Journal of Personality and Social Psychology*, *64*(3), 479–490. doi:10.1037/0022-3514.64.3.479 PMID:8468673

Gebhard, P. (2005, July). ALMA: a layered model of affect. In *Proceedings of the fourth international joint conference on Autonomous agents and multiagent systems* (pp. 29-36). ACM. doi:10.1145/1082473.1082478

Gill & Oberlander, J. (2002). Taking Care of the Linguistic Features of Ex-traversion. *Proc. 24th Ann. Conf. Cognitive Science Soc.*, 363-368.

Gray, J. A. (1987). The neuropsychology of emotion and personality. In S. M. Stahl, S. D. Iversen & E. C. Goodman (Eds.), Cognitive neurochemistry (pp. 171-190). Oxford, UK: Oxford University Press.

Kleinsmith, L. J., & Kaplan, S. (1963). Paired-associate learning as a function of arousal and interpolated interval. *Journal of Experimental Psychology*, *65*(2), 190–193. doi:10.1037/h0040288 PMID:14033436

Kshirsagar, S. (2002, June). A multilayer personality model. In *Proceedings of the 2nd international symposium on Smart graphics* (pp. 107-115). ACM. doi:10.1145/569005.569021

La France, B., Heisel, A., & Beatty, M. (2004). Is There Empirical Evidence for a Nonverbal Profile of Extraversion? A Meta-Analysis and Critique of the Literature. *Communication Monographs*, *71*(1), 28–48. doi:10.1080/03634520410001693148

Laird, J., Newell, A., & Rosenbloom, P. S. (1987). SOAR: An architecture for general intelligence. *Artificial Intelligence*, *33*(1), 1–64. doi:10.1016/0004-3702(87)90050-6

Larsen, R. J., & Ketelaar, T. (1991). Personality and susceptibility to posi-tive and negative emotional states. *Journal of Personality and Social Psychology*, *61*(1), 132–140. doi:10.1037/0022-3514.61.1.132 PMID:1890584

Larsen, R. J., & Shackelford, T. K. (1996). Gaze avoidance: Personality and social judgments of people who avoid direct face-to-face contact. *Personality and Individual Differences*, *21*(6), 907–917. doi:10.1016/S0191-8869(96)00148-1

Lazarus, R. S., & Folkman, S. (1984). Stress. *Appraisal, and Coping*, 456.

Lim, M. Y., Dias, J., Aylett, R., & Paiva, A. (2012). Creating adaptive af-fective autonomous NPCs. *Autonomous Agents and Multi-Agent Systems*, *24*(2), 287–311. doi:10.1007/s10458-010-9161-2

Marinier, R. III, Laird, J., & Lewis, R. (2009). A computational unification of cognitive behaviour and emotion. *Cognitive Systems Research*, *10*(1), 48–69. doi:10.1016/j.cogsys.2008.03.004

Marsella, S. C., & Gratch, J. (2009). EMA: A process model of appraisal dynamics. *Cognitive Systems Research*, *10*(1), 70–90. doi:10.1016/j.cogsys.2008.03.005

Mateas, M. (1999). *An Oz-centric review of interactive drama and believable agents*. Springer Berlin Heidelberg. doi:10.1007/3-540-48317-9_12

McCrae, R. R., & Costa, P. T. (1986). Personality, coping, and coping ef-fectiveness in an adult sample. *Journal of Personality*, *54*(2), 385–404. doi:10.1111/j.1467-6494.1986.tb00401.x

McCrae, R. R., & John, O. P. (1992). An Introduction to the Five-Factor Model and Its Applications. *Journal of Personality*, *60*(2), 175–215. doi:10.1111/j.1467-6494.1992.tb00970.x PMID:1635039

McRorie, M., Sneddon, I., McKeown, G., Bevacqua, E., de Sevin, E., & Pe-lachaud, C. (2012). Evaluation of four designed virtual agent personalities. *Affective Computing. IEEE Transactions on*, *3*(3), 311–322.

Moffat, D. (1997). Personality Parameters and Programs. In R. Trappl & P. Petta (Eds.), *Creating Personalities for Synthetic Actors* (pp. 120–165). New York: Springer Verlag. doi:10.1007/BFb0030575

Napieralski, L. P., Brooks, C. I., & Droney, J. M. (1995). The effect of du-ration of eye contact on American college students' attributions of state, trait, and test anxiety. *The Journal of Social Psychology, 135*(3), 273–280. doi:10.1080/00224545.1995.9713957 PMID:7650931

Neff, M., Toothman, N., Bowmani, R., Tree, J. E. F., & Walker, M. A. (2011, January). Don't scratch! self-adaptors reflect emotional stability. In *Intelligent Virtual Agents* (pp. 398–411). Springer Berlin Heidelberg. doi:10.1007/978-3-642-23974-8_43

Newell, A. (1990). *Unified theories of cognition*. Cambridge, MA: Harvard University Press.

Norris, C. J., Larsen, J. T., & Cacioppo, J. T. (2007). Neuroticism is associated with larger and more prolonged electrodermal responses to emotionally evocative pictures. *Psychophysiology, 44*(5), 823–826. doi:10.1111/j.1469-8986.2007.00551.x PMID:17596178

Northey, M. (2008). *Impact: a guide to business communication*. Toronto: Pearson Prentice Hall.

O'Hair, D., Friedrich, G. W., & Dixon, L. D. (2011). *Strategic communica-tion in business and the professions*. Boston: Allyn and Bacon.

Ortony, A. (2002). On making believable emotional agents believable. Academic Press.

Ortony, A., Norman, D. A., & Revelle, W. (2005). Effective functioning: A three level model of affect, motivation, cognition, and behaviour. In J. Fellous & M. Arbib (Eds.), *Who needs emotions? The brain meets the machine* (pp. 173–202). New York: Oxford Univeristy Press. doi:10.1093/acprof:oso/9780195166194.003.0007

Ostendorf, & Angleitner. (1994). A comparison of different in-struments proposed to measure the Big Five. *European Review of Applied Psychology, 44*(1), 45-55.

Pervin, L. A., & Cervone, D. (2005). *Personality: Theory and Research*. Academic Press.

Poznanski, M., & Thagard, P. (2005). Changing personalities: Towards real-istic virtual characters. *Journal of Experimental & Theoretical Artificial Intelligence, 17*(3), 221–241. doi:10.1080/09528130500112478

Rauthmann, J. F., Seubert, C. T., Sachse, P., & Furtner, M. R. (2012). Eyes as windows to the soul: Gazing behaviour is related to personality. *Journal of Research in Personality*, *46*(2), 147–156. doi:10.1016/j.jrp.2011.12.010

Read, S. J., Monroe, B. M., Brownstein, A. L., Yang, Y., Chopra, G., & Miller, L. C. (2010). A neural network model of the structure and dynamics of human personality. *Psychological Review*, *117*(1), 61–92. doi:10.1037/a0018131 PMID:20063964

Russell, J. A. (1980). A circumplex model of affect. *Journal of Personality and Social Psychology*, *39*(6), 1161–1178. doi:10.1037/h0077714

Saberi, M., Bernardet, U., & DiPaola, S. (2014). An Architecture for Personality-based, Nonverbal Behaviour in Affective Virtual Humanoid Character. *Procedia Computer Science*, *41*, 204–211. doi:10.1016/j.procs.2014.11.104

Scherer, K. R. (1999). Appraisal theory. In T. Dalgleish & M. Power (Eds.), Handbook of cognition and emotion, (pp. 637–663). Chichester, UK: John Wiley & Sons. doi:10.1002/0470013494.ch30

Selic, B., Gullekson, G., & Ward, P. T. (1994). *Real-time object-oriented modeling* (Vol. 2). New York: John Wiley & Sons.

Smith, B., Brown, B., Strong, W., & Rencher, A. (1975). Effects of Speech Rate on Personality Perceptions. *Language and Speech*, *18*, 145–152. PMID:1195957

Su, W. P., Pham, B., & Wardhani, A. (2007). Personality and emotion-based high-level control of affective story characters. Visualization and Computer Graphics. *IEEE Transactions on*, *13*(2), 281–293.

Sun, R. (2002). *Duality of the Mind*. Mahwah, NJ: Lawrence Erlbaum Associates.

Tankard, J. W. Jr. (1970). Effects of eye position on person perception. *Perceptual and Motor Skills*, *31*(3), 883–893. doi:10.2466/pms.1970.31.3.883 PMID:5498193

Thiebaux, M., Marsella, S., Marshall, A. N., & Kallmann, M. (2008, May). Smartbody: Behavior realization for embodied conversational agents. In *Proceedings of the 7th international joint conference on Autonomous agents and multiagent systems* (vol. 1, pp. 151-158). International Foundation for Autonomous Agents and Multiagent Systems.

Vandromme, H., Hermans, D., & Spruyt, A. (2011). Indirectly Measured Self-esteem Predicts Gaze Avoidance. *Self and Identity*, *10*(1), 32–43. doi:10.1080/15298860903512149

Wang, F., & Mckenzie, E. (1998). *Virtual life in virtual environments. Uni-versity of Edinburgh*. Computer Systems Group.

Waxer, P. (1977). Nonverbal cues for anxiety: An examination of emotional leakage. *Journal of Abnormal Psychology, 86*(3), 306-314.

Wiggins, S. (2003). *Introduction to Applied Nonlinear Dynamical Systems and Chaos* (2nd ed.). New York: Springer-Verlag.

Zammitto, V., DiPaola, S., & Arya, A. (2008). A methodology for incorpo-rating personality modeling in believable game characters. *Arya, 1*.

Chapter 6
The Contemporary Craft of Creating Characters Meets Today's Cognitive Architectures:
A Case Study in Expressivity

Selmer Bringsjord
Rensselaer Polytechnic Institute, USA

John Licato
Indiana University/Purdue University – Fort Wayne, USA

Alexander Bringsjord
Motalen, Inc., USA

ABSTRACT

What does the contemporary craft of character design (by human authors), which is beyond the reach of foreseeable AI, and which isn't powered by any stunning, speculative, AI-infused technology (immersive or otherwise), but is instead aided by tried-and-true "AI-less" software tools and immemorial techniques that are still routinely taught today, imply with respect to today's computational cognitive architectures? This chapter narrows the scope of this large question, and argues that at present, perhaps only the cognitive architecture CLARION can represent and reason over knowledge at a level of logical expressivity sufficient to capture such characters, along with the robust modeling implied by contemporary story and character design.

DOI: 10.4018/978-1-5225-0454-2.ch006

INTRODUCTION

Ibsen's characters are deep, memorable, and often dark; a contemporary Norwegian writer, Ullmann, is following suit.[1] With such fiction in mind, one might ask:

Q1. Could an AI generate robust and engaging characters like these?
Q2. Could revolutionary AI and sensory-immersion technology augment the ability of human writers to generate robust and engaging characters like these?

Elsewhere, answers, at least provisional ones, have been offered to this pair. In the case of (Q1), S. Bringsjord (2001) has defended a negative (a view not necessarily in concord with those of Licato); in the case of (Q2), Bringsjord & Bringsjord (2009) have defended an affirmative.[2] In the present chapter, the interest is in a very different question, one firmly rooted in the here and now, to wit:

Q3. What does the contemporary craft of character design (by human authors), which is beyond the reach of foreseeable AI, and which isn't powered by any stunning, speculative, AI-infused technology (immersive or otherwise), but is instead aided by tried-and-true "AI-less"[3] software tools and immemorial techniques that are still routinely taught today, imply with respect to today's computational cognitive architectures?

(Q1) is a question for AI researchers and engineers. (Q2) is a question for them as well, and for those working with them, from the field of human-computer interaction. (Q3) however, is not a question for engineers: it is a question for computational cognitive scientists, or more specifically, for computational cognitive modelers. (Q3) is relevant to those who proclaim, today, that they have on hand a "computational cognitive architecture" that captures, in one framework, most, if not all, of the nature and range of human cognition.

(Q3) can be concretized by pinning down the craft in question, and by doing the same for computational cognitive architectures and the modeling made possible by them. For the former, the authors accomplish this by focusing on the craft of character creation and design in the widely used, tried-and-true Movie Outline® system.[4] For the latter, we focus on two cognitive architectures: ACT-R (Anderson, 1993, Anderson & Lebiere, 1998), probably the best-known cognitive architecture today, and one unmistakably aligned with the sanguine view that this architecture is well on its way to capturing the nature and breadth of human cognition (e.g. see the bold Anderson & Lebiere, 2003); and on CLARION (Sun, 2002), a cognitive architecture that uniquely founds its expressive power (among other things) on sub-symbolic processing, in keeping with what Sun (2001) has called "the duality of mind."

The foregoing sets our general context. Since (Q3) is a large question, the authors narrow its scope so as to be able to productively consider it in the space of but the present chapter. The narrowing is accomplished by making three moves. First, the discussion is anchored to a simple, single story – *Double-Minded Man* – populated by only a pair of characters, and crafted in a particular "AI-less" software system taught and used by scriptwriters today. Second, the specific focus on the *expressivity* of a cognitive architecture, in light, specifically, of two expressivity challenges that arise from considering *Double-Minded Man* and its characters.[5] Finally, cognitive architectures are partitioned, with respect to these challenges, into three categories. The first category is composed of those architectures that appear to have no chance of meeting the challenges; the second, of those that perhaps meet the challenges; and the third, of those that clearly meet the challenges. As shall be seen, the cognitive architecture ACT-R falls into the second category, and CLARION falls into the third. Encapsulating, then, here's the core of our answer to (Q3):

(A_{Q3}): While ACT-R and cognitive architectures that have its level of expressivity or less are perhaps able to support cognitive models of characters in narrative, CLARION certainly has a level of expressivity sufficient to capture such characters. While we know of no other cognitive architecture that has this high level of expressivity, any architecture with this level of expressivity would at least be a candidate for the robust modeling implied by contemporary story and character design.

The basis for (A_{Q3}) is a purported demonstration that the nature of characters created in Movie Outline 3 for even a short-short film generates acute challenges. *A fortiori,* full-length works would deliver the same moral.

The structure of the sequel of the present chapter is straightforward: The authors first provide a brief, self-contained overview of the concept of expressivity. Next, in section 3, a synopsis of *Double-Minded Man* is provided, and by doing so, its only two characters are introduced: Harriet and Joseph; a summary is made of the demands made by Movie Outline 3 on the author who wishes to create rich characters; and a pair of specific expressivity challenges arising from the composition of *Double-Minded Man* in Movie Outline 3 is presented. The chapter then briefly discusses the category of cognitive architecture that: have no chance to meet the two expressivity challenges; have perhaps a chance of meeting the two challenges; and can without question meet the two challenges. As an exemplar of architectures in the second of these categories, the authors use ACT/ACT-R (Anderson & Lebiere, 2003), and as an exemplar of architectures in the third category, the authors use CLARION (Sun 2003). Next, the authors address a number of concerns and objections. The chapter ends with a brief section devoted to summing up the moral of the authors' investigation, and to pointing toward where its future lies.

WHAT IS EXPRESSIVITY?

Perhaps the best way to approach the topic of expressivity is to treat English declarative sentences as input to an algorithm R that outputs the formalized meaning of the input sentences,6 in the form of representations of those sentences, where these representations are expressed as formulae in the formal language of some logic placed within some received continuum.7 For instance, there is the well-known continuum that begins with zero-order logic (ZOL), then moves to first-order logic (FOL), then to second-order logic (SOL), then to third-order (TOL), and so on; each of these is more expressive than its predecessors.8 One can then view R as computing a function, and accordingly write R: $\sigma \rightarrow \varphi\sigma L$, where σ is an English declarative sentence, and $\varphi\sigma L$ is a formula in the particular logic L that's in play. Even for readers unfamiliar with the continuum just alluded to, it's easy to get the hang of it in general, and easy to get the hang of how it works in connection with R. Here's how.

Consider first the following sentence, which is true in Double-Minded Man.

(A) Harriet loves Joseph.

We can easily represent this sentence in ZOL (which subsumes the propositional calculus). ZOL allows us to use relation symbols to represent relations (or properties), constants to represent individual objects, and to then build formulae that ascribe these relation symbols to these constants. This logic also includes the familiar quintet of truth-functional connectives: $\wedge$ ("and"), $\vee$ ("or"), $\rightarrow$ ("if _ then _"), and $\leftrightarrow$ ("_if and only if _"). No quantification is available in ZOL, and there are no variables (but identity, =, is included; with = comes inference schemas for reasoning with identity). For handling (A), a relation symbol is needed to represent the two-place property of one thing loving another, and two constants are needed, one to denote Harriet and one to denote Joseph. With these ingredients and some straightforward punctuation, one can specify R[(A)]:

(A') $L(h, j)$

Now here's an example of a more expressive English sentence that is also true with respect to the story *Double-Minded Man:*

(*B*) Joseph owns a BMW sportbike.

To represent (*B*), we can use the following formula in FOL:[9]

(*B*') $\exists x(B(x) \wedge S(x) \wedge O(j, x))$

The symbol "∃" is the well-known *existential quantifier*, and when paired with variable *x*, as in "∃*x*," can be read as "there exists an *x* such that." As to *B, S,* and *O* in (*A*'), they of course are relation symbols that represent, resp., the properties *being BMW-manufactured, being a sportbike,* and *being a thing _ that owns a thing* _. FOL, as most readers will doubtless recall, also includes the *universal quantifier*, ∀, which allows us to capture such sentences as:

(*C*) Every spouse of Joseph owns more than one Bible.

which is also true of Harriet in *Double-Minded Man,*[10] and would by ***R*** yield

(*C*') $\forall x[Sp\,(x, j) \rightarrow (\exists x \exists y \exists z(x \neq y \wedge Bi\,(x) \wedge Bi\,(y) \wedge O(z, x) \wedge O(z, y)))]$

To be clear about the aforementioned continuum: ZOL was invoked by regimenting the ascription of properties to particular individual things picked out by constants, then the move from ZOL to FOL was made by allowing quantification over lower-case variables *x, y,*... – that are placeholders for individual objects. To next move from FOL to SOL, simply extend quantification so that it ranges over not only individual objects, but over relation symbols (and hence ultimately over properties). To quickly get a sense of what this specifically buys us, consider an English sentence that communicates another truth about *Double-Minded Man:*

(*D*) There are attributes Joseph has that Harriet's mental model of him doesn't.

Let's denote Harriet's mental model of Joseph by *m*. Then we can represent (*D*) in SOL as follows:

(*D*') $\exists X\,(X(j) \wedge \neg X\,(m))$

A move, at this point, to TOL would consist in a direct analogue of what got us started in the first place when we built ZOL: viz., we allow ascription of a new class of relation symbols to our original set of relation symbols. So far, relation symbols have been pulled from the set of upper-case Roman letters, and applied to individual objects only. For our new class of relation symbols, which allow us to ascribe properties to properties, we can avail ourselves of *F, R, S, ...*; we put these symbols to use below, when issuing the first of our two promised expressivity challenges.

The second of those challenges requires the deployment of machinery from a different class of logics: *epistemic* logics.[11] Whereas the FOL-to-TOL sequence, as well as the logics that follow in the continuum, are well-suited to modeling non-cognitive domains like number theory, epistemic logics are marked by the addition of an entirely new device: *operators.* The two key operators in epistemic logic are **K** and **B,** which represent resp., "knows" and "believes." In order to add these operators to our ZOL-to-TOL sequence systematically, we have only to affirm the following grammatical rule:

(R) If φ is a well-formed formula, then so are $\mathbf{B}_c\varphi$ and $\mathbf{K}_c\varphi$ where c is a constant denoting a member of a reserved set of constant symbols used to denote agents.

Note that operators can be iterated. In Bringsjord and Ferrucci (2000), an example from rudimentary mystery fiction is used to show that, where *d* denotes the detective, *v* the villain the detective is seeking to find and bring to justice, and *V* is a relation symbol representing the property *being a villain,* a formula like the following one represents something that writers of such fiction routinely know in the course of their writing:

(V) $\mathbf{B}_d\mathbf{B}_v\mathbf{B}_d V(v)$

DOUBLE-MINDED MAN, AND EXPRESSIVITY CHALLENGES THEREFROM

Gist of Double-Minded Man

Double-Minded Man is a two-character, three-scene short-short film script written by S. Brinsgjord and A. Bringsjord (1992) in and through *Movie Outline 3,* in a way that follows the approach to creative writing described (Bringsjord, 1992). The term "Double-Minded" in the title refers to the mind of Joseph, by all accounts for years a rather boring and morally upright financial advisor working out of his own small suburban firm, but who is in reality a hard-drinking, cocaine-and-heroin-using, sportbike-driving, decidedly non-monogamous chap. Here's the current version of the first scene:

TWIRL – DAY

68-year-old Harriet Smith sits with two wrinkled hands firmly on the wheel of her rust-eaten Subaru wagon, staring straight ahead through the top level of bifocals as she waits serenely at a red light.
Harriet is alone in the car. To her right is another vehicle, also waiting, in this case to make a right turn; it's a sleek, low-slung, black Camaro.
We are inside the cabin with Harriet. The Subaru's sound system softly plays choral music. Harriet's lips move slightly as she internally sings along, mouthing a slow aria. Her head weaves slightly side to side, in rhythm with the music.
Things are calm as can be here inside the car with Harriet. There are a pair of well-worn Bibles on the empty passenger seat beside her, one with a gold-lettered "Harriet" on its leather front cover, the other with a matching "Joseph" on its front cover.
Harriet's eyes swivel up to the light: still red. We wait with her.
Suddenly there is a piercing SCREECH outside. Harriet jerks her head to the right and we follow her line of sight.
A sleek motorcycle has swerved out of its lane and is now streaking straight for the right side of the Camaro beside Harriet's car.
The bike slams with CLANG into the side of the Camaro. Its rider is flung up and forward into the air, twirling passed Harriet's windshield.
We now watch from Harriet's POV, in slow motion. The black-leather-clad motorcyclist sails by Harriet's windshield, airborne. We see a man's face, clearly: His elephant-hide skin tells us that he is well beyond middle-age. Yet thick, black curls of youthful hair emerge from under his helmet. The face has only one half of a black, bushy, swept-out, waxed mustache. His eyes are weary and grey, and appear to lock with Harriet's for an instant.
We return to normal speed. The body is now lying on the incoming lane to the left of Harriet's Subaru, perfectly still on the blacktop, the head twisted into an impossible angle. Blood seeps from a nostril. Beside the lifeless head, a BMW medallion lies on the pavement, glinting in the sunlight.

Joseph, Harriet's 69-year-old husband, is thus dead on the spot. As we learn in the next scene, set in a morgue, Joseph's real hair consists of but anemic, white fuzz, and the second half of his fake mustache has been removed. The third and final scene is Harriet's visit to Joseph's secret lair, a small and dingy storage unit where he parked his sleek and super-fast motorcycle in secret. She there comes upon a number of things that reveal the dark aspects of Joseph's secret life, of which, hitherto, she had not the slightest inkling.

BASIC DEMANDS OF MOVIE OUTLINE 3

Movie Outline 3, like its predecessors 1 and 2, is a system that makes instructively serious demands of the creative writer who would use it. This is especially true with respect to the *characters* in a story created and crafted by the creative writer. For example, the creation of a character requires that 16 initial data-fields be completed. One of these fields is Education, and in Joseph's case it's recorded that he has a BS in Math, an MBA in Finance, and is a Chartered Financial Planner (CFP); but none of these details appear in the script itself. Notice that we here say *initial* data: Movie Outline 3 asks for no less than 68 deep questions in order to evince and record rich information about a given character; this is the so-called Interview component of character construction in the system. There are other aspects of character design encouraged by Movie Outline 3, but what we say here is sufficient, we trust, to allow our readers to understand that the pair of expressivity challenges to which we now turn are not in the least exotic. It would be easy, and fair given what Movie Outline 3 requests, to present much more demanding expressivity challenges. Parallel propositions would arise in connection with the creation in Movie Outline 3 of *any* significant characters in *any* substantive narrative.[12]

HARRIET, JOSEPH, AND A PAIR OF EXPRESSIVITY CHALLENGES

In *Double-Minded Man,* the following sentence holds:

1. Joseph knows at some time before his death that his wife Harriet doesn't know (at that same time) that Joseph has a racy lair.

At a minimum, when **R** is applied to (1), we obtain:

(1') $\mathbf{K}_j\,(t < d \wedge \neg\mathbf{K}_h\,(t, R(j)))$

As to TOL, there is the following, necessarily known by the authors of *DMM:*

2. There are attributes Joseph has that are dark and racy.

For (2), notice that we need to quantify over relations (SOL), and we reach TOL because one or more of these relations has the property of being dark-and-racy:

(2') $\exists X\,(X(j) \wedge DR(X))$

These, then, are the two expressivity challenges we had promised to issue.

THE "NO WAY" CATEGORY AND THE EXPRESSIVITY CHALLENGES

Some cognitive architectures, as far as anyone can honestly tell, clearly are incapable of meeting expressivity challenges like (1) and (2). Since it's likely to be somewhat distressing for proponents and developers of the cognitive architectures in this category to learn that the architectures they presumably think highly of are in fact (in this regard) lowly, we tread very lightly by saying very little about membership in this category. Unfortunately, it seems to us that a number of cognitive architectures still under active development fall into the "no way" category. To give at least one example, we note that RCS (Albus, 2000), which to our knowledge is not under active development, appears to be unable to represent (1)/(1') and (2)/(2'). The reason is simple. The parts of RCS designed to hold representations of the external world in a knowledge-base, to use terms Albus (2000) employs, the parts of RCS that include "frames, objects, classes, and rules," are not associated with any such machinery as complex quantification, application of properties to properties, and operators added to the mix as well.

THE "MAYBE" CATEGORY, ACT-R, AND THE EXPRESSIVITY CHALLENGES

Turning now to ACT-R, even under the most charitable assumptions, it appears incapable of expressing (1') and (2'), and hence appears incapable of representing the cognition that arises from writing or reading and understanding *Double-Minded Man.* This is so for the simple reason that at most ACT-R (as a descendant of ACT) has the expressivity of FOL, as shown in Anderson (1976). We say "at most" because for example the process described by Anderson (1976) for converting a formula φ_σ in FOL to an expression in ACT passes through a phase where $\varphi_\sigma L$ (where L = FOL) is replaced by a corresponding formula $\varphi^{SNF}{}_{\sigma L}$ in Skolem Normal Form. But it's a well-known theorem that the meaning of $\varphi^{snf}{}_{\sigma L}$ doesn't in the arbitrary case correspond to the meaning of $\varphi_\sigma L$. That is, it's not the case that any model **M** satisfying $\varphi_\sigma L$ also satisfies $\varphi^{SNF}{}_{\sigma L,}$ and *vice versa.*

Other work devoted to providing a "formal semantics" for ACT-R since 1976 has not covered expressivity in our sense, and has nothing to do with formal semantics as operationalized in the logician's and linguist's sense of ***R***. For example, Gall and Frühwirth (2014) give, as they explain, merely an *operational* formal semantics for ACT-R. Theirs is an investigation of the computational processes associated with production-system-processing, not of the expressivity of ACT/ACT-R with respect to formulae as the formal bearers of the meaning of declarative sentences in natural languages like English.

Note that it's very hard to see how identity in the semantics provided for ACT in Anderson (1976) provides the power of FOL. For example, the first prime number greater than 3 is identical to the sum of 4 and 1. This fact is easily captured in FOL by for example:

$$\exists x\,[P(x) \land x > 3 \land \neg\exists y\,(P(y) \land y < x) \land x = +(4,1)]$$

It doesn't seem possible for simple arithmetical facts like this to be represented directly in ACT/ACT-R, in such a way that meaning is preserved, even leaving aside the apparently fatal problem we have already cited. This is so because the way identity works in FOL doesn't seem to be carried over to ACT/ACT-R. In FOL, for that matter in ZOL, if we have that $R(a)$ and that $a = b$, the upshot is that exactly the same models render $R(b)$ true as render $R(a)$ true.[13] Yet Anderson (1976) writes that in ACT, even given that $a = b$, $R(a)$ and $R(b)$ don't have the same meaning. Yet these formula *do* have the same meaning in ZOL and FOL.

Furthermore, the sense of expressivity used by the present paper should be distinguished from the idea of *knowledge level* (Newell, 1981). Soar, for example, can justifiably boast of its ability to represent and reason over a wide variety of types of knowledge that vary qualitatively: procedural, episodic, declarative, and so on (Rosenbloom, Newell, & Laird, 2014). But like ACT/ACT-R, little to no work has been done in demonstrating that these representations have a logical expressivity equivalent to FOL, much less second- or any higher-order logics. In short, then, the metric that we are concerned to apply and discuss herein is one that hasn't been addressed. We don't know how Soar would measure on this metric, but one can of course view the present chapter as, in the end, a call for a wider application of the rigorous metric that is our concern. Put another way, since our metric is standard fare in measuring the relative expressivity of precise ways of expressing declarative information (e.g. see Lindström's Theorems in Ebbinghaus, Flum, & Thomas, 1984), it would appear that the Soar community hasn't yet measured the capacity of Soar to represent declarative information.

Finally, note that while (as we have just noted) Anderson (1976) discusses expressions having *believes* within them, this string is treated no differently than any other predicate, and the referential opacity of belief in English isn't formalized.[14] There would be two ways to carry out such formalization. The first way would be through some machinery tailor-made for the purpose, and beyond standard model theory (or a variant) for ordinary first-order logic; for instance, through a possible-worlds semantics (which is often used for epistemic logic; e.g., see Fagin, Halpern, Moses, & Vardi, 2004). But we read:

Development of a possible worlds semantics may also be a better way to deal with the opacity of propositions just discussed. However, I do not feel technically competent to develop a possible worlds semantics (Anderson, 1976, p. 225).

Neither is an alternative pursued. For instance, one might pursue a proof-theoretic semantics (Bringsjord, Govindarajulu, Ellis, McCarty, & Licato, 2014), but that isn't done by Anderson (1976).

Despite the foregoing problems, we rest content out of charity to make only the circumspect claim that it certainly *seems* to be the case that ACT-R cannot meet our two expressivity challenges. After all, the thrust of Anderson (1976) is that ACT has the expressivity of FOL, not TOL, and certainly not TOL in conjunction with operators.

We turn now to CLARION, and its capacity to meet our pair of challenges.

CLARION AND THE EXPRESSIVITY CHALLENGES

Overview of CLARION

CLARION is an integrative cognitive architecture that has a dual-process structure consisting of two levels: explicit (top level) and implicit (bottom level) (Sun, 2002). CLARION has been able to model a wide variety of cognitive phenomena while maintaining psychologically plausible data structures and algorithms; this makes it an ideal choice for our purposes. By showing that structured reasoning can emerge from no more than the mechanisms in CLARION which previous literature have already shown to be psychologically plausible, we provide a strong foundation for showing that these new structures are psychologically plausible as well.

The architecture is further divided into four *subsystems,* each with explicit and implicit levels, which specialize in different aspects of cognition: The Motivational Subsystem (MS), the Metacognitive Subsystem (MCS), the Action-Centered Subsystem (ACS), and the Non-Action-Centered Subsystem (NACS). We will be focusing on the NACS in the present essay.

NACS: THE NON-ACTION-CENTERED SUBSYSTEM

The NACS contains general knowledge about the world that is not contained in the ACS. Whereas the ACS is meant to capture the knowledge that directly causes decision-making which interacting with the world, the knowledge in the NACS is often more deliberative and is used for making inferences. The top level of the NACS is called the General Knowledge Store (GKS), and it contains localist chunks that can be linked to each other using Associative Rules (ARs).

The bottom level of the NACS is called the AMN, or the Associative Memory Network, and it contains implicit associative knowledge encoded as dimension-value pairs (DV pairs). Each GKS chunk is connected to a set of DV pairs in the AMN with some weight that can be adjusted over time. This unique structure gives CLARION the ability to define a *directed* similarity measure between two chunks c_1 and c_2, which is derived from the amount of overlap between the DV pairs connected to the two chunks (Sun, 1995, Tversky, 1977, Sun & Zhang, 2004):

$$S_{c_1 \to c_2} = \frac{\sum_{i \in c_2 \cap c_1} W_i^{c_2} \times A_i}{f\left(\sum_{i \in c_2} W_i^{c_2} \times A_i\right)} \tag{1}$$

where $f(x) = x^{1.0001}$. Sun and Zhang (2004) define A_i as the strength of activation of the values of dimension i in chunk c_1 and $W_i^{c_2}$ as the weights of the DV pairs specified with respect to c_2. However, in this chapter we simplify things by settling all A and W values to 1, which reduces Equation 1 to a function of the number of DV pairs connected to c_1 and c_2:

$$S_{c_1 \to c_2} = \frac{\left|c_1 \cap c_2\right|}{\left|c_2\right|^{1.0001}} \tag{2}$$

Note that it is possible for the denominator in Equation 2 to be zero, in which case the entire equation is given the default value of 1.

The associative rules (ARs) link groups of chunks to other chunks in the GKS, and consist of a set of condition chunks c_1, c_2, … and a single conclusion chunk d. For any given AR, each condition chunk i has a weight W_i such that $\Sigma_i W_i = 1$. We write out a single associative rule in the following format:

$$(c_1, c_2, \ldots, c_n) \Rightarrow d$$

The chunks in the GKS and DV pairs in the AMN have activation levels, which can be set by CLARION's other subsystems. Activations can also spread through the NACS using the chunk-DV pair connections and the top-level ARs. The manner in which this activation spreads can be restricted: other subsystems can temporarily disable Rule-Based Reasoning (activation spreading through ARs) or Similarity-Based Reasoning (activation spreading through chunk similarity), or perform activation propagation as some weighted combination of both of these reasoning types. These abilities are detailed further in Sun and Zhang (2004, 2006), in which these mechanisms are shown to be psychologically plausible by using them to closely emulate the results of psychological studies. We use no more than these mechanisms to construct the knowledge structures in this chapter.

THE EXPRESSIVITY OF CLARION

Representing Structured Knowledge in General

The associative rules linking NACS chunks already seem to impart a kind of weak structure to the GKS, but they do not constitute structural knowledge in the sense described by, for example, (Hummel & Holyoak, 1997, Holyoak & Hummel, 2000, Licato, 2015). In order to impart structure and to allow CLARION to represent knowledge with high levels of expressivity, (Licato, Sun, & Bringsjord, 2014a) introduced *types* of chunks. The scope of these types, however, holds only within the context of a complete structure, for reasons we will explain shortly. It is convenient to start by basing our structures on the well-established S-expressions, which are perhaps most notable for their use in the programming language LISP. An S-expression is of the form:

$(P\ o_1\ o_2 \ldots o_n)$

where P is a predicate, and each o_i is either an object or another S-expression. Our first chunk type, then, is the object chunk. We also define a proposition chunk, which is both a marker of the relationship between the object chunks, and a placeholder for the proposition's predicate symbol. The proposition and object chunks are pictured in Figure 1 as oval-shaped objects.

Of course, something needs to link these chunks together, and that is where Cognitively Distinguished Chunks (CDCs) come in. Given that all neurobiologically normal adult humans are capable of performing structured reasoning, we should assume that there are some common cognitive abilities that are either innate or develop very early in life which allow for structured knowledge to emerge. CDCs

Figure 1. A knowledge structure representing the proposition CHASES (DOG, CAT). On the right is the simplified version, which omits the CDCs and many of the ARs, though they are there (just not pictured). (Things are kept generic here via reference to the proverbial dog-cat scenario. Building structures specifically tailored to Double-Minded Man we judge to be otiose in the present context – but see our return to this issue in the final section of the present chapter.)

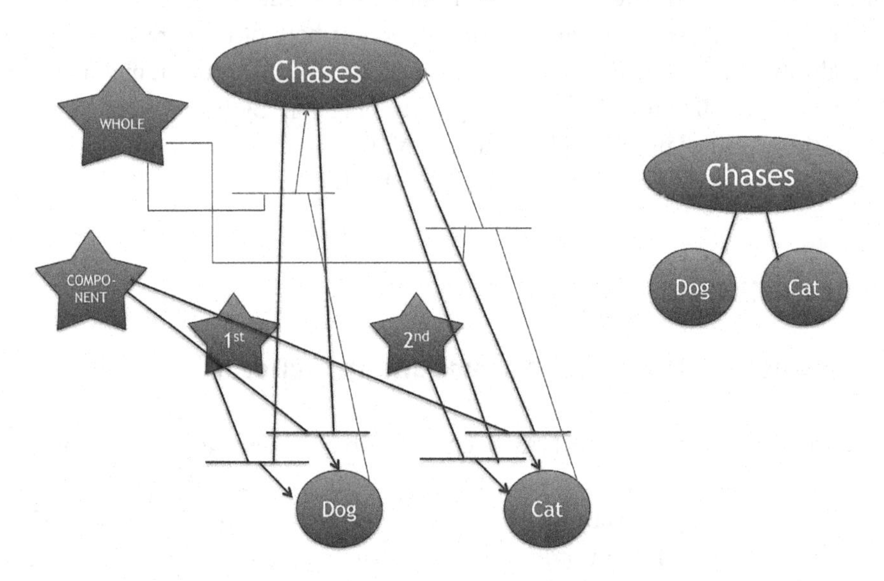

are meant to reflect these abilities, and we maintain psychological plausibility by placing the following restrictions on them: Firstly, CDCs are fixed – we do not define any algorithms that create or destroy CDCs. Secondly, CDCs are known to basic reasoning algorithms. For example, the algorithm used to perform analogical reasoning (Licato, Sun & Bringsjord, 2014a) can refer to certain CDCs directly, under the assumption that these are basic features of structured knowledge. Finally, if there is a function that can be easily performed using a CDC, then that function is assumed to be a basic ability of any neurobiologically normal adult human reasoner. This final point will be elaborated shortly.

CDCs are depicted as star-shaped (Figure 1). Associative rules link the CDCs to the chunks in the structure. For example, the *WHOLE* CDC links object nodes to proposition nodes. In Figure 1, which depicts the proposition *CHASES (DOG, CAT),* the *WHOLE* CDC is part of two ARs (depicted in the Figure as an arrow with multiple tails and one head):

(DOG, WHOLE) $\Rightarrow$ *CHASES*

(CAT, WHOLE) $\Rightarrow$ CHASES

Each object chunk in the condition of these ARs has a weight of 0.5. In fact, for all ARs we mention in the present chapter, the weight is distributed evenly amongst all objects in the AR's condition unless otherwise mentioned. If an AR contains a CDC as one of its condition chunks, then we call that AR a *CDC-based AR*. As we mentioned earlier, one limitation on CDCs is that CDC-based ARs should correspond to basic abilities of neurobiologically normal adult human reasoners. This means that given some chunks in a knowledge structure, if it is easy to traverse CDC-based ARs and retrieve some other part of that structure, then that retrieval should also be something humans can do easily. The "basic ability" that the above two ARs correspond to is the ability to recall propositions involving an object, given nothing but that object. Imagine being asked to recall some fact about dogs. Among others, one of the facts that likely would be recalled is that dogs chase cats (assuming, of course, that the reasoner in question is aware of this fact). That is modeled here by activating the *WHOLE* and *DOG* chunks. The activation would spread through any ARs which contain those two chunks in their conditions, and the resulting proposition nodes would be activated.

A *COMPONENT* CDC is also defined to introduce some redundancy into the structure, such that for every rule involving a *WHOLE* CDC, a complementary rule going in the other direction is created with a *COMPONENT* CDC. Whole chunks are always pictured above component chunks. To preserve argument order, we introduce *Ordinal CDCs,* which are also pictured in Figure 1 as 1st, 2nd, etc. Ordinal CDCs simply preserve the roles objects play within propositions in a general way that does not name the roles specifically (contrast this with the LISA model used in Hummel and Holyoak (2003), which has distinct role units for every type of role).

The basic proposition structure we have been describing can also be nested, so that instead of an object chunk a proposition chunk can have another proposition chunk as a component. A proposition chunk can even have a single object chunk as a component multiple times, as would be necessary in the proposition $P(a, X, a)$.

FOL-LEVEL EXPRESSIVITY IN CLARION

The work described in this chapter was designed in part to be at least as expressive as first-order logic (FOL), the touchstone for assessing the expressivity of extensional logics (Ebbinghaus, Flum, & Thomas, 1994).[15] In this section we show that a major part of the goal to at least reach FOL has been met. We do this by first showing how the full syntax of FOL can be represented in our knowledge structures.

We spare the reader a straightforward proof by induction over terms and formulae in first-order logic, in which one demonstrates that every formula in a countably infinite progression of all first-order formulae can be represented in the scheme we have introduced.

In order to represent FOL formulae, we adopt a structure that directly maps to human-readable syntax. We represent the universal and existential quantifiers as if they were higher-level predicates, using identity links to connect quantified variables to instantiations within the variable's range. This can be seen in Figure 2, along with an example of negation, which is similarly treated as a kind of single-place predicate.

Such structures can exist in this format for easy recall by the reasoner. When they are to be used in active reasoning processes, however, the structures can be recalled from long-term memory and transformed into a form more amenable to reasoning processes (e.g., to the structures in Figure 2), perhaps by some procedure which originates in one of CLARION's other subsystems. Such a process is a bit more involved than the examples we demonstrate here. For example, this process needs to distinguish between the chunks corresponding to the universal and existential quantifiers in order to treat their corresponding structures differently in inferences.

Allowing for a native representation that contains quantifiers is a first step in simulating a so-called "natural" reasoning process, that is, a set of mechanisms that are known to better correspond to how humans, as opposed to machines, reason (e.g., something akin to *natural deduction,* introduced in Jaskowski (1934)). Aiming at natural reasoning may seem an odd choice, considering that modern automated theorem provers tend to prefer methods such as first-order resolution, but we remind the reader that our goal here is to model reasoning in a psychologically plausible way.

Figure 2. First-order-logic formulas $\forall x\ P(x)$ and $\exists x\ \neg P(x)$

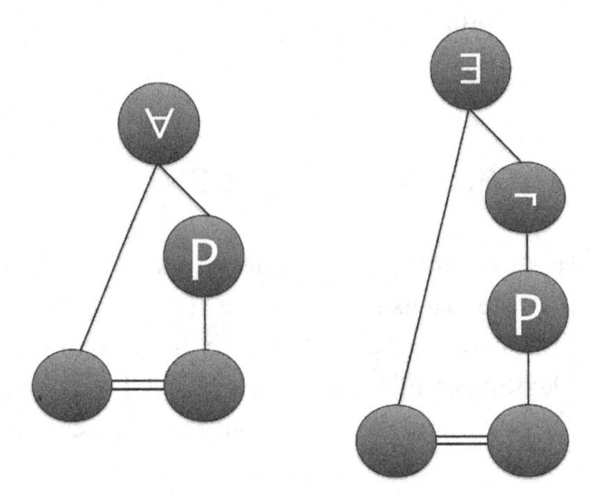

CHALLENGES (1) AND (2) MET

By expanding on the structuring method used to represent FOL formulae in the NACS, we can represent formulae in SOL and TOL (Figure 3). For SOL, the expansion requires that the first argument under the quantifier chunk (pictured as the leftmost child of the chunk corresponding to the ∀ quantifier) is connected, via an identity link, to another blank chunk that serves the role of a predicate rather than an object. A similar change makes TOL representation possible.

It should be noted that the difficulty in ensuring cognitive architectures have higher-level expressivity is not in manually creating tree-like structures that resemble formulae in higher-order logics. That can be done rather trivially. Rather, giving a cognitive architecture the expressivity of a higher-order logic is about accurately capturing the *inferential abilities* of that higher-order logic being able to carry out inferences or produce new structures, *if and only if* the formal semantics of the logic allows them (recall our earlier remarks about a failure of deduction using identity in ACT-R).

Ensuring that an inferential system only carries out allowed inferences requires, at minimum, an ability to very carefully reason over deep structural properties of a representation.[16] CLARION's NACS – again, due to work in Licato et al. (2014a) and Licato, Sun, and Bringsjord (2014b) – makes such controlled structure-based inference possible, by making use of what are called *templates,* pictured in Figure 4. A template is simply an organization of NACS structures specifying what properties

Figure 3. The proposed representations of formulae in SOL (left) and TOL (right) are pictured above. These particular structures correspond to the formulae $\forall P\ P(x)$ and $\exists X\ (X(j) \wedge DR(X))$

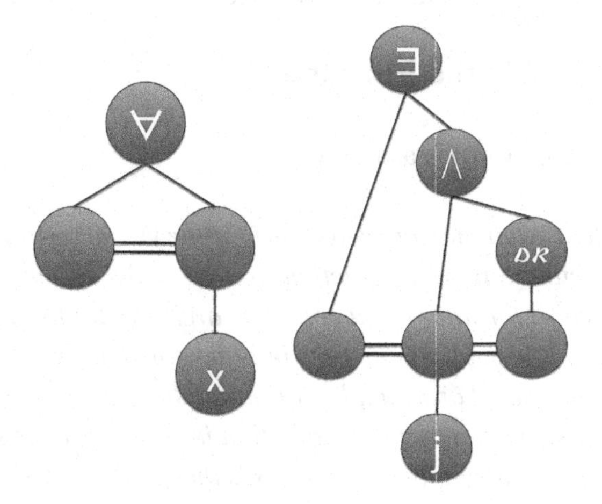

Figure 4. An example of the template structure allowing inferences in CLARION's NACS (from Licato et al. (2014))

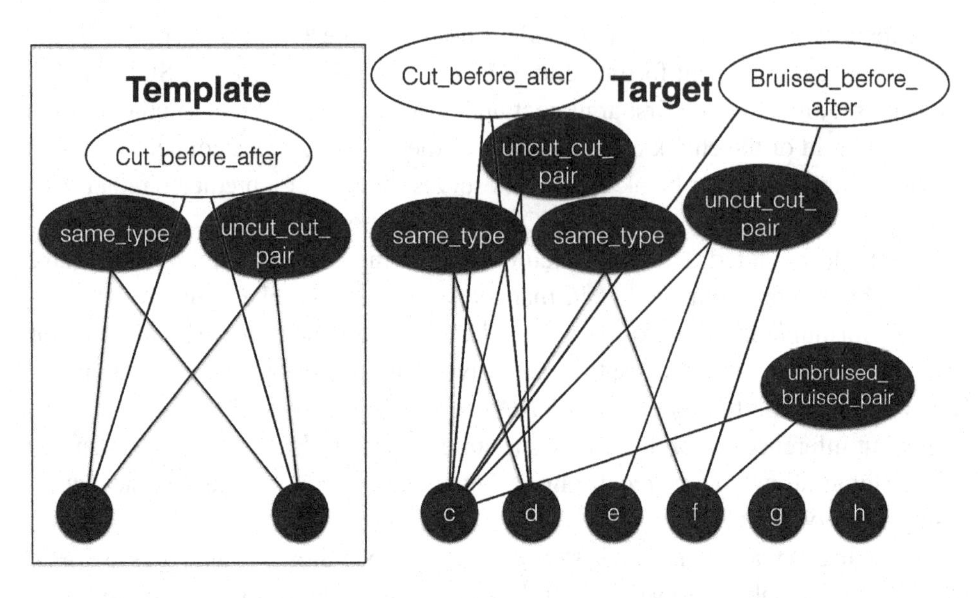

of a structure should hold, and if those properties hold, how inferences should be created. Licato et al. (2014a) used templates to produce basic deductive and analogical inferences, and Licato, Sun, and Bringsjord (2014c) made use of templates to create explanations by generating chains of inferences.

CONCERNS AND OBJECTIONS; REPLIES

In this section we anticipate and address a series of concerns and objections.

Concern 1: "What is the Purpose?"

The first anticipated concern is as follows:

It's not entirely clear what the intended target for the virtual characters in the work is. Is it for social stimulation? Interaction fiction? Video games? It does make a difference in this case, for after all, a cognitive architecture needn't support all of the logic models and devices you discuss in order for that architecture to be suitable for certain purposes. For example, if I'm watching a story being acted out by AI, I don't really care what amount of cognition is going on or is supported by the underlying system; the only thing that matters is the quality of the performance. Do

I believe in the characters and their circumstances? Am I engaged, emotionally, in the proceedings, and so on? How the system in the end works is secondary in this kind of application. In other applications, what is going on behind the scenes can be more important.

We reply as follows. The concern as articulated here appears to ignore the explicitly stated purpose of our investigation, which is to seek to provide a substantive part of an answer to (Q3). Put briefly, (Q3) asks what the craft of character creation and design, when carried out in systems dedicated to framing and facilitating such creation and design, implies with respect to cognitive architectures and the science thereof. We are not concerned with the possibility of creating, designing, and deploying (in engaging narrative) characters in ways that entirely obviate the programmatic nature of the science of cognitive architectures.[17] Rather, as (Q3) makes concrete and specific, we are concerned to see how the scientific program of seeking an adequate cognitive architecture, that is, a cognitive architecture that can enable the computational modeling of human cognitive activity, fares when measured against the cognitive activity of creating, designing, and deploying characters (via sophisticated tools like Movie Outline 3).

It's important to realize that our interest in (Q3) is a particular instance of a more general question. (Q3) has any number of close relatives, for domains other than character design and narrative. For instance, there's a version of (Q3) for the craft of creating not narrative, but introductory (differential and integral) calculus books; viz.,

$(Q3)_{calc}$ What does the contemporary craft of designing (by human authors) introductory (differential and integral) textbooks, an activity which is beyond the reach of foreseeable AI, and which isn't powered by any stunning, speculative, AI-infused technology (immersive or otherwise), but is instead aided by tried-and-true "AI-less" software tools and immemorial techniques that are still routinely taught today, imply with respect to today's computational cognitive architectures?

In the case of $(Q3)_{calc,}$ as in the case of (Q3), expressivity is a key issue. One reason is that a central concept that must be understood by any human engaging in the craft of designing the textbooks in question that of a limit, defined in Bolzano-Cauchy-Weierstrass (B-C-W) style. Confirmation of this route is easy to obtain, by empirical means. For example, in the 10th edition of the best-selling *Calculus* by Larson and Edwards (2014), we read, at the outset:

So, one way to answer the question "What is calculus?" is to say that calculus is a "limit machine" that involves three stages. The first stage is precalculus mathemat-

ics, such as the slope of a line or the area of a rectangle. The second stage is the limit process, and the third stage is a new calculus formulation, such as a derivative or integral (p. 42).

The very same conception can be found in every single other widely used calculus textbook.[18] Yet it is a logico-mathematical fact that writing the definition of a B-C-W limit requires writing out (and presumably understanding) extremely robust levels of quantification. In fact, even basic arithmetic at the level of what is learned in elementary and grade school requires robust quantification.[19]

Concern 2: "What Exactly are these 'Challenges'?"

The second concern we consider is related to the first, but is more specific; it can be expressed as follows:

Your discussion introduces two 'challenges' for cognitive architectures to pass, but the terms and conditions of the challenges are never made clear. For example, just because a system does not contain complex quantification, this does not mean it is not capable of supporting or meeting the challenges given in the chapter. For instance, a system does not have to be able to fully express. 'There are attributes Joseph has that are dark and racy' for it to assign attributes to Joseph that are dark and racy, and to induce Joseph to act in that fashion. If the challenge is to have characters behave according to some kind of rich character definition, more systems than just CLARION will be able to rise to that challenge. If supporting complex quantification is in itself the challenge, then of course many systems would not meet that challenge, but then we would need to debate the relevancy or necessity of such a challenge in the first place.

Given our reply to the previous concern, our response to the second concern is straightforward: On the contrary, the two challenges have been specified, and they cannot be dodged by any cognitive-architecture aspiring to complete the scientific program of modeling human cognition that is part and parcel of creating, designing, and deploying characters, at the level of what is requested by the likes of Movie Outline, and such as are seen in the simple *Double-Minded Man* (let alone in the dramas of Ibsen and the modern novels of Ullmann).[20] An exactly parallel point, as we have noted, can be asserted with respect to the cognition that is part and parcel of creating not fiction, but technical non-fiction (regarding calculus and arithmetic).

Concern 3: "Can CLARION Handle *Double-Minded Man*?"

Here's the third anticipated concern:

Your chapter reasonably establishes that CLARION can express the various logic systems discussed earlier. But what does that show exactly? That earlier discussion of yours is surely not the same as using CLARION to create an enactment of Double-Minded Man, with characters acting independently and autonomously according to their natures. And isn't that more important?

This is a cogent concern. In keeping with our avowed purpose and plan, we rest content with having only demonstrated that CLARION *can* be used to represent the characters in *Double-Minded Man,* and that CLARION can thereby support an enactment. We haven't built the specific representations and carried out the enactment. We have shown that this additional work can be carried out, given more time and space.

Concern 4: "What about the Character-Centric AI of Mateas et al.?"

The fourth concern we anticipate is as follows:

Doubtless it's unintentional on your parts, but the foregoing will give some readers the distinct impression that the connection you forge between deep characterization and AI is unprecedented. Is there no related work in this space? Or no related work worth mentioning?

In response, we first note that our firm, explicit focus on the nexus between the craft of character creation, design, and deployment on the one hand, and formal metrics applied to the science of cognitive modeling (via application to cognitive architectures), does indeed seem to be without precedent (for better or worse). That said, there certainly is some very impressive and important work that relates to AI, characters, and art; and this work is in part relevant to our present essay. For example, there is a form of AI invented (as far as we know) by Mateas (2001): *Expressive AI.* As Mateas puts it, Expressive AI is the marriage of "art practice" with "AI-based practice," in order to build "cultural artifacts." Mateas says that "Expressive AI changes the focus from an AI system as a thing in itself (presumably demonstrating some essential feature of intelligence), to the communication between author and audience" (Mateas, 2001, p. 153). Since ultimately at least one of the main purposes in building three-dimensional characters is to produce narrative that moves audiences in profound ways, Expressive AI is obviously relevant to our purposes herein – yet only to a degree. As we have pointed out, it's mathematically possible for a moving narrative, replete with prose that brings characters to life for readers, to be randomly generated, independent of any cognition whatsoever. Hence it follows that it's mathematically possible that cognitive modeling via a cognitive architecture of an author's mind and notebooks/files is entirely superfluous. But our assumption herein is that the standard practice on the part of human authors is worth examining; and when it's examined (with help from the cognition that Movie Outline and other such tools demand from authors), it becomes clear that, at least with respect to the metric of expressivity and algorithm ***R***, cognitive architectures face a tall order.[21]

CONCLUSION

The authors conclude that the contemporary first-rate craft of character design, if it measures up to the level of thoroughness called for by sophisticated but AI-less software designed to facilitate such design, generates expressivity challenges that proponents of cognitive architectures should find quite sobering. Some architectures are clearly not up to these challenges; some perhaps have at least some elements that are somewhat promising (ACT, and presumably descendant ACT-R); and some have a very impressive capacity to handle even natural language (and *a fortiori* background facts) that requires third-order logic (TOL) and quantified epistemic logic. The authors advise the cognitive modeling community to ascertain and announce which of these three categories their favorite cognitive architecture falls under, and why.

As to future work in the authors' own case, they intend to raise the classification of cognitive architectures, with respect to expressivity (and, subsequently, with respect to equally important measures in the relevant mathematics[22]) to a level of greater rigor, so that such classification is expressed and confirmed by relevant meta-theorems. This future for cognitive modeling is directly in line with the "theorem-anchored" one called for in Bringsjord (2008).

ACKNOWLEDGMENT

We are indebted to two anonymous referees, and editor Jeremy Turner, for insightful suggestions and objections. Any remaining deficiencies are due to our own failings.

REFERENCES

Albus, J. S. (2000). 4-D/RCS reference model architecture for unmanned ground vehicles. In *Proceedings of the 2000 IEEE International Conference on Robotics and Automation*. San Francisco, CA: IEEE. doi:10.1109/ROBOT.2000.845165

Anderson, J. (1976). Formal semantics of ACT representations. In Language, Memory, and Thought (pp. 220-251). Hillsdale, NJ: Psychology Press.

Anderson, J., & Lebiere, C. (2003). The Newell Test for a theory of cognition. *Behavioral and Brain Sciences*, *26*(05), 587–640. doi:10.1017/S0140525X0300013X PMID:15179936

Anderson, J. R. (1993). *Rules of mind*. Hillsdale, NJ: Lawrence Erlbaum Associates.

Anderson, J. R., & Lebiere, C. (1998). *The atomic components of thought*. Mahwah, NJ: Lawrence Erlbaum Associates.

Andrews, P. (2002). *An introduction to mathematical logic and type theory: To truth through proof*. New York, NY: Springer. doi:10.1007/978-94-015-9934-4

Barwise, J., & Etchemendy, J. (1999). *Language, proof, and logic*. New York, NY: Seven Bridges.

Blackburn, P., & Bos, J. (2005). *Representation and inference for natural language: A first course in computational semantics*. Stanford, CA: CSLI.

Bringsjord, S. (1992). CINEWRITE: An algorithm-sketch for writing novels cinematically and two mysteries therein. In M. Sharples (Ed.), *Computers and writing: State of the art. Kluwer*. Dordrecht, The Netherlands: Springer. doi:10.1007/978-94-011-2674-8_12

Bringsjord, S. (1995a). Could, how could we tell if, and why should—androids have inner lives? In K. Ford, C. Glymour, & P. Hayes (Eds.), *Android Epistemology* (pp. 93–122). Cambridge, MA: MIT Press.

Bringsjord, S. (1995b). Pourquoi Hendrik Ibsen Est-II une menace pour la littérature générée par ordinateur? In A. Vuillemin (Ed.), *Littérature et informatique la littérature générée par orinateur*. Arras, France: Artois Presses Universite.

Bringsjord, S. (1995c). Pourquoi Hendrik Ibsen Est-II une menace pour la littérature générée par ordinateur? In A. Vuillemin (Ed.), *Littérature et informatique la littérature générée par orinateur* (pp. 135–144). Arras, France: Artois Presses Universite.

Bringsjord, S. (2001). Is it possible to build dramatically compelling interactive digital entertainment (in the form, e.g., of computer games)? *Game Studies, 1*(1). http://www.gamestudies.org

Bringsjord, S. (2008). Declarative/logic-based cognitive modeling. In R. Sun (Ed.), *The handbook of computational psychology* (pp. 127–169). Cambridge, UK: Cambridge University Press. doi:10.1017/CBO9780511816772.008

Bringsjord, S., & Bringsjord, A. (2009). Synthetic worlds and characters and the future of creative writing. In C. A. P. Smith, K. Kisiel, & J. Morrison (Eds.), *Working through synthetic worlds* (pp. 235–255). Surrey, UK: Ashgate.

Bringsjord, S., & Ferrucci, D. (2000). *Artificial intelligence and literary creativity: Inside the mind of Brutus, a storytelling machine*. Mahwah, NJ: Lawrence Erlbaum Associates.

Bringsjord, S., Govindarajulu, N., Ellis, S., McCarty, E., & Licato, J. (2014). Nuclear deterrence and the logic of deliberative mindreading. *Cognitive Systems Research, 28*, 20–43. doi:10.1016/j.cogsys.2013.08.001

Charniak, E., & McDermott, D. (1985). *Introduction to artificial intelligence*. Reading, MA: Addison- Wesley.

Cohen, L. H. (2014). Suddenly that summer. *The New York Times Sunday Book Review,* p. BR15

Ebbinghaus, H. D., Flum, J., & Thomas, W. (1984). *Mathematical logic*. New York, NY: Springer- Verlag.

Ebbinghaus, H. D., Flum, J., & Thomas, W. (1984). *Mathematical logic* (2nd ed.). New York, NY: Springer-Verlag.

Fagin, R., Halpern, J., Moses, Y., & Vardi, M. (2004). *Reasoning about knowledge*. Cambridge, MA: MIT Press.

Gall, D., & Frühwirth, T. (2014). A formal semantics for the cognitive architecture ACT-R. In M. Proietti & H. Seki (Eds.), Lecture notes in computer science: Vol. 8981. Logic-based program synthesis and transformation (pp. 74–91). Cham, Switzerland: Springer. Retrieved from http://www.informatik.uniulm.de/pm/fileadmin/pm/home/fruehwirth/drafts/act_r_semantics.pdf

Goble, L. (Ed.). (2001). *The Blackwell guide to philosophical logic*. Oxford, UK: Blackwell Publishing. doi:10.1111/b.9780631206934.2001.x

Holyoak, K. J., & Hummel, J. E. (2000). The proper treatment of symbols in a connectionist architecture. In E. Deitrich & A. Markman (Eds.), *Cognitive dynamics: Conceptual change in humans and machines*. Cambridge, MA: MIT Press.

Hummel, J. E., & Holyoak, K. J. (1997). Distributed representations of structure: A theory of analogical access and mapping. *Psychological Review*, *104*(3), 427–466. doi:10.1037/0033-295X.104.3.427

Hummel, J. E., & Holyoak, K. J. (2003). Relational reasoning in a neurally-plausible cognitive architecture: An overview of the LISA project. *Cognitive Studies: Bulletin of the Japanese Cognitive Science Society*, *10*, 58–75.

Jaskowski, S. (1934). On the rules of suppositions in formal logic. *Studia Logica*, *1*, 5–32.

Keisler, H. J. (1986). *Elementary calculus: An infinitesimal approach.* Amsterdam, The Netherlands: Prindle, Weber and Schmidt.

Larson, R., & Edwards, B. (2014). *Calculus*. Boston, MA: Brooks and Cole.

Licato, J. (2015). *Analogical constructivism: The emergence of reasoning through analogy and action schemas.* (PhD thesis). Retrieved from Rensselaer Polytechnic Institute, Troy, NY.

Licato, J., Sun, R., & Bringsjord, S. (2014a). Structural representation and reasoning in a hybrid cognitive architecture. In *Proceedings of the 2014 International Joint Conference on Neural Networks (IJCNN)*. doi:10.1109/IJCNN.2014.6889895

Licato, J., Sun, R., & Bringsjord, S. (2014b). Using a hybrid cognitive architecture to model children's errors in an analogy task. In *Proceedings of CogSci 2014*.

Licato, J., Sun, R., & Bringsjord, S. (2014c). Using meta-cognition for regulating explanatory quality through a cognitive architecture. In *Proceedings of the 2nd International Workshop on Artificial Intelligence and Cognition*.

Mateas, M. (1997). *Computational subjectivity in virtual world avatars*. Technical Report FS-97-02. AAAI. Retrieved from http://www.aaai.org/Papers/Symposia/Fall/1997/FS-97-02/FS97-02-021.pdf

Mateas, M. (2001). Expressive AI: A hybrid art and science practice. *Leonardo: Journal of the International Society for Arts, Sciences, and Technology*, *34*(2), 147–153. doi:10.1162/002409401750184717

Newell, A. (1981). The knowledge level. *Artificial Intelligence*, *18*, 81–132.

Partee, B., Meulen, A., & Wall, R. (1990). *Mathematical methods in linguistics*. Dordrecht, The Netherlands: Kluwer.

Rosenbloom, P., Newell, A., & Laird, J. (2014). Toward the knowledge level in SOAR: The role of the architecture in the use of knowledge. In K. van Lehn (Ed.), *Architectures for intelligence: The 22nd Carnegie Mellon Symposium on Cognition*. Psychology Press.

Russell, S., & Norvig, P. (2009). *Artificial intelligence: A modern approach* (3rd ed.). Upper Saddle River, NJ: Prentice Hall.

Sullivan, K. (1974). *The teaching of elementary calculus: An approach using infinitesimals*. (PhD thesis). University of Wisconsin.

Sun, R. (1995). Robust reasoning: Integrating rule-based and similarity-based reasoning. *Artificial Intelligence*, *75*(2), 241–295. doi:10.1016/0004-3702(94)00028-Y

Sun, R. (2001). *Duality of the mind*. Mahwah, NJ: Lawrence Erlbaum Associates.

Sun, R. (2002). *Duality of the mind: A bottom up approach toward cognition*. Mahwah, NJ: Lawrence Erlbaum Associates.

Sun, R. (2003). A tutorial on CLARION 5.0. In Cognitive Science Department Technical Reports. Rensselaer Polytechnic Institute. Retrieved from http://www.cogsci.rpi.edu/~rsun/sun.tutorial.pdf

Sun, R., & Zhang, X. (2004). Accounting for similarity-based reasoning within a cognitive architecture. In *Proceedings of the 26th Annual Conference of the Cognitive Science Society*. Mahwah, NJ: Lawrence Erlbaum Associates.

Tversky, A. (1977). Features of similarity. *Psychological Review*, *84*(4), 327–352. doi:10.1037/0033-295X.84.4.327

Ullmann, L. (2014). *The cold song*. New York, NY: Other Press.

KEY TERMS AND DEFINITIONS

ACS: Action-Centered Subsystem; a subsystem of CLARION containing procedural knowledge, directly having to do with the actions of an agent.

CLARION: A cognitive architecture, described by Sun (2002), built primarily on two orthogonal dichotomies: action-centered- versus non-action-centered-knowledge, and explicit versus implicit knowledge.

FOL: First-Order Logic; a logic that extends ZOL by allowing quantification over objects.

NACS: Non-Action-Centered Subsystem; a subsystem of CLARION containing general declarative knowledge about the world not directly having to do with an agent's actions.

SOL: Second-Order Logic; a logic that extends FOL by allowing quantification over predicate / relation symbols.

TOL: Third-Order Logic; a logic that extends SOL by allowing higher-order formulae which treat predicate / relation symbols as objects.

ZOL: Zero-Order Logic; a logic consisting of the truth-functional connectives, predicate / relation symbols, identity, and objects. ZOL and the propositional calculus are equally expressive.

ENDNOTES

1. In the former case, examples — such as Hakon Werle! — abound e.g. in The Wild Duck, readily available online; in the latter, see The Cold Song (Ullmann 2014), which features the endlessly fascinating Alma,

   the novels finest and thorniest creation. [...] she is by turns endearingly wistful and frighteningly willful. Love and loathing entwine in her like a double helix, and her greatest transgressions are part and parcel with her greatest strength: a fierce (possibly compulsive) insistence on telling the truth. (Cohen 2014)

   These are just two examples picked from countless others. They are selected mainly because of our prior thinking about Ibsen and deep characters; e.g., see (Bringsjord 1995b) (and see also additional prior Bringsjordian work cited below). We recognize that some readers will be unfamiliar with the universal view that Ibsen's characters are seminally deep and robust. We regret that we don't have the space to substantiate this view from the standpoint of literary theory and criticism, and direct readers to (Bringsjord 1995c) as a starting place. The fact is that what Movie Outline 3 requires of scriptwriters (see below) in connection with fleshing out characters is something that Ibsen did before penning a single word of a drama.

2. There is also of course the related question: *(Q3') Could an AI generate belletristic fiction (which requires robust and engaging characters)?* A negative answer to this question is given in (Bringsjord & Ferrucci 2000); but that same work provides evidence in support of an affirmative response to a version of this question formed by supplanting 'belletristic' with 'formulaic.'

[3] By use of this adjective we mean to simply report that the kind of *intelligent agents* taken to distinguish the field of AI (Russell & Norvig 2009) are not present in the software in question.

[4] Available (including in trial form) at http://www.movieoutline.com. For the record, we have no stake in, nor do we benefit in any way financially from purchases of, this software. As mentioned above, another excellent system, which we could have used for the present paper, and for composing Double-Minded Man, is Dramatica (http://dramatica.com; also available in trial form).

[5] The notion of high expressivity as discussed in the present chapter is, we cheerfully concede, ultimately a single metric for the examination of a cognitive architecture, and this metric is only one piece of a bigger puzzle — a puzzle the fuller exploration of which is beyond our scope and available space.

[6] A future algorithm ***R2*** might take as input some formulae Φ, which may be produced by ***R***, and determine whether the formulae in Φ are validly inferentially linked. For example, our formula (A') is validly inferentially linked to the formula $L(h,j) \vee L(j,j)$.

[7] The algorithm ***R*** can be applied to any declarative sentence, irrespective of the role that that sentence might have. In cognitive models arising from processing in cognitive architectures, some information in declarative memory might be believed or acted on without any deeper reasoning or decision-making. Algorithm ***R*** takes no account of such shades. The same would go for the practice of mathematics: Some declarative sentences involved in the practice of mathematics are largely ignored; others receive a lot of attention. Yet ***R*** can be applied to 1+1=2 just the same as to some great unsolved conjecture, such as that $\mathbf{P} \neq \mathbf{NP}$.

[8] The continuum is e.g. presented cogently in (Andrews 2002).

[9] Cognoscenti in pursuit of building ***R*** would call for pairing FOL with e.g. the additional machinery of the λ-calculus, as e.g. in (Blackburn & Bos 2005). But we can leave this aside. Readers interested in the formal tools in linguistics needed for intelligent pursuit of R should begin their study with (Partee, Meulen & Wall 1990). By our lights, such material should be in every toolbox of every cognitive modeler. After all, the human ability to understand the meaning of natural language is a key part of what it is to be us, cognitively speaking.

[10] This fact isn't revealed in DMM, but is something which, in keeping with the authorial craft of creating robust characters, is in the author's mind when writing. For more on such techniques and Ibsen, see (Bringsjord 1995c).

[11] Coverage of such logics in connection with cognitive modeling is provided in (Bringsjord 2008). For an introduction to epistemic logic and other logics with operators, see (Goble 2001).

12 We offer here no formal definition of 'substantive narrative.' There is a longstanding tradition in AI to count such things as 'Bill went to the store and bought a candy bar' as a story (e.g. see Charniak & McDermott 1985). But clearly such a thing isn't a *substantive* story.

13 See any standard textbook for confirmation. E.g., (Barwise & Etchemendy 1999).

14 In fact, the idiosyncratic treatment of identify that we noted in the previous paragraph is declared to be necessary to block such inferences as that if someone believes $R(a)$, and $a = b$, that same someone would then have to believe $R(b)$. (Given that Smith believes that John Le Carré is the wealthy author of *The Spy Who Came in From the Cold*, and that JL = David Corwell (which is true), it would follow that Smith believes DC to be wealthy. But Smith may not know that 'Le Carré' is a pen name.) But this unwanted consequence only arrives if belief is formalized naïvely, instead of in an operator-based logic like epistemic logic.

15 Extensional logics do not permit non-extensional operators like believes, which e.g. can be applied to propositions φ whose semantic values do not predictably generate a semantic value for the "outer" proposition. This is why, earlier, we stepped outside the ZOL-to-TOL sequence to epistemic logic. E.g., even if we know that φ is false, we cannot infer that 'Jones believes φ' is false (or, for that matter, that it's true). For introductory discussion of this phenomenon in the context of computational cognitive modeling, see (Bringsjord 2008).

16 E.g., an algorithm such as ***R2*** (see footnote 6)

17 In light of fundamental formal facts, we cheerfully concede that there may be any number of ways to create, design, and deploy characters that completely dodge the expressivity challenge, and for that matter completely dodge all such challenges. For example, the random generation of characters and or pixels could produce compelling narrative, replete with robust three-dimensional characters. As explained elsewhere (Bringsjord 1995a), for any empirical, finite stimuli that a human viewer would be impressed with, there is a completely serendipitous path to producing that stimuli.

18 The proviso 'widely used,' note, is crucial. This is so because e.g. a teacher of calculus determined to teach via infinitesimals can of course either use his/her own content, or use some published material. In the latter case, some excellent Leibnizian-Robinsonian options are available: (Keisler 1986, Sullivan 1974).

19 The Peano Axioms for arithmetic specifically require either an infinite number of quantifier-rich formulae at the level of FOL, or a finite number of quantifier-rich formulae in SOL. For details, see (Ebbinghaus et al. 1994).

20 Again, in the case of ACT-R, its scientific program includes what is explicitly set out in (Anderson & Lebiere 2003).

21 The literary cultural artifacts that Mateas and colleagues have themselves created, by the way, could to a significant degree play the roles we have assigned herein to the characters in Double-Minded Man, and those in the drama of Ibsen and fiction of Ullman. However, it's worth noting that there are no formulae corresponding to those at the crux of the expressivity challenge we articulate herein, to be found. E.g., the representations in (Mateas 1997) at the code level are thin, and wouldn't serve to flesh out the requirements in Movie Outline 3 for declarative information about and constitutive of a character (this isn't in the least a criticism).

22 Again, see algorithm ***R2*** in note 6.

Chapter 7
Virtual Soar-Agent Implementations:
Examples, Issues, and Speculations

Jeremy Owen Turner
Simon Fraser University, Canada

ABSTRACT

This chapter provides a brief overview of those virtual agent implementations directly inspired by the cognitive architecture: Soar. This chapter will take a qualitative approach to discussing examples of virtual Soar-agents. Finally, this chapter will speculate on the future of Soar virtual characters. The goals of this chapter are sixfold. The first goal is to explain why cognitive architectures are becoming increasingly important to virtual agent design(s). The second goal is to convey why this chapter focuses exclusively on virtual agents that utilize the Soar architecture. The third goal is to explore some of Soar's technical details. The fourth goal is to showcase a few diverse examples where Soar is beginning to have a design impact on virtual agents. The fifth goal addresses Soar's limitations – when applied to agent design in virtual environments. The final goal speculates on ways Soar can be expanded for virtual agent design(s) in the future.

INTRODUCTION

Cognitive architectures have been around for decades.[1] Many of the more canonical architectures such as CLARION (CLARION webpage, n.d.) and Soar (Soar webpage, n.d.) have been continually updated over the years for agent implementation

DOI: 10.4018/978-1-5225-0454-2.ch007

by external developers, academics and hobbyists. However, most virtual agents (e.g. bots, Non-player-characters [NPCs]) created up to the present day, still do not directly utilize cognitive architectures for interaction with the agent's embodied environment. Instead, such agents often depend on cognition-less systems that may include the use of: finite-state machines, subsumption-level (i.e. reactive) algorithms (Brooks, 1991), hierarchical search interaction trees, heuristic search, and brute-force techniques, or some combination thereof. In recent years, computing power has exponentially increased and this has allowed for a serious reconsideration of cognitive architecture implementations. This new technological climate is beginning to encourage tractable innovations in artificial general intelligence (AGI) generally, and experimentation with cognitive architectures in particular. Academics and developers are increasingly interfacing cognitive architectures with virtual agents for more deliberation and cognitive potential in their interaction design.

This chapter provides an opportunity to qualitatively discuss and tally some of the state-of-the-art of those virtual agents that employ a particular canonical cognitive architecture, Soar (State Operator And Result)[2] (invented by Laird, Newell & Rosenbloom, ca. 1983). There are numerous cognitive architectures existing today (e.g. ACT-R, CLARION, RASCALS etc.) but due to size and other limitations explained below, Soar will be the focus of this chapter.

The thematic goals of this chapter are sixfold. The first goal is to explain what a cognitive architecture is and why they are becoming increasingly important to virtual agent design(s). The second goal is to convey why this particular chapter focuses exclusively on virtual agents that utilize the Soar architecture. The third goal is to explore some technical details that are uniquely characteristic of Soar's handling of cognition. The fourth goal is to showcase - via a few diverse examples - the continued legacy and design impact Soar is beginning to have on virtual agents. The fifth goal is to address Soar's limitations – when applied to NPC/bot design in virtual environments. The final goal is to speculate on additional ways Soar can be expanded for virtual agent design in the near future.

CAVEAT: VIRTUAL AGENTS MIGHT NOT EVEN REQUIRE COGNITIVE ARCHITECTURES

Design motivations behind whether or not to deploy a cognitive architecture often varies according to whether the virtual agent is meant to academically evaluate embodied and/or extended cognition or instead, to implement a virtual character (NPC) for social and/or ludic entertainment purposes. With these contextual contingencies in mind, most contemporary virtual agents are not required to be cognizant of the visual inputs beyond the raw numerical data underlying each object and path.

Basically, a virtual agent can successfully interact and navigate its environment without being directly aware of it in any introspective way. An agent can react to numbers alone. For example, a bit-mapped image or any other virtual object can be represented simply as a stream of numeric representations (e.g. binary digits). These numbers provide the agent with coordinates, proximity-relations, and other manually coded identification data that allow the agent to respond in a completely reactive way. In many video-games, for example, an NPC can appear to be intelligent simply by mindlessly reacting to the changes in numerical inputs afforded by the dynamic changes occurring via game-play interaction. However, cognitive architectural enhancement might be useful for any agent-based cognitive task involving reasoning and mnemonic assistance. Otherwise, a cognitive architecture may only be useful in this numerically-reduced case if we wish to imbue the agent with the ability to contemplate its environment as an "[...] idealized version of the underlying reality" (Best et al., 2006, p. 186). This underlying numerical reality might be modeled to enable embodied, extended, situated, and distributed (EESD) cognition so that virtual agents can actively "[...] structure or manipulate their environments" to "simplify, sequence, or otherwise, support cognitive processes" (Smart & Sycara, 2015, p. 3837).

The Need for Integrating Cognitive Architectures with Virtual Agents

Most virtual agents to date – whether embodied in video-games, interaction simulators or social virtual worlds - are not enabled with any cognitive architectures or algorithms modeling deeper cognitive processes. Cognitive representation of some kind is crucial if designers and developers wish to create artificially intelligent agents capable of some manner of introspective thought and deliberation – before executing an action. Due to the simplicity of construction and the current cost of available computational resources, behavioral regulation comes instead in the form of finite-state machines, and/or purely reactive systems where a particular input immediately corresponds to an (observable) output. Some designers see the behavioral affordances of finite-state machine-driven characters as "brittle" (Best et al., 2006, p. 190). Further, as the number of states (and state-transitions) increase in more behaviorally realistic games, the more states the programmers have to manually account for. Further, each of these states would require the programmer to account for an increasing number of pre-programmed responses while entering, transitioning to, or experiencing these additional states. At worst, "[...] human players often learn to game the finite state machine and take advantage of the idiosyncrasies of the opponents" (Best et al., 2006, p. 190). If virtual agents are one day intended to be perceived by gamers and other interacting avatars as "believable characters", then it seems that one should expect an increasing integration in the near-term of cognitive architectures as well as cognition-inspired algorithms and models in a virtual agent's deliberative and behavioral design.

COGNITIVE ARCHITECTURES

A cognitive architecture is semantically distinct from any other system or algorithmic structure that parses and models representations of cognition. To qualify this distinction, one must first clarify those categories that are not considered to be cognitive architectures. Firstly, an agent imbued with cognitive potential might utilize a cognitive architecture as one of many possible modular components for deliberation and (inter)action. However, the agent itself, is not architectural. Cognitive architectures are ontologically distinct from cognitive agents in that they resemble computer architectures. Like computer architectures, cognitive architectures specialize in parsing: "memories, processing and control components, representations of data, and input/output devices" (Laird, 2012, p. 5-6). A cognitive agent using any cognitive architecture ideally involves Artificial General Intelligence (AGI) not for computation's sake, but only for "[symbolic] representation, acquisition, and use of knowledge to pursue goals." (Laird et al., 2012, p. 6). Secondly, a cognitive model resembles a cognitive architecture however, "[...] a number of task-specific cognitive models can be developed that all draw on the same underlying (cognitive) mechanisms [which would comprise an architecture]" (Smart et al., 2014a, p. 3). A cognitive (infra)architecture can summarize a "common set of underlying mechanisms" to unify and explain "[...] a number of otherwise disparate behaviors" (Smart et al., 2014a, p.3).

A cognitive architecture is fundamentally, more general-purpose in that it can be applied towards a host of different agents and multi-agent systems. An agent uses a portion of an architectures' "fixed and task independent" (Smart et al., 2014a, p. 6) attributes. These attributes are customarily identified as: memories, natural-language patterns, data-structures and algorithms. These attributes learn and encode specific knowledge configurations, problem-specifications, rational deliberations, and actions for goal-directed intelligent performance in a stochastic environment (e.g. virtual world) (Smart et al., 2014a, p. 8). Cognitive architects usually understand that their cognitive architecture cannot be so general purpose as to handle every possible contingency and action. Due to the stochastic nature of an external environment,cogntive agents are assumed to "exist in an environment and [...] pursue tasks similar to those we find in the world we inhabit". Therefore, cognitive agents – regardless of architectural type - must be custom tailored to handling situations unique to their task domain (Smart et al., 2014a, p. 28). Some cognitive architectures assert their cognitive influence on a single agent by focusing on introspection and metareasoning capabilities. To clarify, metareasoning is the "process of [recursively] reasoning about [its own] reasoning cycle" (Cox & Raja, 2011, p. 4).The strong AI community traditionally employs a robust cognitive architecture (e.g. Soar) to model an agent's reasoning or metareasoning capabilities in the form of a "decision cycle within an action-perception loop [...] (Cox & Raja, 2011, p. 4).

Why Discuss Soar and Not Other Cognitive Architectures?

Other than Soar, there are a few canonical cognitive architectures that can be (and are being used) for tractable virtual agent implementations. The three cognitive architectures - briefly discussed below in chronological order - are by no means representative of all cognitive architectures but are sufficient for encapsulating three basic architectural types as belonging to one canonical corpus.

The most established cognitive architecture for virtual agents is known as ACT-R [Adaptive Control of Thought-Rational] (Anderson ca. 1990)]. ACT-R was an early model for representing brain-like cognition and was re-iterated from previous models known as "HAM" (1973) and "ACT-*" (1983).[3] ACT-R is one of the first cognitive architectures to explore and implement the relationships between deliberation, conceptualization memorization, and action. Further, ACT-R is an early instantiation of a hybrid architecture that dynamically draws from: symbolic, rule-based, and statistically tunable sub-symbolic representations of memory and cognition (Best et al., 2006, p. 188).

ACT-R has been implemented primarily for use by NPCs (non-player characters)[4] in first-person shooter video games for pathfinding, maze navigation, and strategic team communication (Best et al., 2006). Additionally, ACT-R has been occasionally deployed for more cerebral social simulations modeling human cognition and/or human-surrogate attributes.[5] For example, some virtual agents use a centralized intelligence network to send and receive news items in order to semantically interpret and predict terrorist attacks (Smart et al., 2014). The bulk of these particular ACT-R projects are military affiliated and/or directly funded. As mentioned previously, ACT-R is perhaps the most widely used cognitive architecture for virtual agent implementations. Therefore, it was decided that this chapter's contribution would be a focus on a cognitive architecture that is just as canonical as ACT-R but was not quite as popular and whose virtual agent implementations had not yet been summarized into one text.

Another canonical cognitive architecture is CLARION (Connectionist Learning with Adaptive Rule Induction ON-Line) (Sun et al., 2006). CLARION, like ACT-R, is a hybrid architecture that is focused on distinguishing between implicit and explicit cognition by using means of distributed representation.[6] For implicit cognition, CLARION can also parse subconscious processes, sub-symbolic knowledge (Sun, 2006, p. 82-83) and employ tacit knowledge. This implicit knowledge is effectively a priori knowledge which, like ACT-R, is represented and managed using bottom-level functional approximations of neural-nets. CLARION's neural nets are quite elaborate as they can support: back-propagation, feed-forward behaviors (Sun, 2006 p. 82), and Q-learning techniques (Watkins, 1989 in Sun, 2006., p. 83, 85). For explicit cognition, CLARION focuses on levels of: conscious awareness, active introspection, and intentional deliberation. CLARION frequently generates elements of focused top-down knowledge from bottom-up intuition in the form of: deductions, inductions, abductions, hypothesis testing, and explicit symbolic representation (Sun, 2006, p. 83-84). CLARION is also one of the few cognitive architectures capable of representing meta-cognition (Sun, 2006, p. 92-94). Of all the canonical cognitive architectures, CLARION theoretically has the most potential for implementing virtual agents with cognitive faculties for self-awareness, innate motivations and drives, complex strategic deliberation, and even learning from reactive behavior. However, CLARION's issue is the reverse of ACT-R in that there are very few known virtual agent implementations of CLARION. Because of this under-representation, the main active developer of CLARION has even decided to translate its entire code-base from C# into Python in order to entice character designers to use CLARION for their virtual agents.[7] Therefore, a discussion of CLARION will be reserved for a future paper on virtual agent implementations. It is hoped that tractable CLARION examples will become more plentiful over time.

There are many other cognitive architectures. Some of these architectures have emerged from fringe beginnings. However, due to interesting implementations with virtual agents, the occasional architecture had the potential to become canonical in its own right. Take the example of RASCALS (Rensselaer Advanced Synthetic Character Architecture for "Living" Systems) (Bringsjord et al., 2005). The RASCALS cognitive architecture was developed to explore the mathematical modeling of morality in AGI-enabled agents. RASCALS "[...] uses simple logical systems (first-order ones) for low-level (perception & action) and mid-level cognition, and advanced logical systems (e.g., epistemic and deontic logics) for more abstract cognition" (Bringsjord et al., 2005, p. 31). RASCALS' primary virtual agent was a personified embodiment of moral evil existing in the social virtual world Second Life ("Second Life", n.d.)[8] - aptly named, "E". RASCALS has been shown to be technically distinguished from CLARION due to the additional aforementioned emphasis of natural language processing capabilities (Bringsjord et al., 2005, p. 31). RASCALS can also be distinguished from earlier versions of other canonical cognitive architectures based on production rules such as Soar. Once again, RASCALS, like CLARION, has not included many virtual agent examples from which to examine. In fact, the RASCALS project is no longer formally maintained by

Bringjord's research team. Therefore, the rest of this discussion will focus primarily on Soar. Soar is an architecture with enough examples to examine and also deserves a higher-profile for its under-represented virtual agent implementations.

Soar

Soar ("State Operator And Result") is a production rule-based AGI cognitive architecture with military and entertainment clientele. However, Soar's creators claim that Soar-agent developers do not need previous programming nor computer-science experience to understand its architectural syntax. Soar is also appealing as a cognitive architecture because it is generalizable enough to handle different: problem-space, tasks, states, and operators. Ultimately, Soar is known for handling multi-faceted problem-domain approaches to a particular task (Laird, 2012, p. 4).

Canonical Features of Soar

Soar is highly scalable to support a large number of rules. In recent versions, Soar has also provided reinforcement learning capabilities. The most recent versions of Soar even have found a way to represent affective states and mental imagery processing (Lathrop & Laird, 2009). Unlike other Rule Based Systems (RBS), Soar-agents have the ability to dynamically update their beliefs about the external environment via "computationally inexpensive truth maintenance algorithms"("SoarTech Corporate Brochure", n.d.). Having a flexible belief system allows a Soar-agent to adapt to multiple ontological commitment opportunities within its particular task-domain. Soar's growing number of memory banks are completely associative, and currently store procedural, semantic, and episodic memories. There is even active speculation that Soar might include a predictive prospective memory bank in the near future (Li & Laird, 2013).

Soar's chunking mechanism is also rule-based and is a form of "explanation based learning" (Laird, 2012, p. 8) that is consolidated in episodic memory. Rule-based chunking allows a Soar-agent to compress a batch of pre-conditions from a specific episodic experience into a single generalized pre-condition for later possible activation of an operator that would be governed by a production rule. Soar's chunking procedure governs all learning from input through deliberation on the object-level and back to output via proposing, then selecting an operator for future action(s) in the external environment. Any novel or unfamiliar real-time episodic event perceived by the agent "returns a result to [the agent's] supergoal" (Laird, 2012, p. 8). The agent's actions and interpretations from this event are immediately "chunked" into a production rule. This chunked rule contains new conditional statements that relate directly to the episodic experience just witnessed and interpreted by the agent. Just like Soar's operators and states, this rule-chunk is stored associatively between the agent's short-term episodic memory and its long-term procedural memory banks for cross-retrieval. These episodic "interpretations" are based on a priori or previously acquired knowledge relating to apparently similar situations. For future interaction scenarios where a similar episodic event matches some of the conditions in that rule-chunk, the agent can retrieve suggested preferences for the most appropriate operators or actions to use in that new situation. Over time, Soar's reinforcement learning feature "[...] tunes the actions of rules that create preferences for operator selection" by rewarding the agent for acting out "good" preferred operators and actions (Laird, 2012, p. 18).

Soar also employs "automatic subgoaling" (an extension of the chunking mechanism). After recognizing a conflict (impasse) during a novel decision-making scenario, the Soar-agent generates a new state known as a "subgoal". As this new state is based on rule-based knowledge, Soar's freshly generated subgoal can oscillate between superstate and substate designations. Soar's corporate literature claims that other Rule Based Systems (RBS) can only deal with one fixed state at a time. The Soar agent automatically generates monitoring substates that "decompose complex operators into similar operators" (Laird, 2012, p. 66). Therefore, Soar's assignment of a monitoring role to these substates implies that Soar also offers rudimentary on-demand metareasoning or self-reflection capabilities (Laird, 2012, p. 66).

Soar's characteristic feature is its decision cycle. Soar's decision cycle is a "fixed processing mechanism that works in five phases: input, elaboration, decision, application, and output" (Lehman, Laird, & Rosenbloom, 2006, p.17). The Soar-agent must be able to parse continuous percepts (e.g. observations, perceptions) from a stochastic environment in order to make informed real-time decisions on which operators to select next. The Soar-agent handles this perpetually updating percept stream of new symbols by temporarily storing them first in a local "perceptual short-term memory" before adding them to a more global and action-driven working memory (Laird, 2012, p. 18). Next, the Soar-agent reflects on recently occurred episodes stored in its short-term working memory by elaborating on the elements of these events as conditions and suggested actions.

The Soar-agent elaborates on these learned episodes by comparing the history of its operator-preferences (thoughts) and operations (actions) it made at the time with conditional statements it had acquired from the rules it had chunked and stored from previous experience into its long-term memory. The Soar-agent then has the opportunity to decide on a course of action. If any rules generated (learned) from a recent episode match with any other remembered rules, those matching rules will fire in parallel. Once these matching combination of rules fire, the Soar-agent will have updated to its next state which contains new "suggestions, or preferences, for selecting the current operator" (Lehman, Laird, Rosenbloom, 2006, p.17). From a list of highly rated and preferred operators, the Soar-agent can then propose, select, and apply the operator to be used for the next action. The programmer also has the luxury of manually preferring operators by adding a "+" symbol to a favored operator. With the reinforcement learning procedure, the Soar-agent updates an operator with a "+" symbol and a corresponding numerical rating value automatically when rewarded for an action that was produced by that operator.

Soar's architecture is known mostly for its ability to simultaneously fire an extremely large number of hard-coded production rules without having these rules initially undergo a matching procedure. Soar's operators and states are also rule-driven. In fact, Soar differentiates itself from other Rule Based Systems (RBS) in that it is most concerned with selecting the best operator that will impact the agent's next state. In other RBS, the focus is on hierarchically rating each rule and sequentially determining which single rule should fire next (Laird, 2012, p. 4). Soar instead determines the best combination of rules to be fired in parallel, given the "knowledge and structure'" of a particular task-context (Laird, 2012, p. 4-5). These operators, states, and rules are completely modular atomic elements and can be recycled from one memorized context to another for action-selection. This means that Soar saves on rule-firing speed and resolves any conflicting logical contradiction or decision impasse by being highly associative with different elements of each of its parallel long and short term memory banks. However, this convenient association and modularity among memorized rules comes at a structural cost. For example, Soar can retrieve any other chunk (concept with data and a form) from one memory bank and automatically turn it into a sub-goal for use in episodic context. This cross-retrieval process is executed without requiring any knowledge of the previous associative context. Consequently, Soar is task-agnostic as it only focuses on how the parallel optimization of a large number of hard-coded rules result in operational changes "[...] to the features and values" in the agent's state (Lehman, Laird, & Rosenbloom, 2006, p. 17).

Soar's Technical Integration with Virtual Agents

This chapter is focused on historical virtual agent implementation examples rather than on their technical interfacing protocols between agent, cognitive architecture and virtual environment. However, it can be stated in passing that generally speaking, integration between Soar and character-driven virtual environments are contingent on the technical affordances of a particular software application. For example, NPCs residing in gaming engines such as Unreal Tournament can be interfaced with Soar using a control methodology and data specifications that are packaged by the company SoarTech as a "Human Behavior Model Interface Standard" (HBMIS, van Lent et al., 2004). It is entirely feasible that this interface standard can be migrated across similar gaming engines such as Unity. The most established gaming engine implementation treats each human behavioral model as an asynchronous external software module that (for now) communicates with a proprietary "UnRealScript" function interface (van Lent et al., 2004, p. 3). Each model exclusively processes Unreal's simulation data through the Soar General Input/Output mechanism (SGIO, Laird et al., 2002) before passing this data through its own decision cycle (van Lent et al., 2004, p. 3). Fortunately, these models are designed so that their processing cycles do not interfere with the game engine's own processing cycles (van Lent et al., 2004, p. 4).

SoarTech has made further high-level modifications to this Unreal/SGIO interface to account for state initialization and representation, (multiple) independent procedural threading, and group action sequences (van Lent et al., 2004, p. 3-4). For data control, SoarTech has also created their own low-latency DLL file that is executed once the Unreal engine initializes (van Lent et al., 2004, p. 4). In order to account for Unreal's real-time rendering capabilities, the SGIO mechanism must be able to execute and return game loops at a rapid speed. In fact, deliberative latency might even compromise the virtual environment's frame-refresh rate (i.e. below 30 frames per second, van Lent et al., 2004, p. 4). Therefore, Soar's once impressive ability to refresh a new decision cycle every 50 milliseconds might still prove to be too temporally cumbersome for cognition to believably function at the speed of contemporary game-play (van Lent et al., 2004, p. 4). One of the agent designers mentioned in this chapter has developed a plausible workaround to this interface issue. In this case, the high-level interface virtually encapsulates three selectable lower-level interfaces that operate at run-time instead of real-time (Magero et al., 2004, p. 879). Magerko's research team has optimised the performance capabilities of SGIO by "embedding [...] multiple copies [of] Soar within the UT process using a C API" (Magerko et al., 2004, p. 879).

Gaming engines in general (esp. VBS2, Unity3D and Unreal Tournament) can now be interfaced with Soar using SoarTech's "PoL" ("Patterns of Life") behavior modeling infrastructure (SoarTech, "Entity-Level Patterns of Life for Virtual Environments, promotional brochure, 2013). This PoL infrastructure comes bundled with sample virtual agents for evaluation. These agents feature parameterized and schedulable patterns that correspond to self-monitoring behaviors, beliefs, desires and intentions (SoarTech, "Entity-Level Patterns of Life for Virtual Environments, promotional brochure, 2013).

Outside of the gaming context, other Soar-agent designers have created novel integration solutions such as implementing an elaborate messaging interface that links various technical components and software application components. These disparate software components can even run on different machines, if required. For example, the proprietary components that created the cognition for the first virtual agent example (STEVE, Rickel & Johnson, 2000, p. 11) passed messages between the VIVIDS simulation engine (USC Behavioral Technologies Laboratory, 1997), Vista Viewer (for visual interfacing, Lockheed Martin, 1995), and Entropic's Text-to-Speech software. These messages were dispatched using another piece of proprietary software known as, ToolTalk (SunSoft, 1994).

Soar Examples

These four Soar implementation examples were chosen to showcase the various ways a rule-based cognitive architecture can enhance the interactive and deliberative potential of a virtual agent. These examples are presented chronologically in historical order from 2000-2013. These Soar-agent examples address military (e.g. industrial and cultural sensitivity training), entertainment (e.g. video-games), and fine-arts (e.g. dance) contexts.

The Soar Training Expert for Virtual Environments (STEVE)

The first example involves a US Navy-funded "Soar Training (Virtual) Expert for Virtual Environments" known colloquially by "his" personified acronym, "STEVE" (Rickel & Johnson, 2000). STEVE is a virtual humanoid who interacts with naval officers wearing a head-mounted display (HMD) in a virtual reality (VR) environment (Rickel & Johnson, 2000, p. 101-105). STEVE can respond to text-to-speech queries as well as reason about the motivations and goals for enacting the correct operational procedures for maintaining has turbine engines on naval ships[9] (Rickel & Johnson, 2000, p. 96). In order to carry out these training procedures in a believable manner, STEVE can orient his avatar body, walk around, and point towards salient virtual objects within the turbine-engine simulation while identifying each object by name (Rickel & Johnson, 2000, p. 97). Further, STEVE can shift his gaze from the inquiring officer to the queried virtual objects that he is currently identifying for pedagogical purposes.

STEVE uses a perception module to monitor as well as semantically interpret all of the appropriate messages and events that result in updating the (virtual) world-state (Rickel & Johnson, 2000, p. 106). STEVE constantly relays the current world-state in a way that is continuously interpretable by the STEVE's cognition module (ie. where Soar resides) (Rickel & Johnson, 2000, p. 106). This cognition module is then responsible for deciding what STEVE's next action-sequence should be. Once one decision cycle has transpired, STEVE transmits motor-commands to the motor control module which allows STEVE's voice to speak and avatar body to move (Rickel & Johnson, 2000, p. 107). STEVE's cognition module carries out serial procedures independently from the other modules while the perception and motor control modules work in parallel (Rickel & Johnson, 2000, p. 107). STEVE is unique from many other virtual agent implementations because Soar in this case, does not supply STEVE's virtual mind with higher-level cognitive processes (e.g. cerebral deliberation and meta-cognition). By contrast, Soar represents STEVE's lowest level of cognition (Rickel & Johnson, 2000, p. 106-107). Soar was only modeled for lower-level cognition because the original Soar architecture was never optimized to assist with those demonstration and conversational activities more familiar to the perceptual and actuational (motor-control) modes of operation (Rickel & Johnson, 2000, p. 106-107)

STEVE repeatedly deployed his cognition module to choose the next appropriate action (Rickel & Johnson, 2000, p. 114). STEVE engaged the cognitive module by perpetually looping Soar's trademark decision cycle. Based on the hardware specifications circa 2000, STEVE used to complete up to ten decision cycles per second (Rickel & Johnson, 2000, p. 106-107). Even back then, STEVE appeared to be very responsive to his environment (Rickel & Johnson, 2000, p. 115). From each percept received by the perception module, STEVE responds to a trainee's question or demonstrates a new procedure by executing a behavioral operator. These

Figure 1. STEVE giving a demonstration of the turbine engine simulator
Permission granted by the author (Rickel & Johnson, 2000)

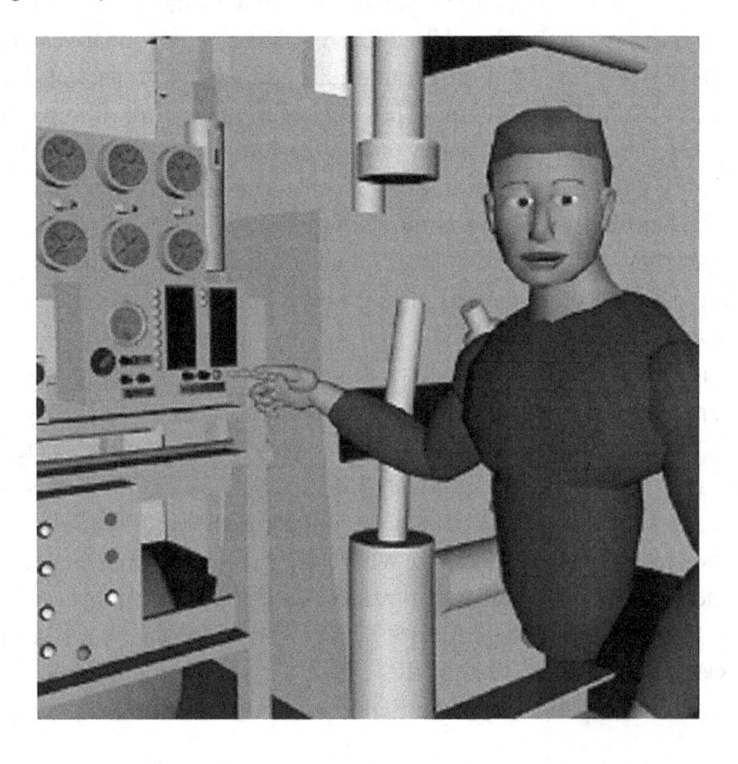

operators are represented by production rules that effect a change in the world-state and might even influence the next sequence of proposed operators themselves (Rickel & Johnson, 2000, p. 116). Once an operator has been applied, STEVE then determines which preferred operator to select and apply for the next time-step. These operators conform to action-categories related to: object manipulation, object-movement, self-proximity, and condition checking while the actions themselves correspond with STEVE's communicative faculty (Rickel & Johnson, 2000, p. 116). STEVE hierarchically compares, prefers, selects and then applies operators based on their contextual priority. STEVE's first priority is usually to answer the trainee's queries. However, if there are no queries, STEVE might instead attempt to steer the conversation towards focusing on more of his own verbal explanations and demonstrations of the training course (Rickel & Johnson, 2000, p. 114-115). In other cases, production rules might prompt STEVE to move to the next appropriate demonstration location or compel him to point at a particular object of immediate importance (Rickel & Johnson, 2000, p. 114-115). STEVE's communicative acts are in the form of Soar production-rules but ultimately have the restricted interactive characteristics of finite-state machines (Rickel & Johnson, 2000, p. 115).

Soar-NPCs in Unreal Tournament

These virtual Soar-agents are NPC (Non-Player Character) humans that use the Soar General Input/Output (SGIO, Laird et al., 2012) interface in order to communicate with Soar's architecture and the Unreal custom video-game environment (Magerko et al., 2004, p. 879). In this video game known as "Haunt 2", the player is the ghost and the player's goal is to uncover the mystery behind his/her death by subtly haunting the rule-driven NPCs using a limited ethereal mode of non-verbal communication. The player also cannot directly manipulate objects in the game and so, can only psychologically influence the haunted human NPCs to manipulate objects on behalf of the ghost's otherworldly will (Magerko et al., 2004, p. 878). The player-as-ghost can either scare the NPC into acting on the ghost's behalf or temporarily possess a more psychologically receptive NPC. When an NPC is in a possessed state, the ghost-player can actually observe and directly manipulate the NPC's Soar-mediated cognitive processes as they occur (Magerko et al., 2004, p. 878). However, each NPC has its own idiosyncratic psychological tolerance, bias, goals, motivations, needs (e.g. comfort), drives (e.g. hunger, thirst etc.), and a unique knowledge-representation of the background context (Magerko et al., 2004, p. 879).

Plot points in Haunt 2 are divided into preconditions and postconditions that can be optimally mediated by Soar operators to address both general and specific narrative outcomes. Preconditions describe "[...] what has to be in true in the world in order for this plot point to be relevant". In contrast, postconditions describe "[...]

Figure 2. A screenshot of the player haunting Soar-NPCs as a ghost in the video-game HAUNT 2 Permission for this image granted by the author. Accessed online from http://ai.eecs.umich.edu/~Soar/sitemaker/projects/haunt/images/blackboard.jpg on March 16, 2015
(Magerko et al,2004).

what actions should take place once these conditions are met" (Magerko et al., 2004, p. 880-881). In Haunt 2, there is no explicit mention of in-conditions which would describe what should be continually happening in the world to retain the truth of that particular condition. The closest thing to in-conditions in Haunt 2 is a "timing constraint" that provides situations when certain plot points lose their relevance over time (Magerko et al., 2004, p. 881). In addition to the player-character (PC) and NPC, Haunt 2 has a "Director" that runs a simulation modeling the player's actions in order to hypothesize over "[...] what declarative knowledge [the ghost] is acquiring through his [event triggered or time-lapsed] experiences [by keeping] an up-to-date internal structure describing the state of the world" (Magerko et al., 2004, p. 882). For example, the Director can alter the NPC's communication style (e.g. to be more outwardly "social"), and/or control the NPC's motor-functions and available text-strings for dialogue with other NPCs (Magerko et al., 2004, p. 880). The Director determines the success or failure of a world-state hypothesis depending on whether the player advances the story's plot points towards the expected narrative conclusion or instead, thwarts the activation of a plot-point by ignoring some particular pre-condition (Magerko et al., 2004, p. 880).

Virtual Soar-Agents using the Cultural Cognitive Architecture (CCA)

Some Virtual Soar-agents have been designed to interpret and communicate the complex ethnographic nuances of behavioral protocols, values, and social norms from other cultures. The military contractors known as "SoarTech" ("SoarTech", n.d.) have designed a Cultural Cognitive Architecture (CCA) within SOAR[10] in order to cognitively enhance conversational virtual agents for cross-cultural training purposes (Taylor & Sims, 2009, p.1). Soar in particular was chosen for its: "scalable long-term memory, [...] associative memory activation [functions], mixed automatic and deliberative reasoning [capabilities], goal-directed, problem-solving behavior, ability to incorporate multiple reasoning methods, [...], and graph-based working memory, which maps well to Scripts/Frames approaches to schema representation" (Taylor et al., 2007, p. 2; Minsky, 1975; Schank & Abelson, 1975).

Built upon the Cultural Schema Theory (CST), the CCA draws from an ontology that contains a growing corpus of embodied ethnographic knowledge (Taylor et al., 2007, p. 4). Outside of virtual environments, this ethnographic knowledge is usually acquired when human agents are embedded within a particular culture for a sufficient length of time (Taylor & Sims, 2009, p. 1). The cultural schema itself encodes the social hierarchies, known protocols, predicted social outcomes, culturally salient features of the environment as well as the particular contextually-sensitive goals that are likely to drive characters from particular ethnographic backgrounds (Taylor & Sims, 2009, p. 5). A culturally instantiated schema is subject to "[...] a process of activation, decay, and deactivation based on relevance to events perceived in the environment" (Taylor et al., 2007, p. 4). SoarTech claims that the CCA and CST are generalizable enough to be applied to understanding various cultures[11]. If any cultural differences cannot be captured with generality, the CCA generates leaf nodes to represent any noticeable cultural idiosyncrasies (Taylor & Sims, 2009, p. 5).

The Soar-CCA implements its virtual agents by balancing an emotional system dealing with an event-dependent appraisal mechanism with a cognitive system that establishes higher-level goals and generated behaviors (Taylor & Sims, 2009, p. 5-6). These conversational virtual agents exist in the 3D animation engine and behavioral simulator, "Vcommunicator". Vcommunicator has access to "[...] over 60 culturally diverse virtual human models as well as 40 facial expressions and 500 gestures, and can automatically lip-sync to over 22 mouth shapes that map to over 100 speech sounds of the International Phonetic Alphabet." (Taylor & Sims, 2009,

Figure 3. This image depicts a virtual soldier attempting cross-cultural negotiations with a Middle-Eastern Soar-CCA Virtual Agent. Permission to use this image granted by the author
(Taylor & Sims, 2009, SoarTech).

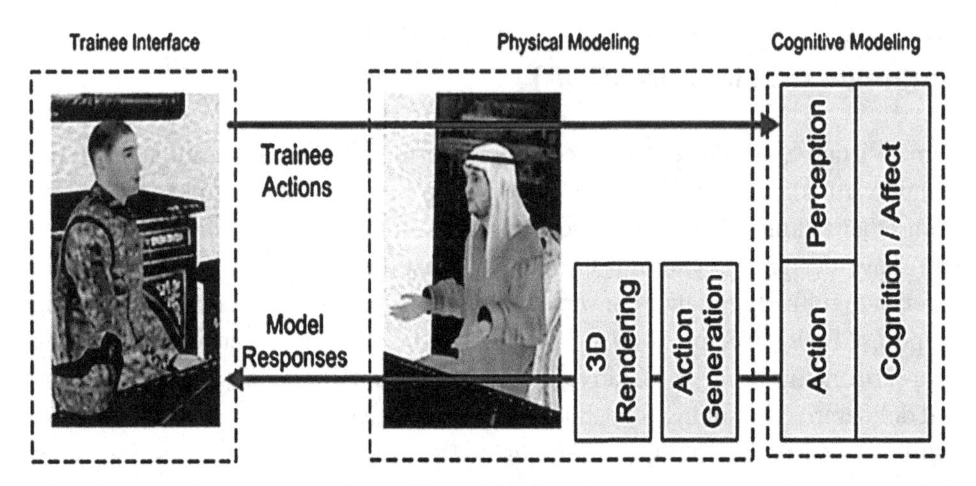

p. 4). A culturally successful agent will be able to ascertain the relevance of goals, and the benchmark set for societal norms while assessing the relative novelty of a particular ethnographic ritual (Taylor & Sims, 2009, p. 4). SoarTech claims that a virtual agent drawing from the Soar-CCA can dynamically assess the most typical emotional state (e.g. relative standards of "rudeness" and "politeness") based on the episodic sum of the outputs from the emotional system and the appraisal value collected from a specific cultural encounter (Taylor & Sims, 2009, p. 4). Overall, each virtual character receives both the recommended action-sequence and the expected emotional state while having parameterized control over situations where the relative emphasis of certain culturally appropriate gestures is required (Taylor & Sims, p. 2009, p. 7).

The SoarTech team has conceded to some challenges regarding virtual character implementations of the CCA. For one, it was difficult for the team to determine the ideal granularity for cognitive and physical actions. Dealing with the granularity at the atomic level might be well-tailored to very specific cultural instances but would likely not be generalizable enough to be interoperable with multiple cultural contexts. More generalized models are easier to engineer. However, if the models are too generalized, then this would create a culture-agnostic character which might not be seen as believable or realistic (Taylor & Sims, 2009, p. 9).

In theory, but not entirely realized in practice, the CCA-agents are potentially scalable to handle rich data models and a fluent understanding of various natural languages. However, with increasing complexity and size comes an equal increase in available computational resources. Within the virtual world context, this poses no real obstacle because the "[...] fidelity of the data would be only as good as the fidelity of the interactions available in the virtual world" and most virtual world data is limited to culturally unrealistic gestures and inauthentic chat-responses anyway (Taylor & Sims, 2009, p. 9). To date, however, there does not appear to be any firm evidence of Soar-agents (including CCA-agents and any other AI-enhanced agents) that have sufficiently transcended the limitations of finite-state organization to handle natural language understanding and generation at an adult human level. At the very least, some noteworthy attempts (see Lindes et al., 2015) have been made to use Soar for grounded language acquisition and linguistic mapping. In one particular case, a Soar-agent can "[...] associate the sentence 'put a large red block in the table' with an instantiated operator which will achieve the intended goal" (Mohan et al., 2012., p. 124).

CAA-agents are similar to the Director in Haunt 2 as it can select goals that will encourage following the currently activated narrative script towards its ritualistic conclusion (Taylor et al., 2007, p. 4). Soar's impasse threshold-breaking mechanism is mirrored in CAA's ability to prefer, apply, and fire those current script-relevant operators and rules from working memory that already have the highest activation potential (Taylor et al., 2007, p. 3-4). The issue here is that the CAA-agent will always act based on what is determined to be the most culturally appropriate action rather than in its own self-interest (Taylor et al., 2007, p. 3-4).

A Virtual Soar-Agent using Augmented Reality for Improvised Choreography

VAI (Viewpoints Artificial Intelligence, Jacob et al., 2013) is an augmented reality virtual Soar-agent designed for interactive improvised choreography in real-time. VAI is projected onto a 2D surface in order to interact with human-participants in a turn-taking manner (Jacob et al.,, 2013, p. 4). VAI is influenced by the collision physics of the human participant's shadow and can learn movement sequences that are scanned from a human's skeletal vertex points acquired via a Microsoft Kinect system ("Kinect for Windows", n.d.). Cognitively, VAI accesses "perception, reasoning and action modules" in order to calculate relative vertex values of: joint position, tempo (pacing), energy[12] levels, and velocity. These values are conveniently represented for Soar as "predicates" (Jacob et al., 2013). These Soar-ready predicates convert VAI's ambient movements into color values when gestures are being made at low-acceleration rates (Jacob et al., 2013, p. 4). VAI stores all input and output gestures into an "internal interaction history" (Jacob et al., 2013, p. 4) which likely encompasses various stages of Soar's working memory. VAI either randomly accesses these working memory elements – composed of various dance-sequence durations - in order to determine the best movement operator to propose, select, and apply, or ignores the interaction history entirely (Jacob et al., 2013, p. 4). In response, VAI can choose between refusing to apply an operator, repeat the input gesture (i.e. mimicking the dancing human participant), or perform some novel choreographic variation on said gesture. VAI's creators have asserted their desire to enable the possibility for VAI to propose and apply operators based on instances of similarity based reasoning principles (i.e. the similarity between "source and target [...] predicates", Jacob et al., 2013, p. 5). VAI also has a heuristic for showing an awareness of the human participant's presence and dance-movement choices while also showing the capacity for improvisation and creative autonomy. This heuristic allows VAI to follow a human participant's repeated gesture twice (consecutively) before performing an improvised gestural variation of its own devising (Jacob et al., 2013, p. 5).

VAI's limitations were more the result of its perceived levels of interaction and not so much due to any cognition parsing issues produced by Soar. For example, some people were too impatient with the turn-taking approach to VAI's choreographed interactions (Jacob et al., 2013. p. 4-5). Otherwise, some human participants either found VAI's interactions to be arbitrary (random), or merely copying the human movements. In fact, some human participants were deluded into thinking

Figure 4. A snapshot of VAI taking turns improvising a dance routine with a human participant. Permission for this image granted by the author
(Jacob et al., 2013)

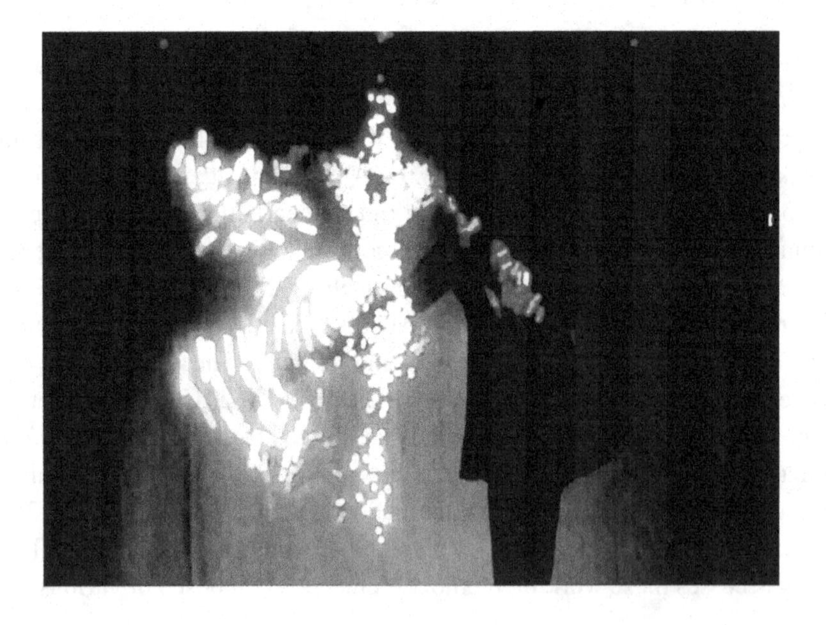

the VAI agent possessed more deliberative capabilities than were actually present under-the-hood (Jacob et al., 2013, p. 4-5). The latter can be seen as beneficial for those designers wishing to elicit conceivable rather than accessible interactive states in the mind of the interactor.[13]

Criticism of Soar-Implementations

Current versions of Soar – although centered on rules – work best with a robust knowledge base containing a large number of hand-coded productions. Ultimately, Soar works best for purely deterministic simulated environments but has been criticized for not being able to solve even rudimentary logic problems based on hidden, partially observable or incomplete empirical knowledge (Bringsjord et al., 2005, p. 35). For many virtual agent implementations, cognitive architectures such as Soar do not usually require full scale logic operations. This is because they are usually "[…] too slow for real-time perception-and-action-heavy computation" (Bringsjord et al., 2005, p. 35) anyway. Soar might be scalable while keeping the computational cost relatively low but eventually, Soar's cognition lags when a large number of production-rules need to successfully interpret a dynamically changing virtual environment. Basically, Soar's chunking mechanism can optimize limited computational resources but a stochastic simulation modeled after the "real world" will cause the automatic-subgoaling mechanism to gradually tax the Soar's working and long-term memory capabilities.

The general issue of abstracting out semantically-appropriate information from a stochastic virtual environment and "specifying the [symbolically represented] conditions under which productions can apply" (Best et al., 2006, p. 212) provides ongoing engineering challenges to those particular cognitive architectures that are production rule-driven (esp. Soar). At best, a production-oriented cognitive architecture can cope with generalized symbolic approximations of environmental conditions through some rudimentary form of case-based reasoning process. In this instance, a cognitive agent can focus on a single "canonical case" that can be applied to most particular interactive scenarios where conditional decision making applies (Best et al., 2005, p. 213). Culturally speaking, cognitive elements representing a "canonical case" – and its associated natural language structures - might not even be sufficiently decomposable into production rules and rule-chunks.

In virtual environments optimized for multi-agent system interactions, cognitive architectures are usually under-represented as they are criticized for neglecting multi-agent implementations. This is because cognitive architectures were originally to focus on the egocentric cognitive concerns of a single agent instead of allocentrically mapping the overall cognitive climate of interacting agents. According to Smart et al., developers of these multi-agent swarms, usually "[...] assume that individual agents have very rudimentary cognitive processing capabilities, and in some cases, aspects of agent cognition (e.g., beliefs or attitudes) are represented by simple numerical values that do not respect the constraints imposed by the human cognitive system" (Smart et al. 2014a., p. 1).

Speculations for Future Virtual Soar-Agent Implementations

Overall, the above examples are diverse enough from each other to address the range of Soar's potential and limitations when cognitively interfacing with a virtual agent. It is up to each of the above researchers whether or not they wish to re-iterate their virtual agents with newer Soar-related features. Regardless, there are some plausible research directions these and other researchers could take to develop the next-generation virtual Soar-agents. For example, STEVE could utilize Soar's reinforcement learning capability to emergently discover how to train naval officers with equipment operating in unknown or hostile virtual simulations (e.g. a war-zone). In addition, the NPCs in Haunt 2 could be modified to explicitly model separate "in-conditions" from pre and post conditions. Further, CCA-agents could do with a connectionist representation in future versions of Soar – similar to CLARION - where cultural sensitivities can be gradually learned in a supervised manner using tunable weights and/or with reinforcement signals. Perhaps, a virtual-agent can be incrementally rewarded and/or have neural connections solidified (i.e. Hebbian learning activity increased) for producing behaviors (as well as internal thoughts) that seem to match the protocols of other cultures (e.g. scripts for culturally "correct" postures, semantic relationships, and routines). This would be a compelling alternative to classically tuning Soar's rules and operators themselves. Ideally, the issue of a virtual agent's "self-interest" situated within an uncertain cultural context would be more elaborately addressed if Soar was modified to draw from a software ontology (Noy et al., 2001) that contained recursive predicates and slots that dealt with "self-awareness". Monitoring substates is sufficient for finding common ground between complex operators (Laird, 2012., p. 66). However, operators themselves are not necessarily a tractable representation of what one culturally would consider to be a form of self-awareness. As for VAI, it may prove difficult to have Soar's decision cycle operate in a non-serial manner. Turn-taking, however, was an issue for some impatient human participants so VAI's creators might consider either advocating more parallelism in Soar and/or implementing VAI with a different cognitive architecture more suited to parallel decision making.[14]

Generally speaking Soar's current endeavor to develop additional affective-state models and mental imagery (i.e. imagination) representations is encouraging. Soar will also likely be adopted by a newer generation of virtual agent designers once a distinct prospective memory store is finally implemented in future iterations.

Imagination resides in episodic memory[15] as a list of proposed (but not preferred) operators, rules, and actions. These proposed rules, operators and, actions are the atoms of imagination in Soar. These atoms can be further elaborated on via fictional metadata stored in semantic memory, or as plausible (but untested) interactivity algorithms, and/or heuristics stored in procedural memory. Alongside speculations of a prospective memory Soar implementation, a parallel sub-component of episodic memory known as "elaborative memory" or "fantasy memory"[16] would generate imagined novel episodes as myths/fictions. These episodes would propose equally fictional operators, rules, and actions.[17] The elaborative memory would be driven by sub-goals and sub-states. Elaborative memory would determine which sub-goals/ sub-states should be rapidly retrieved from other memory banks. Chunking would involve elaboration by fabricating memories, operators, rules etc. The rule firings themselves could even resemble mirror neurons as they could be simulated internally without actually firing to express cognitive deliberation (Maffre et al., 2001).

Prospective memory on its own would definitely benefit those aforementioned virtual Soar agents that navigate semi-deterministic virtual spaces. The possibility of Soar including a prospective memory as one of its features would be even more impressive if the virtual Soar-agent interface this type of memory with a robust logic component – autonomous from the authority of production rules. Further, a future version of Soar could have its own equivalent representation of CLARION's *rule-extraction refinement algorithm*" (aka. RER, Sun 2006., p. 86). Soar could approximate how CLARION actually connects declarative and procedural knowledge. Like CLARION, Soar's own version of the RER algorithm could be encoded as an explicit knowledge element for use by CLARION's deductive and inductive logic components in the service of actionable inference. In other words, Soar can understand in its own way how CLARION's showcased RER algorithm has enabled a CLARION-agent to learn from sub-symbolic knowledge and find ways to convert this knowledge into logic and/or ontology symbols. If the implicit knowledge component is sufficiently salient and capable of semantic refinement, base empirical experience can even be converted into a Soar production rule. Prospective memory combined with logical inference and a new representation of the RER algorithm may even be able to predict future world-states in a stochastic environment with some actionable degree of confidence.

CONCLUSION

In conclusion, this chapter has showcased the various ways in which Soar can be integrated with virtual agents representing multi-faceted roles such as: military trainers, video-game NPCs, culturally sensitive interpreters, and improvised choreographers.

This chapter has discussed the relevance for using a cognitive architecture in the first place and has contextualized the strengths and weaknesses of implementing virtual agents with Soar.

Soar's legacy remains as an established scalable and stable production-rule architecture still maintained by an active development team seeking further iteration and improvement to its fundamental design. The speculations mentioned in this chapter merely serve to inspire cognitive architects to push the boundaries of interactions driven by mental-processes even farther than anticipated.

REFERENCES

Anderson, J. R. (1996). ACT: A simple theory of complex cognition. *The American Psychologist*, *51*(4), 355–365. doi:10.1037/0003-066X.51.4.355

Anonymous. (n.d.). *CLARION Cognitive Architecture*. Retrieved February 27, 2015, from https://sites.google.com/site/clarioncognitivearchitecture

Anonymous. (n.d.). *Kinect for Windows*. Retrieved March 15, 2015, from http://www.microsoft.com/en-us/kinectforwindows/

Anonymous. (n.d.). *LSL access portal*. Retrieved March 16, 2015, from http://wiki.secondlife.com/wiki/LSL_Portal

Anonymous. (n.d.). *Second Life. LSL access portal*. Retrieved March 16, 2015, from http://secondlife.com

Anonymous. (n.d.). *Soar Home Page*. Retrieved February 27, 2015, from http://Soar.eecs.umich.edu/

Anonymous. (n.d.). *SoarTech Corporate Brochure*. Retrieved February 28, 2015, from http://ai.eecs.umich.edu/Soar/sitemaker/docs/misc/SoarRBSComparison.pdf

Anonymous. (n.d.). *SoarTech* webpage. Retrieved March 15, 2015, from http://www.Soartech.com/about/

Best, B. J., & Lebiere, C. (2006). Cognitive agents interacting in real and virtual worlds. In R. Sun (Ed.), Cognition and Multi-Agent Interaction: From Cognitive Modeling to Social Interaction. New York: Cambridge University Press, 2006.

Bringsjord, S., Khemlani, S., Arkoudas, K., McEvoy, C., Destefano, M., & Daigle, M. (2005). Advanced synthetic characters, evil, and E. *Game-On*, *6*, 31–39.

Brooks, R. (1991). Intelligence Without Representation. *Artificial Intelligence*, *47*(1-3), 139–159. doi:10.1016/0004-3702(91)90053-M

Cox, M. T., & Raja, A. (2011). Metareasoning: An Introduction. In M. T. Cox & A. Rajam (Eds.), *Metareasoning: Thinking About Thinking* (pp. 3–14). Cambridge, MA: MIT Press. doi:10.7551/mitpress/9780262014809.003.0001

Crawford, C. (2003). *The Art of Interactive Design: A euphonius and illuminating guide to building successful software* (pp. 77–90). San Francisco: No Starch Press.

Isbister, K. (2006). *Better Game Characters By Design: A Psychological Approach.* San Francisco: Elsevier.

Jacob, M., Coisne, G., Gupta, A., Sysoev, I., Verma, G., & Magerko, B. (2013). Viewpoints AI.*Proceedings of the Ninth Annual AAAI Conference on Artificial Intelligence and Interactive Digital Entertainment (AIIDE).*

Laird, J. (2012). *The Soar Cognitive Architecture.* Cambridge, MA: MIT Press.

Laird, J., Hucka, M., Huffman, S., & Rosenbloom, P. (1991). An analysis of Soar as an integrated architecture. *ACM SIGART Bulletin*, *2*(4), 98–103. doi:10.1145/122344.122364

Laird, J. E., Assanie, M., Bachelor, B., Benninghoff, N., Enam, S., Jones, B., . . . Wallace, S. (2002). A Test Bed for Developing Intelligent Synthetic Characters. *AAAI 2002 Spring Symposium Series: Artificial Intelligence and Interactive Entertainment.*

Laird, J. E., Kinkade, K. R., Mohan, S., & Xu, J. Z. (2012). Cognitive robotics using the Soar cognitive architecture. *Cognitive Robotics AAAI Technical Report*, *WS-12*, 6.

Laird, J. E., & Rosenbloom, P. S. (1993). Integrating execution, planning, and learning in Soar for external environments. Academic Press.

Laird, R., Newell, A., & Rosenbloom, P. S. (1987). Soar: An Architecture for General Intelligence. *Artificial Intelligence*, *33*(1), 1–64. doi:10.1016/0004-3702(87)90050-6

Lathrop, S. D., & Laird, J. E. (2009, March). Extending cognitive architectures with mental imagery. *Proceedings of the second conference on artificial general intelligence.*

Lehman, J. F., Laird, J. E., & Rosenbloom, P. S. (1996). A gentle introduction to Soar, an architecture for human cognition. *Invitation to Cognitive Science, 4,* 212-249.

Leonard, A. (1997). *Bots: The Origin of New Species.* San Francisco: HardWired.

Li, J., & Laird, J. E. (2013). The Computational Problem of Prospective Memory Retrieval. *Proceedings of the 17th International Conference on Cognitive Modeling.*

Lim M.Y., Dias, J., Aylett, R., & Paiva, A. (2010). *Creating Adaptive Affective Autonomous NPCs. Autonomous Agent Multi-Agent Systems.* DOI: 10.1007/s10458-010-9161-2

Maffre, E., Tisseau, J., & Parenthoen, M. (2001). Virtual Agent's Self-Perception in Story Telling. In *Virtual Storytelling: Using Virtual Reality Technologies for Storytelling.* Berlin: Springer-Verlag.

Magerko, B., Laird, J., Assanie, M., Kerfoot, A., & Stokes, D. (2004). AI characters and directors for interactive computer games. In *IAAI'04 Proceedings of the 16th conference on Innovative applications of artifical intelligence.*

Mateas, M. (1997). *Computational Subjectivity in Virtual World Avatars.* AAAI Technical Report FS-97-02, 1-6.

Minsky, M. (1975). A Framework for Representing Knowledge. In The Psychology of Computer Vision (pp. 211-277). New York: McGraw Hill.

Moon, J., & Anderson, J. R. (2012). Modeling Millisecond Time Interval Estimation in Space Fortress Game.*34th Annual Conference of the Cognitive Science Society*, Sapporo, Japan.

Noy, N. F., & McGuinness, D. L. (2001). *Ontology Development 101: A Guide to Creating Your First Ontology.* Stanford Knowledge Systems Laboratory Technical Report KSL-01-05 and Stanford Medical Informatics Technical Report SMI-2001-0880, March 2001.

Reidl, M. O., & Stern, A. (2006). Failing Believably: Toward Drama Management with Autonomous Actors in Interactive Narratives. In S. Göbel, R. Malkewitz, & I. Iurgel (Eds.), *TIDSE 2006, LNCS 4326* (pp. 195–206). Berlin: Springer-Verlag. doi:10.1007/11944577_21

Rickel, J., & Johnson, L. W. (2000). Task-Oriented Collaboration with Embodied Agents in Virtual Worlds. In J. Cassell, J. Sullivan, & S. Prevost (Eds.), *Embodied Conversational Agents* (pp. 95–122). Cambridge, MA: MIT Press.

Ritter, F., Kim, J. W., & Sanford, J. P. (2012). *Soar Frequently Asked Questions List.* Retrieved January 30, 2016, from http://acs.ist.psu.edu/projects/soar-faq/soar-faq.html#G3

Schank, R. C., & Abelson, R. P. (1975). *Scripts, plans, and knowledge.* Yale University.

Smart, P. R., Richardson, D. P., Sycara, K., & Tang, Y. (2014a). Towards a cognitively realistic computational model of team problem solving using ACT-R agents and the ELICIT experimentation framework.19th ICCRTS C2 Agility: Lessons Learned from Research and Operations. Southhampton University School of Electronics and Computer Science.

Smart, P. R., & Sycara, K. (2015, July) Situating Cognition within the Virtual World.*6th International Conference on Applied Human Factors and Ergonomics*, Las Vegas, NV.

Smart, P. R., Sycara, K., & Lebiere, C. (2014b, September). Cognitive Architectures and Virtual Worlds: Integrating ACT-R with the XNA Framework.*Annual Fall Meeting of the International Technology Alliance*, Cardiff, UK.

Sun, R. (2006). The CLARION cognitive architecture: Extending cognitive modeling to social simulation. In Cognition and multi-agent interaction, (pp. 79-99). New York, NY: Cambridge University Press.

Taylor, G., Quist, M., Furtwangler, S., & Knudsen, K. (2007, May). Toward a hybrid cultural cognitive architecture. *CogSci Workshop on Culture and Cognition.* SOARTech.

Taylor, G., & Sims, E. (2009). Developing Believable Interactive Cultural Characters for Cross-Cultural Training. In Online Communities and Social Computing (pp. 282-291). Springer Berlin Heidelberg.

van Lent, M., McAlinden, R., Probst, P., Silverman, B. G., O'Brien, K., & Cornwell, J. (2004). Enhancing the behaviorial fidelity of synthetic entities with human behavior models. Departmental Papers (ESE), 300.

ADDITIONAL READING

Anstey, J., & Pape, D. (2002, October). Scripting the Interactor: An Approach to VR Drama.*IC&C '02, October 16-26, 2002. ACM 1-58113-465-7/02/0010.* Pp. 150-156.

Anstey, J., Pape, D., & Sandin, D. (2000). *The Thing Growing: Autonomous Characters in Virtual Reality Interactive Fiction. Virtual Reality Proceedings* (pp. 71–78). New Brunswick, New Jersey: IEEE; doi:10.1109/VR.2000.840366

Ayiter, E., Glasauer, S., & Moswitzer, M. (2013). LPDT2: La plissure du texte 2. In D. Harrison (Ed.), *Digital Media and Technologies for Virtual Artistic Spaces* (pp. 75–90). Hershey, PA: IGI Global. doi:10.4018/978-1-4666-2961-5.ch006

Bates, J. (1994). The Role of Emotion in Believable Agents. *Communications of the ACM, 37*(7), 122–125. doi:10.1145/176789.176803

Blumberg, B., & Galyean, T. (1995, August). Multi-Level Direction of Autonomous Creatures for Real-Time Virtual Environments.*Proceedings of SIGGRAPH 95, ACM SIGGRAPH, Los Angeles, CA, August 1995.* pp. 47-54. doi:10.1145/218380.218405

Brahnam, S. (2001). Creating Physical Personalities for Agents with Faces: Modeling Trait Impressions of the Face.*Proceedings of the Workshop on "Attitudes, Personality and Emotions in User-Adapted Interaction", held in conjunction with the Conference on User Modeling 2001.* Pp. 1-7.

Bringsjord, S. (1992). *What Robots can and can't be*. Norwell, MA: Kluwer Academic Publishers. doi:10.1007/978-94-011-2464-5

Bringsjord, S. (2008). Declarative/Logic-Based Computational Cognitive Modeling. In R. Sun (Ed.), *The Cambridge Handbook of Computational Psychology* (pp. 127–169). Cambridge, UK: Cambridge University Press. doi:10.1017/CBO9780511816772.008

Bringsjord, S., Shilliday, A., Taylor, J., Werner, D., Clark, M., Charpentier, E., & Bringsjord, A. (2008). Toward logic-based cognitively robust synthetic characters in digital environments. *Frontiers in Artitifical Intelligence and Applications, 171*, 87–98.

Burden, D. J. H. (2009, October). Deploying Embodied AI into Virtual Worlds. *28th SGAI International Conference on Artificial Intelligence. Knowledge-Based Systems, 22*(7), 540–544. doi:10.1016/j.knosys.2008.10.001

Chong, H.-Q., Tan, A.-H., & Ng, G.-W. (2007). Integrated cognitive architectures: A survey. *Artificial Intelligence Review, 28*(2), 103–130. doi:10.1007/s10462-009-9094-9

Colton, S. (2008, Spring). Creativity versus the Perception of Creativity in Computational Systems.*Proceedings of the AAAI Spring Symposium on Creative Systems.*

El-Nasr, M., Yen, J., & Ioerger, T. R. (2000). FLAME - Fuzzy Logic Adaptive Model of Emotions.[Netherlands.]. *Autonomous Agents and Multi-Agent Systems, 3*(3), 219–257. doi:10.1023/A:1010030809960

Friedenberg, J., & Silverman, G. (2014). *Cognitive Science: An Introduction to the Study of Mind* (2nd ed.). New York: SAGE publications.

Funge, J., Tu, X., & Terzopoulos, D. (1999, August). Cognitive Modeling: Knowledge, Reasoning and Planning for Intelligent Characters.*Proceedings of the 26th Annual Conference on Computer Graphics and Interactive Techniques (SIGGRAPH '99)*, pp. 29-38, Los Angeles, CA, August 1999. doi:10.1145/311535.311538

Goertzel, B. (2009, June). OpenCogPrime: A cognitive synergy based architecture for artificial general intelligence. In *Cognitive Informatics, 2009. ICCI'09.8th IEEE International Conference* (pp. 60-68). IEEE. doi:10.1109/COGINF.2009.5250807

Goertzel, B. (2014). GOLEM: Towards an AGI meta-architecture enabling both goal preservation and radical self-improvement. *Journal of Experimental & Theoretical Artificial Intelligence*, 1–13.

Goertzel, B., & Pennachin, C. (2007). *Artificial General Intelligence*. New York: Springer. doi:10.1007/978-3-540-68677-4

Gordon, A. S., Hobbs, J. R., & Cox, M. T. (2011). Anthropomorphic Self-Models for Metareasoning Agents. In Metareasoning: Thinking About Thinking (pp. 295-305). MIT Press: Cambridge. 335 pages. doi:10.7551/mitpress/9780262014809.003.0019

Hart, D., & Goertzel, B. (2008). OpenCog: A Software Framework for Integrative Artificial General Intelligence. *Artificial General Intelligence, 2008:Proceedings of the First AGI Conference* (Vol. 171, p. 468-472). IOS Press.

Hebb, D. O. (1949). *The Organization of Behavior*. New York: Wiley & Sons.

Hibbard, B. (2012). Model-based utility functions. *Journal of Artificial General Intelligence, 3*(1), 1–24. doi:10.2478/v10229-011-0013-5

Holland, O. (2004). The Future of Embodied Artificial Intelligence: Machine Consciousness? Lecture Notes in Computer Science, 2004, Vol. 3139/2004, 629, DOI: . Pp. 74-85. doi:10.1007/978-3-540-27833-7_3

Kluwer, T., Adolphs, P., Xu, F., Uszkoreit, H., & Cheng, X. (2010). Talking NPCs in a Virtual Game World. *ACLDemos '10 Proceedings of the ACL 2010 System Demonstrations. Association for Computational Linguistics (Stroudsburg).*

Legg, S., & Hutter, M. (2005, July). A universal measure of intelligence for artificial agents. *International Joint Conference on Artificial Intelligence(*Vol. 19*)* p. 1509. Marwah, NJ: Lawrence Erlbaum Associates Ltd.

Lekavy, M., & Navrat, P. (2007, May/June). Expressivity of STRIPS-like and HTN-like Planning. Lecture Notes in Artificial Intelligence, Vol. 4496 Agent and multi-agent Systems. *Technologies and applications. 1st KES International Symposium, KES-AMSTA 2007, Wroclaw, Poland, May/June 2007*. - Germany, Springer-Verlag Berlin Heidelberg, 2007. pp. 121-130

Lim, M. Y., Dias, J., Aylett, R., & Paiva, A. (2010). *Creating Adaptive Affective Autonomous NPCs. Autonomous Agent Multi-Agent Systems* (pp. 287–311). Berlin: Springer.

Lindes, P., Lonsdale, D. W., & Embley, D. W. (2015, October). Ontology-Based Information Extraction with a Cognitive Agent.*Twenty-Ninth AAAI Conference on Artificial Intelligence*. Pp. 558-564.

Loyall, A. B., Reilly, W. S. N., Bates, J., & Weyhrauch, P. (2004). System for Authoring Highly Interactive Personality-rich Interactive Characters". *Eurographics/ ACM SIGGRAPH Symposium on Computer Animation*.

Marques, H. G., & Holland, O. (2009). Architectures for Functional Imagination. *Neurocomputing*, *72*(4-6), 743–759. doi:10.1016/j.neucom.2008.06.016

Marr, D. (1982). *Vision: A Computational Approach*. San Francisco: Freeman & Co.

McCormack, J., Romero, J., & Machado, P. (2007). Facing the Future: Evolutionary Possibilities for Human-Machine Creativity. In J. McCormack, J. Romero, & P. Machado (Eds.), *The Art of Artificial Evolution: A Handbook on Evolutionary Art and Music* (pp. 417–451). Berlin: Springer.

Minsky, M. (2006). *The Emotion Machine: Commonsense Thinking, Artificial Intelligence and the Future of the Human Mind*. New York: Simon & Schuster.

Mueller, E. T., & Dyer, M. G. (1985). Towards a Computational Theory of Daydreaming.*Proceedings of the Seventh Annual Conference of the Cognitive Science Society*, Irvine, CA, pp. 120-135.

Nareyek, A. (2004, February). AI in Computer Games. *Queue*, *1*(10), 58–65. doi:10.1145/971564.971593

Nau, D. S., Au, T. C., Ilghami, O., Kuter, U., Muñoz-Avila, H., Murdock, J. W., & Yaman, F. et al. (2005, March-April). Applications of SHOP and SHOP2. *IEEE Intelligent Systems*, *20*(2), 34–41. doi:10.1109/MIS.2005.20

Nivel, E., Thórisson, K. R., Steunebrink, B. R., Dindo, H., Pezzulo, G., Rodriguez, M., . . . Jonsson, G. K. (2013). *Bounded Recursive Self-Improvement*. arXiv preprint arXiv:1312.6764.

Nowak, K.L. (2004, January). The Influence of Anthropomorphism and Agency on Social Judgment in Virtual Environments. *Journal of Computer-Mediated Communication. Vol. 19, Issue No.2* Pp. 2-30.

Orkin, J. (2006, March). Three states and a plan: the AI of FEAR.*Game Developers Conference(*Vol. 2006, *p.*4*).* 18 pages.

Perlis, D. (2011). There's No "Me" in "Meta" – Or Is There? In Cox, M.T.m Raja, A. (Eds.), Metareasoning: Thinking About Thinking (pp. 15-26). Cambridge: MIT Press: Cambridge. 335 pages.

Rabin, S. (2009). [Rabin, S., Ed.] Introduction to Game Development, Second Edition. Chapter 5.3: Artificial Intelligence: Agents, Architecture and Techniques. Newton Center, MA: Charles River Media. Pp. 538-574.

Ranathunga, S., Cranefield, S., & Purvis, M. (2011, May). Interfacing a Cognitive Agent Platform with a Virtual World: A Case Study using Second Life (Extended Abstract). *Proceedings of the 10th International Conference on Autonomous Agents and Multiagent Systems (AAMAS 2011).* Turner, Yolum, Sonenberg and Stone (eds.). May 2-6, 2011, Taipei, Taiwan, pp. 1181-1182.

Robinett, W. (1998). Interactivity and Individual Viewpoint in Shared Virtual Worlds. In Dodsworth Jr., C. (Ed.), Digital Illusion: Entertaining the Future with High Technology, Vol. 28, Issue 2. New York, NY: ACM Press, Ch. 21, pp. 331-342.

Roman, M., Sandu, I., & Buraga, S. C. (2011). Owl-Based Modeling of RPG Games. *Informatica, LVI*(3), 83–90.

Russell, S., & Norvig, P. (2010). Artificial Intelligence: A Modern Approach. Third Edition. Chps: Intro, 3, 12, 7-17. Upper Saddle River, NJ: Prentice Hall.

SoarTech. (2013). *Entity-Level Patterns of Life for Virtual Environments*. Promotional Brochure.

Strannegård, C. (2005). *Anthropomorphic Artificial Intelligence. Philosophical Communications #33* (pp. 169–181). Department of Philosophy, Gothenburg University.

Sutton, R. S., & Barto, A. G. (1998). Reinforcement learning: An introduction: Vol. 1. *No. 1*. Cambridge, MA: MIT press.

Thórisson, K., & Helgasson, H. P. (2012). Cognitive architectures and autonomy: A comparative review. *Journal of Artificial General Intelligence*, *3*(2), 1–30. doi:10.2478/v10229-011-0015-3

Ventrella, J. (2010). *Virtual Body Language: Avatars, Smileys and Digital Puppets: Evolving the Future of Nonverbal Expression on the Internet. Beta Edition*. Petaluma, CA: Eyebrain Books.

Waterworth, E.A. & J.A. (2003). The Illusion of Being Creative. In Riva. G. Et al (Eds.), *Being There: Concepts, Effects and Measurements of User Presence in Synthetic Environments* (pp. 223-236), Chapter 15, pp. 223-236. Amsterdam, NL: IOS Press.

Whitelaw, M. (2004). *Metacreation: Art and Artificial Life*. Cambridge, MA: MIT Press.

Zhao, R., & Szafron, D. (2012). *Generating Believable Virtual Characters Using Behavior Capture and Hidden Markov Models. Advances in Computer Games* (pp. 342–353). Berlin: Springer.

Zilberstein, S. (2011). Metareasoning and Bounded Rationality. In Cox., M.T., Raja, A. 2011 [Eds.], Metareasoning: Thinking About Thinking (pp. 27-40). Cambridge, MA: MIT Press 335 pages. doi:10.7551/mitpress/9780262014809.003.0003

KEY TERMS AND DEFINITIONS

Bot: Social virtual world slang for a Virtual Agent.

Chunk: Episodic associations compressed into a singular generalized concept.

Cognitive Architecture: A unified collection of cognitive models used to regulate mental processes.

Declarative Knowledge: Memorized Information that pertains to factual information.

NPC: Video-game slang for a "Non-Player Character" (aka. a Virtual Agent).

Production: A type of rule that is dynamically governed by alterable conditional statements.

Virtual Agent: The academic name for an artificially intelligent entity in a virtual environment.

ENDNOTES

1. Since at least 1971. For one of the earliest cognitive-architecture flow-charts, see page 3b & 5b of Atkinson, R. C., & Shiffrin, R. M. (1971). The control

processes of short-term memory. Technical report no. 173. Institute for Mathematical Studies in the Social Sciences, Stanford University. 23 pages.

2 Soar was originally treated as an acronym. Therefore, Soar would normally be capitalized as SOAR. However, there is now evidence to suggest that the Soar community no longer uses the acronym convention to describe this cognitive architecture (Ritter, Kim, and Sanford, 2012).

3 For more historical contextualization, read Anderson 1996 & Best et al. 2006 (in Sun, 2006, p. 187-189).

4 Of note, the creator of ACT-R has interfaced his architecture a 2D virtual space-ship that is the main player-avatar of the videogame "Space Fortress" (Bothell, 2010). However, to limit the scope of this paper, a space-ship will, in in this case, not be considered as a "character". See Moon & Anderson, 2012.

5 Smart et al. (2014) make the distinction that their abELICIT agents designed by Smart et al. (2014) for ACT-R are not really "[…] based on a cognitive architecture that is modeled after the human cognitive system" (Smart et al. 2014a, p. 5).

6 For more information, see van Gelder, T. (2001). "Distributed vs. Local Representation" (pp. 236-*238).* In R.A. Wilson., F.C. Keil., [Eds.]. The MIT Encyclopedia of Cognitive Science (MITECS). Cambridge, MA: MIT Press. 964 pages.

7 Based on email correspondence with the lead CLARION developer, Michael Lynch on March 04, 2015.

8 Second Life is the most popular social virtual world with its own proprietary "Linden Scripting Language" (LSL) for virtual agent AI implementations.

9 Apparently, STEVE can draw from different declarations related to other domains that support virtual worlds and associated objects, if needed (Rickel & Johnson, 2000, p. 96). Also, STEVE can re-tailor his training demonstration by to adjusting his plan to world-state changes and "failure modes" induced by human-interactors in the VR world (Rickel & Johnson, 2000., p. 97).

10 To clarify, SoarTech developed a "[…] hybrid symbolic-numeric model" that is part of more comprehensive sub-architecture built entirely within Soar. (Taylor et al., 2007, p. 8).

11 This research was intended to analyze and interpret Middle-Eastern (esp. Arabic) cultural protocols.

12 Energy in this context, represents the "amount of movement" (Jacob et al., 2013, p. 4).

13 An inverse of the "common-sense" idea of privileging accessible over conceivable states (advocated by Crawford, 2003).

14 With Soar, only the decision cycle runs serially. The rules themselves can fire in parallel.

[15] In future Soar versions, the imagination will likely also reside in prospective memory.

[16] These are my terms to describe what should be future memory components in Soar.

[17] For example, Soar's syntax might include "elaborate-operator" as well as "propose-operator".

Chapter 8
Towards Truly Autonomous Synthetic Characters with the Sigma Cognitive Architecture

Volkan Ustun
University of Southern California, USA

Paul S. Rosenbloom
University of Southern California, USA

ABSTRACT

Realism is required not only for how synthetic characters look but also for how they behave. Many applications, such as simulations, virtual worlds, and video games, require computational models of intelligence that generate realistic and credible behavior for the participating synthetic characters. Sigma (Σ) is being built as a computational model of general intelligence with a long-term goal of understanding and replicating the architecture of the mind; i.e., the fixed structure underlying intelligent behavior. Sigma leverages probabilistic graphical models towards a uniform grand unification of not only traditional cognitive capabilities but also key non-cognitive aspects, creating unique opportunities for the construction of new kinds of non-modular behavioral models. These ambitions strive for the complete control of synthetic characters that behave as humanly as possible. In this paper, Sigma is introduced along with two disparate proof-of-concept virtual humans – one conversational and the other a pair of ambulatory agents – that demonstrate its diverse capabilities.

DOI: 10.4018/978-1-5225-0454-2.ch008

1. INTRODUCTION

Twenty years ago, Tambe et al. (1995) discussed the generation of human-like synthetic characters that can interact with each other, as well as with humans, within the emerging domain of highly interactive simulations. Many of these simulations strove to create environments that looked realistic, and synthetic characters that looked and behaved as real people to the extent possible. The behavioral models in these simulations extensively utilized *cognitive architectures* (Langley, Laird, & Rogers, 2009) – models of the fixed structure underlying intelligent behavior in natural and/or artificial systems – as the underlying driver for human-like intelligent behavior. Twenty years later, developments in computer graphics and animation have allowed for extremely realistic-looking interactive simulation environments; it is now possible to create almost photo-real synthetic characters with realistic gaits and gestures. However, progress in behavior generation has been more mixed. Mainstream cognitive architectures, including Soar and ACT-R, originated as production systems and are fairly capable of modeling the reactive, knowledge-intensive, and goal-driven aspects of human behavior. For example, Tambe et al.'s (1995) work in the air-combat simulation domain utilized Soar (Laird, 2012) to model the behavior of pilots. These cognitive architectures are also capable of working in real time and, in ACT-R's case, with explicit models of human reaction times and limitations. However, they have not yet been able to successfully incorporate all the capabilities that are required for human-like intelligence.

As Swartout (2010) has pointed out, behaving like real people requires synthetic characters to, among other things: (1) use their perceptual capabilities to observe their environment and other virtual/real humans in it; (2) act autonomously in their environment based on what they know and perceive, e.g. reacting and appropriately responding to the events around them; (3) interact in a natural way with both real and other virtual humans using verbal and nonverbal communication; (4) possess a *Theory of Mind* (*ToM*) to model their own mind and the minds of others; (5) understand and exhibit appropriate emotions and associated behaviors; and (6) adapt their behavior through experience. The Soar and ACT-R communities worked toward addressing these six capabilities for synthetic characters (referred to as the *capability list* hereafter) but some items were simply not feasible within the core architecture. For example, external modules were required for acceptable perceptual and communication capabilities. Likewise, most of the emotion models were also outside the core. More importantly, they haven't been able to fully capture the advances that have been made in recent years in behavioral adaptation, or in other words, learning. A number of aspects of learning were successfully incorporated but generality in statistical machine learning, for example, has eluded them.

Probabilistic graphical models (Koller & Friedman, 2009) provide a general tool that combines graph theory and probability theory to enable efficient probabilistic reasoning and learning in ways that haven't been possible with traditional cognitive architectures based on production systems. The machine learning community employs such models as one of its primary tools, yielding state-of-the-art results for at least four of the listed capabilities that have challenged traditional cognitive architectures: perception, autonomy, interaction and adaptation. However, most of these improvements have been achieved independently, as examples of narrow Artificial Intelligent (AI) systems, with little effort toward cross-integration.

One of the main reasons for the relative lack of integration efforts is the inherent difficulty of general intelligence research: not only is it challenging to integrate all of the requisite capabilities, but it is also hard to measure the resulting incremental progress toward human-like intelligence (Goertzel, 2014). In contrast, many forms of narrow AI systems can easily track incremental progress as they try to improve "intelligent" behaviors in very specific contexts. Therefore, it is easier to assess the merit of these systems through simple comparisons. This strategy has helped narrow AI approaches dominate the research in the last two decades.

The objectives for generally intelligent systems, however, are quite different from narrow AI systems. Generally intelligent systems are not necessarily focused on the extremes of performance in limited contexts but rather on the ability to achieve a wide range of goals and to carry out a variety of tasks in many different contexts and environments, as stated by Goertzel (2014). Such systems require the traits of the capability list to be integrated together and work coherently. As pointed out by Swartout (2010), this integration can be quite difficult, but it can potentially yield more than the sum of its parts. Mainstream cognitive architectures, as natural candidates for the desired integration, have attempted to add some of these capabilities without actually updating their cores (Ji, Gray, Guhe, & Schoelles, 2004). This approach has improved their overall functionality but still has not been enough to capture many critical developments, especially in perception and learning.

This paper introduces a nascent cognitive architecture/system, Sigma (Rosenbloom, 2013), that conflates traditional cognitive architectures and probabilistic graphical models. The long-term goal is to understand and replicate the fixed structure underlying intelligent behavior in *both* natural and artificial systems. Sigma is motivated by the original grand goal of AI, as well as by the more recent reformulation of this goal within artificial general intelligence (AGI) (Goertzel, 2014), in an attempt to create a full cognitive system that can implement the qualities in the capability list and more. As such, it is intended ultimately to support the real-time needs of truly autonomous characters, whether robots or synthetic characters (virtual humans).

Sigma's development is guided by a quartet of desiderata: (1) *grand unification*, uniting the requisite cognitive and non-cognitive aspects of embodied intelligence for intelligent behavior in complex real worlds; (2) *generic cognition*, spanning both natural and artificial cognition at a suitable level of abstraction; (3) *functional elegance*, yielding broad cognitive (and sub-cognitive) functionality from interactions among a small general set of mechanisms; and (4) *sufficient efficiency*, executing rapidly enough for anticipated applications. The first, second and last desiderata are directly relevant to the construction of broadly capable, real-time virtual humans (VHs), while the third implies a rather unique path towards them, where instead of a disparate assembly of modules, all of the required capabilities are constructed and integrated together on a simple elegant base.

Most of the work to date on Sigma has explored various capabilities individually: learning (Rosenbloom, 2012, 2014; Rosenbloom, Demski, Han, & Ustun, 2013), memory and knowledge (Rosenbloom, 2010, 2014; Ustun, Rosenbloom, Sagae, & Demski, 2014), decision making and problem solving (Chen et al., 2011; Rosenbloom, 2011b), perception (Chen et al., 2011), speech (Joshi, Rosenbloom, & Ustun 2014), ToM (Pynadath, Rosenbloom, Marsella, & Li, 2013; Pynadath, Rosenbloom, & Marsella, 2014), and emotions (Rosenbloom, Gratch, & Ustun 2015). But, even more importantly, *Sigma's* non-modular, *hybrid* (discrete + continuous) *mixed* (symbolic + probabilistic) character also supports attempts at a deep form of integration across the capability list, straddling the traditional boundary between symbolic cognitive processing and numeric sub-cognitive processing. Sigma provides the ability to exhibit this combination of capabilities in a unified manner because of its grounding in a *graphical architecture* that is built from graphical models (Koller & Friedman 2009) (in particular, factor graphs and the summary product algorithm (Kschischang, Frey, & Loeliger 2001)), n-dimensional piecewise linear functions (Rosenbloom, 2011a), and gradient descent learning (Rosenbloom, Demski, Han, & Ustun, 2013). The required VH capabilities emerge in a functionally elegant manner from both interactions among this small but general set of mechanisms and knowledge captured in Sigma models. For example, reinforcement learning (RL) arises from the interactions between gradient descent learning and particular forms of both domain-specific and domain-independent knowledge (Rosenbloom, 2012). Truly autonomous characters would require a computational intelligent behavior model that pushes the boundaries of how the broad range of traits in the capability list may be integrated, with Sigma providing a unique, and potentially powerful, way of doing this.

Still, this is just the beginning of a long journey. The real power of the unique path taken by Sigma is in the emerging synergy between diverse capabilities. Yet the models discussed in this paper are just the first attempts toward developing integrative virtual humans. There is definitely a need to devise more comprehensive virtual humans with more interactions among characters in realistic scenarios and trained across multiple environments in order to assess Sigma's true potential.

Nonetheless, Sigma is maturing day by day, and the diverse progress that has been made so far provides initial evidence that Sigma will be capable of creating truly autonomous VHs as a capstone of its maturation.

There have been a few other recent proposals for the utilization of graphical models in cognitive architectures (Cassimatis, 2010; Danks, 2014; Lee-Urban et al., 2015), but they are mostly just proposals with quite limited implementations. Neural networks also provide a major alternative to graphical models for addressing a number of the items in the capability list that most challenge traditional cognitive architectures, and with the recent focus on deep models provide the current state of the art in statistical machine learning. General intelligence is emerging as a popular research topic in the deep learning community (Mikolov, Joulin, & Baroni 2015; Schmidhuber, 2015; Strannegard, von Haugwitz, Wessberg, & Balkenius 2013), but it remains to be seen what the research in this field will yield. In general, none of the current approaches is yet adequate for comprehensively encompassing the capability list.

In this paper, first a case is made for needing truly autonomous characters that embody human-like intelligence, rather than just characters who may be perceived as intelligent, in interactive simulation environments. Then, Sigma will be introduced and its potential for creating truly autonomous synthetic characters will be discussed along with two disparate proof-of-concept virtual humans – one conversational (Ustun, Rosenbloom, Kim, & Li 2015) and the other a pair of ambulatory agents (Ustun & Rosenbloom, 2015) – that demonstrate its diverse capabilities. Both varieties of virtual human combine cognitive and non-cognitive capabilities, and in the case of the ambulatory agents integrate together diverse forms of learning.

2. TRULY AUTONOMOUS CHARACTERS

Traditional cognitive architectures have controlled virtual characters capably in a number of interactive simulation environments (Tambe et al., 1995; Traum, Marsella, Gratch, Lee, & Hartholt 2008; Zemla et al., 2011). One of the most relevant examples is the use of the Soar cognitive architecture to create a synthetic character with multiple goals and extensive tactics in the computer game Quake II (Laird and van Lent 2001). Laird and van Lent (2001) states: "While the Soar Quakebot explores a level, it creates an internal model of its world and uses this model in its tactics to collect nearby weapons and health, and set ambushes. It also tries to anticipate the actions of human players by putting itself in their shoes (creating an internal model of their situation garnered from its perception of the player) and projecting what it would do if it were the human player" (p. 22). The Soar Quakebot did not have as many rules as a fully developed commercial synthetic character would, nor did it have any probabilistic reasoning or learning capabilities, but it did show some potential to create realistic and believable synthetic characters.

Yet similar efforts never gained much traction due, at least in part, to the reality that major consumers of interactive simulation environments – e.g. training, gaming, and entertainment communities – are not interested in whether these synthetic characters are actually intelligent or not, but only in the perceived intelligence of these characters. This perspective severely limits the resources allocated for efforts in general intelligence. Many of these interactive simulation environments even prefer stricter control on the interactions between synthetic characters and real humans, as such tight control can prevent synthetic characters from exhibiting behaviors that may be potentially perceived as unintelligent. Many simulations use the labor-intensive process of defining hierarchical representations of goals and rules that cover almost all of the simulation's situations. For a higher level of control of these situations, it is a common practice to "cheat" by collecting information on human opponents without actually perceiving them in the gaming environment (Bourg & Seaman 2004; Schaeffer, Bulitki, & Buro, 2008). When these rule-based decision models are combined with very realistic graphics and animations, striking applications can be created with even a relatively simple rule structure (DeVault et al., 2014).

The relative success of applications with extensive rule structures in guiding behavior impedes the penetration of even basic narrow AI approaches into interactive simulation environments. Basically, training by observation is considered to be too time consuming and the resulting behavior unpredictable (Robertson & Watson, 2014; Umarov, Mozgovoy, & Rogers, 2012) when the sole focus is on perceived intelligence. There are even some experts who believe that all that users want is predictability and that it is therefore unnecessary to pursue synthetic characters that learn by experience. On the other hand, there are other experts and researchers who believe that human behavioral models with cognitive capabilities and more human-like intelligence can make simulations more varied and the games more enjoyable (Lucas & Kendall, 2006; Robertson & Watson, 2014). Many interactive simulation models already desire synthetic characters that are not only believable but also indistinguishable from humans – performing coherent sequences of believable behavior sustained across different tasks and environments. Human behavior is still considered a challenge by/for the simulation community (Taylor et al., 2015), where there is a need for creating truly autonomous synthetic characters that actually have generic perception, action and learning capabilities appropriate for a variety of environments. Such characters are almost impossible to generate by strictly controlling their behavior, as this approach would require generating rules for each different task, each different environment, and every possible interaction. These characters need to be trained by observation, and thus they ideally require architectures with general learning capabilities.

Characters that are truly autonomous would be very interesting for interactive simulation environments, as they would be defined by the experiences they have undergone and the environments where they have been, contributing to the personality of each character. Eventually, the hope is that progress in the development of these characters will yield *plug compatibility* between humans and artificial systems (Tambe et al., 1995). Such characters would open doors to uncharted territories and to immense opportunities for the training, gaming, and entertainment communities. One recent example is the deep reinforcement-learning algorithm that has been used to train a synthetic character from high-dimensional sensory inputs to play multiple Atari games (Mnih et al., 2015). The synthetic character in this study received only the game score and pixels as input and was able to achieve a level comparable to humans using the same algorithm, network structure and hyperparameters across games. There was no interaction with other synthetic characters or switching of contexts in this study, but it is still an inspiring achievement in using real perceptions.

Overall, interest in general intelligence is increasing, and many rich applications with truly autonomous synthetic characters are envisioned for the future. Architectures capable of creating such characters are thus very intriguing.

3. SIGMA

The Sigma cognitive architecture is built on *factor graphs* (Kschischang, Frey, & Loeliger 2001) – undirected graphical models (Koller & Friedman 2009) with variable and factor nodes, and functions that are stored in the factor nodes. Graphical models provide a general computational technique for efficient computation with complex multivariate functions – implemented via hybrid mixed *piecewise-linear functions* (Rosenbloom, 2011a) in Sigma – by leveraging forms of independence to: decompose them into products of simpler functions; map these products onto graphs; and solve the graphs via message passing or sampling methods. The *summary product algorithm* (Kschischang, Frey, & Loeliger 2001) is the general inference algorithm in Sigma (Figure 1). Graphical models are particularly attractive as a basis for broadly functional, yet simple and theoretically elegant, cognitive architectures because they provide a single general representation and inference algorithm for processing symbols, probabilities and signals.

The Sigma architecture defines a high-level language of *predicates* and *conditionals* that compiles down into factor graphs. Predicates specify relations over continuous, discrete and/or symbolic arguments. They are defined via a name and a set of typed arguments, with *working memory* (WM) containing predicate instantiations as functions within a WM sub-graph. Predicates may also have perception and/or long-term memory (LTM) functions. For perceptual predicates, factor nodes

Figure 1. Summary product computation over the factor graph for $f(x,y,z) = y^2+yz+2yx+2xz = (2x+y)(y+z) = f_i(x,y)f_2(y,z)$ *of the marginal on y given evidence concerning x and z*

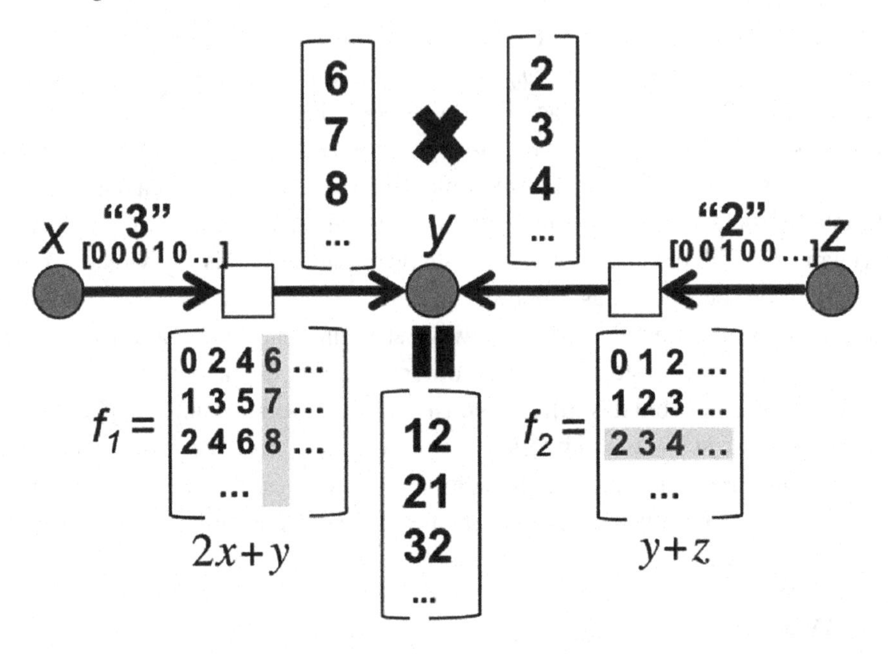

for *perceptual buffers* are connected to the corresponding WM sub-graphs. For example, Observed(object:object visible:boolean) is a perceptual predicate with two arguments: (1) object of type object; and (2) visible of type boolean. This predicate specifies which objects are visible to the agent at any particular time. For memorial predicates, *function factor nodes* (FFNs) are likewise connected to the corresponding WM sub-graphs. Messages into FFNs provide the gradients for learning the nodes' functions. Gradient calculations require identifying parent-child relationships among predicate arguments; for instance, the predicate function for the Object-Location-X(object:object x:location%) predicate defines a distribution over the x coordinate given an object (x is marked as the child variable in this predicate by %).

Conditionals structure LTM and basic reasoning, compiling into more extended sub-graphs that interconnect with the appropriate WM sub-graphs. Conditionals are defined via a set of *predicate patterns* – in which type specifications are replaced by constants and variables – and an optional *function* over pattern variables. *Conditions* and *actions* are predicate patterns that behave like the respective parts of rules, pushing information in one direction from the conditions to the actions. The example conditional in Figure 2 updates the information about which objects have

Figure 2. Conditional for context information

```
CONDITIONAL Seen
    Conditions: Observed(object:o visible:true)
    Actions: Seen-Objects(object:o)
```

been seen so far, based on the information in the Observed predicate. *Condacts* are predicate patterns that support the bidirectional processing that is key to probabilistic reasoning, partial matching, constraint satisfaction and signal processing. Examples of probabilistic networks based on condacts and functions will be seen in the coming sections on virtual humans. Overall, conditionals provide a deep combination of rule systems and probabilistic networks.

The core processing in Sigma is driven by a cognitive cycle that comprises input, graph solution, decisions, learning, and output. Graph solution yields inference via the summary product algorithm, by *product* of the messages coming into each node – including the node's function when it is a factor node – and then *summarization*, via integration or maximum, of unneeded variables from outgoing messages. Perception and action are modeled as sub-symbolic reasoning in Sigma that occurs predominantly within graph solution, rather than within external modules (Joshi, Rosenbloom, & Ustun, 2014). Specifically, perceptual information collected during the input phase is passed to WM, where it is combined with the current contents of the WM and knowledge available within the LTM – such as might define a hidden Markov model (HMM) or a conditional random field (CRF) – yielding new WM content at the end of the graph solution phase.

Decisions in Sigma, in the classical sense of choosing one among the best operators to execute next, are based on an architecturally distinguished selection predicate – Selected (state:state operator:operator) – with typically the operator associated with the highest value, or utility, in the distribution for a state being selected. The learning phase then modifies the long-term functions in the graph via a general form of gradient-descent learning. Finally, in the output phase, selected actions are performed in the outside world.

A single cognitive cycle yields *reactive processing*; a sequence of them yields *deliberative processing*; and *reflective processing* occurs when no operator can be selected, or a selected operator can't be applied, similar to impasses in Soar (Laird, 2012). Activities such as classifying an object or executing a body of rules can all occur reactively, within a single cognitive cycle. Deliberative processing utilizes reactive processing, and thus individual cognitive cycles, as its inner loop in providing sequential, algorithmic behavior that leads to problem solving and reasoning.

When impasses occur, deliberative processing – either in the same or different problem spaces – can be used reflectively to yield the knowledge that would resolve the impasses.

4. THE IMMERSIVE NAVAL OFFICER TRAINING SYSTEM (INOTS)

The first proof-of-concept Sigma Virtual Human discussed in this paper focuses on language and conversation. This Sigma model replicates the mind of a deployed, but cognitively rather simple, virtual human that was originally developed for the Immersive Naval Officer Training System (INOTS) system, a training vehicle for junior leaders in the U.S. Navy in interpersonal communication skills (Campbell et al., 2011). INOTS replaces the live role player in a traditional role-play practice with a virtual human, creating a blended learning environment that merges traditional classroom instruction with a mixed reality training setting. For instance, the virtual human's behavior and dialog are tailored to provide the cues indicative of when to apply specific strategies and skills for the problems on hand. The implementation uses a branching narrative such that each utterance (also known as a challenge in INOTS) by the virtual human is a decision point with a fixed number of possible responses by the trainee. At each challenge, the system displays the possible responses for the trainee to choose. The trainee is expected to speak one of the possible responses that are recognized by the system. Each response is linked to another decision point in the narrative representing the virtual human's reaction and the next set of possible utterances for the trainee. The scripted dialogue of the virtual human and the scripted possible responses for the trainee allow practicing the strategies that an expert leader would use to solve certain types of interpersonal problems (Campbell et al., 2011).

4.1 INOTS Sigma Model

The overall control structure for the INOTS virtual human is based on a branching – directed acyclic – network of states and utterances plus an utterance classifier that is used to determine which choice to make at each point, and thus also what response to make. Two separate tools were combined to make this work, one for the branching structure and the other for the utterance classifier (Campbell et al., 2011). In Sigma these two functionalities were mapped, respectively, onto: (1) deliberative movement in a discourse problem space composed of operators for speaking and listening, and (2) a reactive bag-of-words naïve Bayes utterance classifier (Ustun, Rosenbloom, Kim, & Li 2015).

The basic control structure alternates whose turn it is to talk, with a listen operator selected when it is the human's turn and a speak operator selected when it is the VH's turn. When it is the human's turn, their utterance – which is typed rather than spoken to Sigma – is classified as one from among a small set of standard utterances that have been predefined for the corresponding dialogue state; there are typically 1-3 standard utterances per state. Two distinct types of conditionals realize this naïve Bayes classification: (1) the prior on the utterance being classified (Figure 3), and (2) the conditional probabilities on the words given the utterances, with one conditional per word in the vocabulary (Figure 4). The Prior-Meaning conditional (Figure 3) defines the valid utterances for the current dialog state by having an Utterance(meaning:utterance-number) predicate pattern as a condact, where utterance-number is a discrete numeric type that uniquely identifies the distinct utterances known to the system. The function in the Prior-Meaning conditional, in which only valid utterances for the current dialog state are assigned non-zero probability, captures the prior on the utterances (m is the child variable for this function). In the current implementation, a uniform distribution is used as a prior over the valid utterances for the current dialogue state.

The conditional in Figure 4 demonstrates how conditional probabilities are represented for the word play given the utterance. A conditional probability distribution is defined over the likelihood of the presence of play in a sentence spoken by the trainee for each associated meaning via the function in the Sentence-Word-Play conditional (wp is the child variable for this function). Similarly, for each word in the VH's vocabulary, there is a distinct conditional and function that capture the likelihood of the presence of that word in each utterance. At the end of graph solution, a posterior distribution is yielded for the Utterance predicate and the utterance

Figure 3. Prior on the sentence

```
CONDITIONAL PRIOR-MEANING
     Conditions:  Dialog-State(state-number:sn)
     Condacts:    Utterance(meaning:m)
     Function(sn,m): …
```

Figure 4. Example conditional for the bag-of-words model

```
CONDITIONAL SENTENCE-WORD-PLAY
     Condacts:     Utterance(meaning:m)
                   Word-Play(word-play:wp)
     Function(m,wp): …
```

with the highest probability is selected as the one that is heard. However, if none of the utterances have a probability greater than 0.5, it is assumed that the trainee has entered something unrecognizable, and s/he is asked to reenter the utterance.

The selected standard utterance – as represented in the Heard-Utterance predicate – then feeds the transition conditional (Figure 5), which determines the subsequent dialogue state. When it is the VH's turn to speak, only one possible utterance exists for each dialogue state, so it is simply produced and the next dialogue state is determined as a direct function of the previous one.

This VH mind works in real-time, and is capable of holding a dialogue comparable to the one in the original system. However, it is still rather limited in its capabilities. Even though Sigma is capable of learning the conditional probability distributions for the bag-of-words model, and has done so in a variety of other tasks, for the current implementation these distributions were determined outside of Sigma and then imported into conditional functions. The original goals for implementing this particular VH mind within Sigma were to show that simple such minds could be created simply, and in an integrated fashion, and that they could then be extended to more sophisticated intelligent behavior as required – typically customer requirements in domains like this start simple, but then grow as it is understood what the initial system can and cannot do. The first part of this was demonstrated, but not yet the second part, although such extensions could still be explored in the future.

5. PHYSICAL SECURITY SYSTEM

Chen et al. (2011) discussed the fusion of symbolic and probabilistic reasoning at an earlier stage of the development of Sigma. In that study, initial steps towards grand unification were demonstrated when perception, localization and decision-making were implemented within a single graphical model, with interaction among these capabilities modulated through shared variables. The work here greatly expands on this approach to yield a more significant combination of capabilities, plus a deployment in virtual humans – within a 3D virtual environment (rather than a toy one-dimensional space) – embodied in the SmartBody character animation system

Figure 5. Transition conditional

```
CONDITIONAL Transition
    Conditions:   Heard-Utterance(heard:h)
    Actions:      Dialog-State(state-number:h)
```

(Shapiro, 2011). SmartBody's internal movement, path-finding and collision detection algorithms are used in animating the virtual human's actions, although eventually much of this is to be moved within Sigma. Sigma has no "direct" access to the virtual environment, being limited instead to perceiving and acting on it through a (deliberately) noisy interface.

A number of capabilities within Sigma have been combined to construct an adaptive, interactive virtual human in a virtual environment. The VH is adaptive not only in terms of dynamically deciding what to do based on the immediate circumstances, but also in terms of embodying two distinct forms of relevant learning: (1) the automated acquisition of maps of the environment from experience with it, in the context of the classic robotic capability of Simultaneous Localization and Mapping (SLAM); and (2) reinforcement learning (RL), to improve decision making based on experience with the outcomes of earlier decisions. The VH is interactive both in terms of its (virtual) physical environment – through high-level perception and action – and other participants, although the latter is still quite limited. Speech and language are being investigated in Sigma, but neither is deployed in these virtual humans yet, so social interaction is limited to constructing – actually, learning, with the help of RL – models of the self and others. These forms of adaptivity and interaction are combined together, along with for these initial virtual humans a basic rule-based decision framework, all within Sigma.

5.1 Physical Security System Conceptual Model

Physical security systems are comprised of structures, sensors, protocols, and policies that aim to protect fixed-site facilities against intrusions by external threats, as well as unauthorized acts by insiders. Physical security systems are generally easy to understand but they also allow complex interactions to emerge among the agents. These properties make physical-security-systems simulation a natural candidate as a testbed for developing cognitive models of synthetic characters (Ustun, Yilmaz, & Smith 2006). Similar to the discussion made by Ustun et al. (2006), a physical-security-system scenario in a retail store has been selected as a platform to develop and test Sigma VH models.

In a typical retail-store shoplifting plot, offenders first pick up merchandise in a retail store and then try to leave without getting caught by any of the store's security measures. A simple grab-and-run scenario is considered in this paper (a large number of different scenarios are possible). In this scenario, the intruder needs to locate the desired item in the store, grab it, and then leave the store. The role of security is to detain the intruder before s/he leaves the store. A basic assumption is that it is not possible to tell what the intruder will do until s/he picks up an item and starts running. The security can immediately detect the activity and start pursuing the intruder once the item is picked up (assuming CCTV). If the intruder makes it to the door, it is considered a success for the intruder.

For the basic setup, it is assumed that the intruder does not know the layout of the store and hence it has to learn a map and be able to use it to localize itself in the store. When the intruder locates the item of interest, it grabs the item and leaves the store via one of the exits. In the hypothetical retail store used here (Figure 6), there are shelves (gray rectangles), the item of interest (the blue circle) and two entry/exit doors (red rectangles). The intruder leaves the store via either (1) the door it used to enter or (2) the door closest to the item of interest. The main task for security is to learn about the exit strategies of intruders and use this to effectively detain them.

SmartBody (Shapiro, 2011), a Behavior Markup Language (BML) (Kopp et al., 2006) realization engine, is used as the character animation platform for this study, with communication between

the Sigma VH model and SmartBody handled via BML messages. In the current setup, locomotion and path finding are delegated to the SmartBody engine. Sigma sends commands and queries to SmartBody to perform these tasks and to return perceptual information. Two basic types of perception are utilized by the Sigma VH model: (1) information about the current location of the agent, mimicking the combination of direction, speed and odometry measurements available to a robot; and (2) objects that are in the visual field of the agent, along with their relative distances, mimicking the perception of the environment for a robot. Location information is conveyed to the Sigma VH model with noise added – perfect location information is not available to the model.

Figure 6. Layout of the store and its SmartBody representation

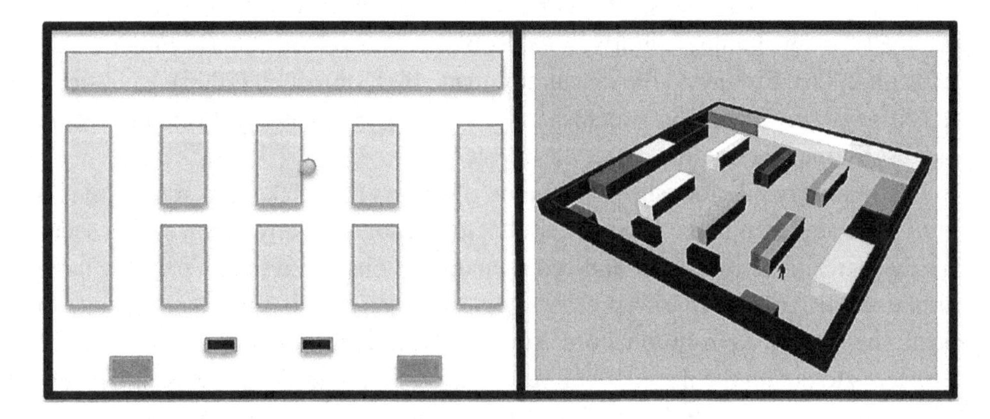

5.2 Physical Security System Sigma Model

In a typical physical security system setting there are intruders (shoplifters) and security personnel. There may also be neutrals, but they are not modeled in this work. In this initial implementation, an intruder and a security agent are modeled as virtual humans. There are two distinct types of learning (and probabilistic inference) in this scenario: (1) The intruder agent does not know the layout of the store in advance, and so it must learn a map of the store while simultaneously localizing itself in the learned map (SLAM model), and (2) The security personnel agent infers the strategy of the intruder agent – i.e., whether it exits through the entry door or the closest door – by first learning a policy for the intruder agent via RL and then using this policy and the intruder agent's actions to determine on each trial the relative likelihoods of the two strategies being used (Section 5.2.1). The intruder also needs a decision framework to dynamically decide what to do based on its immediate circumstances (Section 5.2.2).

5.2.1 Learning in the Physical Security System VH Models

Sigma can effectively utilize different types of learning, a key capability leading towards adaptive and interactive synthetic characters that are truly autonomous. Here, the agents learn the probability distributions in a Bayesian network and Q functions for an RL algorithm using the same basic set of architectural mechanisms, something that would require multiple distinct modules in other architectures if it could be done at all.

In the SLAM model, the intruder has no a priori knowledge about the layout of the retail store (which is as shown in Figure 6). Therefore, it needs to learn a map of the store while simultaneously using the map to localize itself in the store. A 31×31 grid is imposed on the store for map learning. A virtual human only occupies a single grid cell, whereas objects in the environment – such as shelves – can span multiple cells. While performing SLAM, a Dynamic Bayesian Network (DBN) is utilized (Figure 7) (Grisetti, Kummerle, Stachniss, & Burgard 2010). In this representation, l is the location, u captures the odometry readings, p represents perceptions of the environment, and m is the map of the environment.

The Sigma VH model defines two perceptual predicates – Location-X(x:location) and Location-Z(z:location) – to represent the location of the virtual human on the grid; here, the location type is discrete numeric, with a span of 31. Together these two predicates represent the space of 2D cell locations in the grid. They are perceptual predicates, and hence induce perceptual buffers that emulate odometry readings. In particular, the current location of the agent is perceived with noise – the default noise model assumes that any neighboring cell of the correct cell may be perceived

Figure 7. Dynamic Bayesian Network of the SLAM process

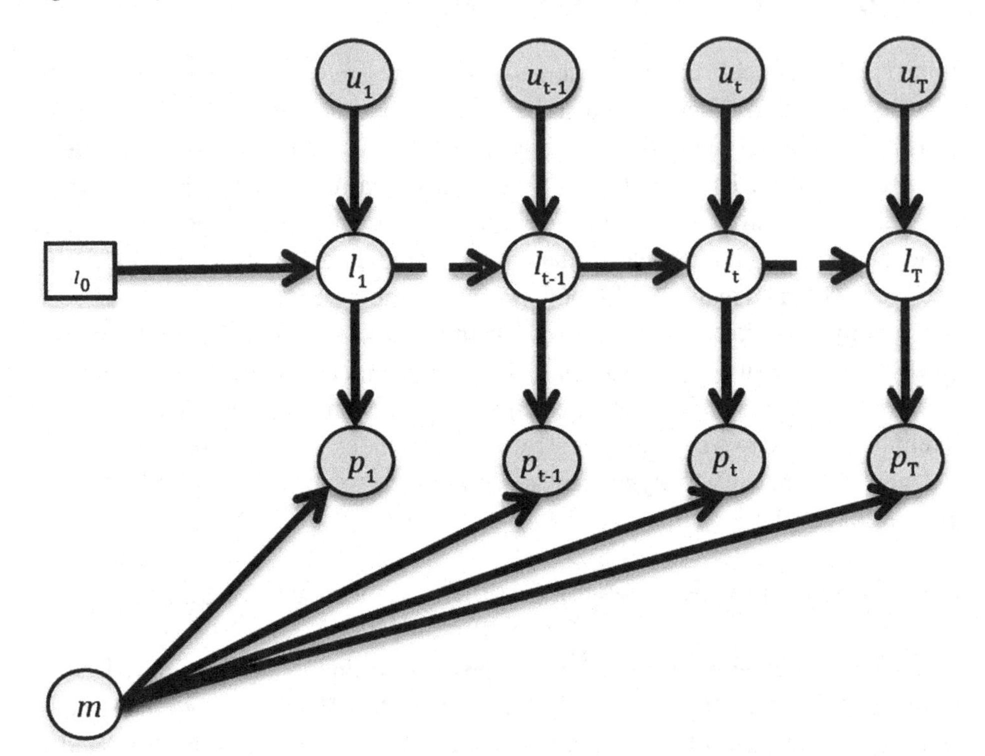

as the agent's current location. In addition, the objects in the visual field are also perceived along with the relative distances of the center of these objects to the agent's location.

The DBN is captured in two almost identical conditionals, one for x (Figure 8) and a similar one for z. These conditionals convert relative locations of objects given the agent to absolute locations in the map. They use Sigma's general capability for affine transforms in visual imagery (Rosenbloom, 2012) to offset the agent's current location by the distance to the object being updated. In Figure 8, the Object-Location-X predicate is a memorial predicate so it has a function that learns the x coordinates in the map via gradient descent. Since both the Location-X and Object-Location-X patterns are condacts, the processing is bidirectional between them; both perception of the VH's location and perception of the object locations have an impact on the posterior for the VH's location. This bidirectional processing forms the basis for SLAM, where the map is learned while it is simultaneously used for localization.

Once it was verified that without noise this implementation of SLAM could acquire appropriate maps, an additional experiment was run on this VH to explore more probabilistic learning under noise, and thus uncertainty. This experiment in-

Figure 8. SLAM conditional for the x coordinate

```
CONDITIONAL SLAM-X
    Conditions:    Observed(object:o visible:true)
                   Object-Distance-X(dist-x:dx object:o)
    Condacts:      Location-X(x:lx)
                   Object-Location-X(object:o x:(lx-dx))
```

vestigated the effect of different noise models on the number of trials required to learn a correct map. A normally distributed noise with mean 0 and varying standard deviations was added to the location perceptions of the VH in this experiment, with the results shown in Figure 9. As expected, the learned map is sensitive to noise, but small amounts of it can be tolerated, even given just a small number of trials (Figure 9 (a)). However, high-variation noise models make it hard to learn correct maps. In Figure 9(c), the correct location of the item of interest is not effectively

Figure 9. Learned distributions over the displacements of the x (left) and z (right) coordinates from the correct values for the item of interest after varying number of trials, under three different noise models

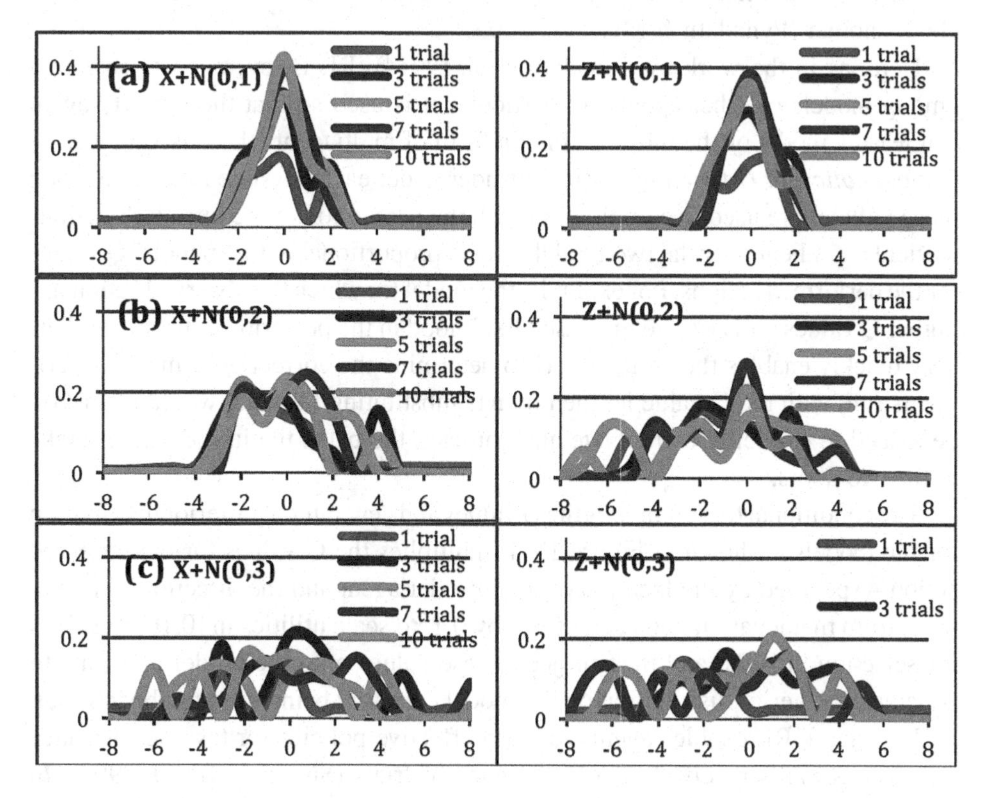

learned even after 10 trials. These experiments provide evidence that general probability distributions are effectively learned in the Sigma cognitive architecture.

In addition to mapping and localizing itself like the intruder, the security agent needs to reason about the intruder agent's actions for effective detainment, and hence it needs to learn about the policies used by the intruder agent. One basic assumption made in this paper is that it is easy to recognize that a grab-and-go scenario has been initiated, by observing the pick-up behavior of the intruder.[1] However, even though security can easily recognize when such a scenario has been initiated, it still needs to intercept the intruder before it leaves the retail store. As there are two exit doors, early anticipation of the intruder's choice increases the chances of a successful detention.

Theory of Mind (ToM) involves formation of models of others and generation of expectations about their behavior based on these models to enable effective decisions in social settings (Whiten, 1991). In decision theoretic approaches to ToM, such models can be represented as reward functions. For the intruder in our scenario there are two possible models, distinguished by whether a reward is received when the agent returns to its door of entry or when it reaches the nearest door from the item of interest. This corresponds to Bayesian approaches to multi-agent modeling that use a distribution over a set of policies to specify the beliefs that one agent has about another (Pynadath & Marsella, 2005).

Here, as in the work by Pynadath et al. (2014), RL is leveraged in selecting among models of other agents; in particular it is used so that the security agent can learn a model of the intruder. First a form of multiagent RL is used to learn a distinct *policy*, or *Q function*, for the intruder under each possible model, and then these policies are used in combination with the perception of the intruder's actions to yield a gradient over the two models that is proportional to the models' Q values for the performed actions. For example, the model for which the observed action has higher Q values will achieve increased likelihood in the posterior distribution. This very quickly enables the security VH to determine the correct door in the experiments run, with the Q value learned by RL substituting for what would otherwise be a need to explicitly extrapolate and compare the paths the intruder might take to the two doors.

The conditional that compares the Q values and generates a posterior distribution for the models is shown in Figure 10. It multiplies the Q values for the observed action – specified by the location of the intruder agent and the direction of movement from that location – in each policy by 0.1, to scale utilities in [0,10] to values for selection in [0,1], and then projects these values onto the model predicate to generate a posterior distribution on the model currently being used by the intruder.

In general, RL enables agents to learn effective policies for task performance based on rewards received over a sequence of trials (Sutton & Barto, 1998). In

Figure 10. Model prediction conditional

```
CONDITIONAL PREDICT-MODEL
     Conditions:  Previous-RL-Loc(location:loc)
                  RL-Direction(direction:d)
                  Q(model:m location:loc direction:d
                    value:[0.1*q])
     Actions:     Model(model:m)
```

Sigma, RL is not a separate architectural learning algorithm, but occurs through a combination of gradient-descent learning and the appropriate knowledge expressed in predicates and conditionals. This also means that, similar to the work by Pynadath et al. (2014), it is possible to move from single-agent to multi-agent RL, and from RL given a single reward function to RL given a range of possible reward functions – i.e., models – by appropriately changing the predicates and conditionals.

5.2.2 Behavior Rules in the Physical Security System VH Models

The objective of the intruder is to grab the item of interest (Figure 6) and to leave the store through one of the exits without being detained. In the current implementation, four basic behaviors, with corresponding Sigma operators, are available to the intruder: (1) *walk towards target object*; (2) *run towards target object*; (3) *pick up target object*; and (4) *walk towards random object*. Target objects play a role in the intruder achieving its goals, while random objects drive exploration of the store.

The intruder is initialized with a sequence of target objects that it needs either to walk towards or to pick-up. Given that the intruder does not have a priori knowledge of the store, the location of the current target object may be unknown to it. In this case it needs to explore the store, mapping it in the process, to locate the target object. The basic operator used for exploration is *walk towards random object*. The intent is that doing this will help the agent discover new objects, and eventually the target object.

This exploratory operator is always available; however, if other more task-relevant behaviors are available, they take precedence. For example, Figure 11 shows a conditional in which the operator *walk towards target object* is suggested for selection with a utility of 0.5 when the virtual human has seen the target object (and hence, there is an estimate of its location). Exploration has a lower utility, so walking to a target object takes precedence if both operators are available.

When the virtual human is within a threshold distance of the target object, a new operator – *pick-up target object* – is selected. The model terminates when the intruder reaches its preferred exit door, which acts as the target object for the *run towards target object* operator.

Figure 11. Conditional suggesting the walk towards target object operator

```
CONDITIONAL WALK-TOWARDS-TARGET
    Conditions:  Target-Object(object:o)
                 Seen-Objects(object:o visible:true)
    Actions:     Selected(operator:walk-target)
    Function:    0.5
```

Combining the short-term adaptivity provided by this rule-based behavior with the long-term adaptivity provided by map and ToM learning yields embodied Sigma-based VHs that exhibit effective social and environmental interaction in the physical security model.

6. CONCLUSION

The two models presented here are important steps in the maturation of Sigma as they are the first uses of Sigma to control virtual characters, demonstrating small but diverse combinations of symbolic and sub-symbolic processing. Each embodies an important subset of the capabilities that virtual humans should support. It is hoped that over the long term such combinations of the capabilities of traditional symbolic cognitive architectures and probabilistic graphical models may assist in the generation of high-fidelity computational behavior models. There are short-term plans for extending these particular models: (1) the INOTS model is to incorporate a Sigma-based continuous speech recognizer, and (2) the physical security model is to incorporate natural language, speech, and affect. Furthermore, the variety of participants involved – intruders, security, and neutrals – makes the physical security model in particular extremely flexible for the generation of scenarios that encompass many different interactions among VHs and between VHs and humans. As the scenarios get more complex, the expectation is that the forms of cognition exhibited will keep up, going beyond simple rule-based reasoning to more involved combinations of reactive, deliberative and reflective processing. Further experiments with these models will help to test the limits and assess Sigma as an architecture for general intelligence.

Broadly capable real-time virtual humans provide ideal test beds for demonstrating and evaluating progress on Sigma's four desiderata, with "broadly capable" challenging the extent of its grand unification; "real-time" challenging its sufficient efficiency (Rosenbloom, 2012); and "virtual humans" with their need to exhibit human-like behavior in artificial systems, challenging its generic cognition. The virtual humans here also leverage Sigma's functionally elegant approach to providing and combining the requisite capabilities, such as rule-based reasoning, language, SLAM, ToM, and RL.

ACKNOWLEDGMENT

This effort has been sponsored by the Office of Naval Research and the U.S. Army. Statements and opinions expressed do not necessarily reflect the position or the policy of the United States Government, and no official endorsement should be inferred. We would also like to thank Ari Shapiro for his overall support with SmartBody.

REFERENCES

Bourg, D. M., & Seaman, G. (2004). *AI for Game Developers*. O'Reilly Media.

Campbell, J. (2011). Developing INOTS to Support Interpersonal Skills Practice. In *Proceedings of the Thirty-second Annual IEEE Aerospace Conference (IEEEAC)*. Washington, DC: IEEE doi:10.1109/AERO.2011.5747535

Cassimatis, N. (2010). An architecture for adaptive algorithmic hybrids. *Systems, Man, and Cybernetics, Part B: Cybernetics. IEEE Transactions on*, *40*(3), 903–914.

Chen, J., Demski, A., Han, T., Morency, L. P., Pynadath, D. V., Rafidi, N., & Rosenbloom, P. S. (2011). Fusing Symbolic and Decision-Theoretic Problem Solving+ Perception in a Graphical Cognitive Architecture. In Biologically Inspired Cognitive Architectures.

Danks, D. (2014). *Unifying the mind: Cognitive representations as graphical models*. MIT Press.

DeVault, D. (2014). SimSensei Kiosk: A virtual human interviewer for healthcare decision support. In *Proceedings of the 2014 International Conference on Autonomous Agents and Multi-agent systems*.

Goertzel, B. (2014). Artificial General Intelligence: Concept, state of the art, and future prospects. *Journal of Artificial General Intelligence.*, *5*, 1–46.

Grisetti, G., Kummerle, R., Stachniss, C., & Burgard, W. (2010). A tutorial on graph-based SLAM. *Intelligent Transportation Systems Magazine, IEEE*, *2*(4), 31–43. doi:10.1109/MITS.2010.939925

Ji, Q., Gray, W. D., Guhe, M., & Schoelles, M. J. (2004). Towards an integrated cognitive architecture for modeling and recognizing user affect. In *AAAI Spring Symposium on Architectures for Modeling emotion: cross-disciplinary foundations*.

Joshi, H., Rosenbloom, P. S., & Ustun, V. (2014). Isolated word recognition in the Sigma cognitive architecture. In Biologically Inspired Cognitive Architectures. doi:10.1016/j.bica.2014.11.001

Koller, D., & Friedman, N. (2009). *Probabilistic Graphical Models: Principles and Techniques*. MIT Press.

Kopp, S., Krenn, B., Marsella, S., Marshall, A. N., Pelachaud, C., Pirker, H., & Vilhjálmsson, H. (2006). Towards a common framework for multimodal generation: The behavior markup language. In Intelligent Virtual Agents.

Kschischang, F. R., Frey, B. J., & Loeliger, H. A. (2001). Factor graphs and the sum-product algorithm. *IEEE Transactions on Information Theory*, *47*(2), 498–519. doi:10.1109/18.910572

Laird, J., & van Lent, M. (2001). Human-level AI's killer application: Interactive computer games. *AI Magazine*, *22*(2), 15.

Laird, J. E. (2012). *The Soar Cognitive Architecture*. Cambridge, MA: MIT Press.

Langley, P., Laird, J. E., & Rogers, S. (2009). Cognitive architectures: Research issues and challenges. *Cognitive Systems Research*, *10*(2), 141–160. doi:10.1016/j.cogsys.2006.07.004

Lee-Urban, S., Trewhitt, E., Bieder, I., Odom, J., Boone, T., & Whitaker, E. (2015). CORA: A Flexible Hybrid Approach to Building Cognitive Systems. In *Proceedings of the Third Annual Conference on Advances in Cognitive Systems Poster Collection* (p. 28).

Lucas, S. M., & Kendall, G. (2006). Evolutionary Computation and Games. *IEEE Computational Intelligence Magazine*, *1*(1), 10–19. doi:10.1109/MCI.2006.1597057

Mikolov, T., Joulin, A., & Baroni, M. (2015). *A roadmap towards machine intelligence*. Retrieved from http://arxiv.org/abs/1511.08130

Mnih, V., Kavukcuoglu, K., Silver, D., Rusu, A. A., Veness, J., Bellemare, M. G., & Hassabis, D. et al. (2015). Human-level control through deep reinforcement learning. *Nature*, *518*(7540), 529–533. doi:10.1038/nature14236 PMID:25719670

Pynadath, D. V., & Marsella, S. C. (2005). PsychSim: Modeling theory of mind with decision-theoretic agents. IJCAI.

Pynadath, D. V., Rosenbloom, P. S., & Marsella, S. C. (2014). Reinforcement Learning for Adaptive Theory of Mind in the Sigma Cognitive Architecture. In Artificial General Intelligence.

Pynadath, D. V., Rosenbloom, P. S., Marsella, S. C., & Li, L. (2013). Modeling two-player games in the Sigma graphical cognitive architecture. In Artificial General Intelligence. doi:10.1007/978-3-642-39521-5_11

Robertson, G., & Watson, I. (2014). A review of real-time strategy game AI. *AI Magazine*, *35*(4), 75–204.

Rosenbloom, P. S. (2010). Combining procedural and declarative knowledge in a graphical architecture. In *Proceedings of the 10th International Conference on Cognitive Modeling*.

Rosenbloom, P. S. (2011a). Bridging dichotomies in cognitive architectures for virtual humans. In *Proceedings of the AAAI Fall Symposium on Advances in Cognitive Systems*.

Rosenbloom, P. S. (2011b). From memory to problem solving: Mechanism reuse in a graphical cognitive architecture. In Artificial General Intelligence.

Rosenbloom, P. S. (2012). Deconstructing reinforcement learning in Sigma. In Artificial General Intelligence. doi:10.1007/978-3-642-35506-6_27

Rosenbloom, P. S. (2012). Towards a 50 msec cognitive cycle in a graphical architecture. In *Proceedings of the 11th international conference on cognitive modeling*.

Rosenbloom, P. S. (2012). Extending mental imagery in Sigma. In *Artificial General Intelligence* (pp. 272–281). Springer Berlin Heidelberg. doi:10.1007/978-3-642-35506-6_28

Rosenbloom, P. S. (2013). The Sigma cognitive architecture and system. *AISB Quarterly, 136*.

Rosenbloom, P. S. (2014). Deconstructing episodic memory and learning in Sigma. In *Proceedings of the 36th Annual Conference of Cognitive Science Society*.

Rosenbloom, P. S., Demski, A., Han, T., & Ustun, V. (2013). Learning via gradient descent in Sigma. In *Proceedings of the 12th International Conference on Cognitive Modeling*.

Rosenbloom, P. S., Gratch, J., & Ustun, V. (2015). *Towards Emotion in Sigma: From Appraisal to Attention*. Artificial General Intelligence.

Schaeffer, J., Bulitki, V., & Buro, M. (2008). Bots Get Smart: Can video games breathe new life into AI research? *IEEE Spectrum*, *45*(12), 48–56. doi:10.1109/MSPEC.2008.4688952

Schmidhuber, J. (2015). *On Learning to Think: Algorithmic Information Theory for Novel Combinations of Reinforcement Learning Controllers and Recurrent Neural World Models*. Retrieved from http://arxiv.org/abs/1511.09249

Shapiro, A. (2011). Building a character animation system. In Motion in Games. doi:10.1007/978-3-642-25090-3_9

Strannegård, C., von Haugwitz, R., Wessberg, J., & Balkenius, C. (2013). A cognitive architecture based on dual process theory. In *Artificial General Intelligence* (pp. 140–149). Springer Berlin Heidelberg. doi:10.1007/978-3-642-39521-5_15

Sutton, R. S., & Barto, A. G. (1998). *Reinforcement Learning: An Introduction*. MIT Press.

Swartout, W. (2010). Lessons learned from virtual humans. *AI Magazine*, *31*(1), 9–20.

Tambe, M., Johnson, W. L., Jones, R. M., Koss, F., Laird, J. E., Rosenbloom, P. S., & Schwamb, K. (1995). Intelligent agents for interactive simulation environments. *AI Magazine*, *16*(1), 15.

Taylor, S. J., Khan, A., Morse, K. L., Tolk, A., Yilmaz, L., Zander, J., & Mosterman, P. J. (2015). Grand challenges for modeling and simulation: simulation everywhere—from cyberinfrastructure to clouds to citizens. *Simulation: Transactions of the Society for Modeling and Simulation*.

Traum, D., Marsella, S. C., Gratch, J., Lee, J., & Hartholt, A. (2008). Multi-party, multi-issue, multi-strategy negotiation for multi-modal virtual agents. In *Intelligent Virtual Agents* (pp. 117–130). Springer Berlin Heidelberg. doi:10.1007/978-3-540-85483-8_12

Umarov, I., Mozgovoy, M., & Rogers, P. C. (2012). Believable and effective AI agents in virtual worlds: Current state and future perspectives. *International Journal of Gaming and Computer-Mediated Simulations*, *4*(2), 37–59. doi:10.4018/jgcms.2012040103

Ustun, V., & Rosenbloom, P. S. (2015). Towards adaptive, interactive virtual humans in Sigma. In *Intelligent Virtual Agents* (pp. 98–108). Springer International Publishing. doi:10.1007/978-3-319-21996-7_10

Ustun, V., Rosenbloom, P. S., Kim, J., & Li, L. (2015). Building High Fidelity Human Behavior Model in the Sigma Cognitive Architecture. In *Proceedings of the 2015 Winter Simulation Conference*. Piscataway, NJ: Institute of Electrical and Electronics Engineers, Inc.

Ustun, V., Rosenbloom, P. S., Sagae, K., & Demski, A. (2014). Distributed vector representations of words in the Sigma cognitive architecture. In Artificial General Intelligence. doi:10.1007/978-3-319-09274-4_19

Ustun, V., Yilmaz, L., & Smith, J. S. (2006). A conceptual model for agent-based simulation of physical security systems. In *Proceedings of the 44th annual Southeast regional conference*. ACM. doi:10.1145/1185448.1185530

Whiten, A. (Ed.). (1991). *Natural Theories of Mind*. Oxford, UK: Basil Blackwell.

Zemla, J. C., Ustun, V., Byrne, M. D., Kirlik, A., Riddle, K., & Alexander, A. L. (2011). An ACT-R model of commercial jetliner taxiing. In *Proceedings of the Human Factors and Ergonomics Society Annual Meeting* (Vol. 55, No. 1, pp. 831-835). Sage Publications. doi:10.1177/1071181311551173

ENDNOTE

1 The details of this recognition process are beyond the scope of this paper but there are a variety of behavioral cues (e.g. posture changes while concealing an item, gait changes under stress etc.) that could be revealing. Exhibiting and detecting such cues is one of a number of intriguing future directions for this work.

Chapter 9
A Universal Architecture for Migrating Cognitive Agents:
A Case Study on Automatic Animation Generation

Kaveh Hassani
University of Ottawa, Canada

Won-Sook Lee
University of Ottawa, Canada

ABSTRACT

In this chapter, the characteristics of a cognitive architecture that can migrate among various embodiments are discussed and the feasibility of designing such architecture is investigated. The migration refers to the ability of an agent to transfer its internal state among different embodiments without altering its underlying cognitive processes. Designing such architecture will address both weak and strong aspects of AI. The authors propose a Universal Migrating Cognitive Agent (UMCA) inspired by onboard autonomous frameworks utilized in interplanetary missions in which the embodiment can be tailored by defining a set of possible actions and perceptions associated with the new body. The proposed architecture is then evaluated within a few virtual environments to analyze the consistency between its deliberative and reactive behaviors. Finally, UMCA is tailored to automatically create computer animations using a natural language interface.

DOI: 10.4018/978-1-5225-0454-2.ch009

INTRODUCTION

After the term artificial intelligence (AI) was coined by John McCarthy in 1955, it was divided into two main categories: weak and strong AI. The weak AI refers to a non-sentient machine intelligence that is domain-dependent and is developed for a specific task. This type of AI is extensively applied in engineering applications and high-tech products such as robotics, search engines, voice recognition, medical expert systems, etc. and has reached a fairly good degree of success (Russell & Norvig, 2009). On the other hand, the strong AI, also known as artificial general intelligence (AGI) aims to create an intelligent machine that can successfully perform cognitive tasks that a human being can do (P. Wang & Goertzel, 2007). Intelligent behavior emerges from cognitive characteristics such as recognition, decision making, perception, situation assessment, prediction, problem solving, planning, reasoning, belief maintenance, execution, interaction and communication, reflection, and learning (Langley et al. 2009).

During the last decades, AGI was dominated by the weak AI. However, there is a new trend in community that is trying to amalgamate the advantages of both types of AI. Research on embodied agents is one of the pioneers in this field.

The embodied intelligence is a prominent topic in the multi-agent systems (MAS) and refers to a coupled mind-body loop in which high-level deliberative processes within the mind that function on a symbolic representation of the world decide the behavior of the agent by controlling a collection of physical or virtual sensors and actuators within the body (Hassani & Lee, 2015). The embodied intelligence mostly follows a dualist perspective which decomposes the agent into a mind and a body. The mind as an abstract layer provides the agent with a set of cognitive functionalities. It receives the perceptions from the body, makes decisions, and then sends the decisions as a set of abstract actions to the body. The body as an embodied layer executes the received actions within the environment and provides the mind with a set of perceptions acquired from its sensors. The continuous interaction between the mind and the body forms a closed perception-cognition-action loop. Although a few architectures such as Censys (Ribeiro et al. 2013) and embodied cognition model (Vala et al. 2012) challenge this strict separation between the mind and the body, the separation enhances the development cycle in real-world applications. The embodiment may also be considered as a situational coupling between an agent and its environment in scenarios such as situated agents (MacLennan, 1996).

The embodied agents have been extensively investigated in both physical world and the virtual environments in terms of robotic systems and intelligent virtual agents (IVA), respectively, and later on have been upgraded to social robotics (Fong et al. 2003) and embodied conversational agents (ECA) (Hassani et al. 2013b) by embedding social context and human interactions. In literature, different classifications are proposed for categorizing the agents. For example, agents can be classified into cognitive, behavioral and hybrid agents from the embodiment point of view (Siegwart & Nourbakhsh, 2004) and logic-based, reactive, belief-desire-intention (BDI) and layered agents from the perspective of internal processes (Wooldridge, 2002).

Designing migrating agents is a recent approach in developing the generic embodied agents that meets some requirements of the AGI. The migration is the ability of an abstract entity to morph from one embodiment into another one and control the new body without altering the internal cognitive processes of the transferred entity. A more specific definition of the migration is introduced within the context of the social companions in which the intrinsic problem is preserving the identity of the companion inhabiting different embodiments (Arent & Kreczmer, 2013). Kriegel et al. (2011) define the companion's identity as a set of persisting features that makes the companion agent unique and recognizable. These features include sets of goals, emotional reaction rules, action tendencies, emotional thresholds, decay rates of the emotions, etc. The migrating companions require by the necessity sufficient levels of abstraction and modularity.

The theory of migrating agents relies on a well-known theory of mind called Functionalism which is the most popular view among philosophers of mind and cognitive scientists. (Mahner & Bunge, 2001). The Functionalism moves beyond both Behaviorism and Identity Theory by assuming: (1) Brain states are not mental states, and (2) Behaviorism cannot account for mental states. According to Functionalism, mental states are functional states identified by what they do rather than by what they are made of and are described by their functional descriptions (i.e. inputs and outputs of a black box model) and causal roles (i.e. the effect that it has on other parts of the system). One magnificent outcome of the Functionalism is that it is possible to transfer mental states among different hardware as long as the hardware can perform the same functions and hence mind-transference is logically possible and nothing will be lost.

The first question about the migrating agents is why they are useful? From an engineering point of view, designing universal migrating agents will result in a mind which can adapt to different digital systems and as a result can alleviate the drawbacks of designing ad-hoc controlling mechanisms. The ultimate goal is to design a body-agnostic mind that can reside in a rover, a satellite, a game character, a Web crawler, etc. without requiring much of customization, a universal ghost that can animate any given body. Furthermore, from man-machine interaction perspective, a migrating agent is an effective method to connect virtual and real worlds (Ogawa & Ono, 2008) that encourages the development of a long-term social relationship of a companion with its user (Koay et al., 2011). These advantages of migrating agents motivate the researchers to attempt to create a unified platform for designing universal migrating agents.

Another important question is about the constraints on the body-to-body migration. The agent's experience depends on its body, which greatly affects its cognitive processes, rendering the migration to a non-similar body to be quite problematic, in spite of similar cognitive abilities. In case that the migration results in drastic changes, it can go through a middle embodiment that has common components with the previous embodiments to smoothen the changes. This process is referred as indirect migration and has been shown to be effective in some context (Arent & Kreczmer, 2013). Moreover, contrary to the definition of migrating agents in the context of social robotics, the authors do not consider the identity of an agent as an important aspect of a universal migrating agent. Indeed an agent will lose its identity if the migration happens between two extreme bodies but it will adapt to the new body and control it if it is provided with a set of possible action-perceptions and operational objectives. Therefore through this chapter the identity preservation is not of an interest. The authors believe that a universal migrating agent can immigrate to a new body only if the new body provides the required computational resources and that is the only constraint considered for the migration. The authors define the migrating agents in a more general sense as follows. A migrating agent is a cognitive agent that can transfer its mind and mind-body interface among different embodiments while only requiring a high-level knowledge of the actions and perceptions associated with the new body.

The embodied agents usually function in a dynamic, uncertain, and uncontrolled environment, and exploiting them is a chaotic and error-prone task which demands high-level behavioral controllers to adapt to the failures at the lower levels of the system (Hassani & Lee, 2014, 2015). The conditions in which space robotic systems such as satellites, spacecraft and rovers operate, inspire by necessity, the development of robust and adaptive control software. These systems can autonomously achieve mission goals and handle unpredicted situations (Hassani & Lee, 2013). The challenges of developing the agent architectures for the onboard autonomy are driven by four major characteristics of the spacecraft as follows: (1) the spacecraft must perform autonomously for long periods of time, (2) it must guarantee the success considering the deadlines and constraints, (3) it must guarantee high reliability considering the high cost of the spacecraft, and (4) it must support concurrent activities among a set of tightly coupled subsystems (Muscettola et al. 1998).

In this chapter, universal migrating cognitive agent (UMCA) architecture is proposed. This agent architecture integrates the autonomous characteristics of the onboard autonomy with the cognitive abilities of the intelligent agents to create an agent architecture that can freely move among different physical or virtual embodiments. To evaluate this architecture, three scenarios are implemented. The first two scenarios are toy worlds in which the agent concurrently performs path finding and obstacle avoidance tasks in both deliberative and reactive manners. In the third scenario, the agent automatically generates 3D animation via a natural language interface. In these scenarios agent receives some instructions in natural language and uses them to manipulate the environment.

BACKGROUND

Agent Architecture

The concept of migrating agents is mostly investigated in the context of social companions and human-robot interactions. In this context, the agent's both high-level characteristics (e.g. emotions and behaviors) and the low-level identities (e.g. voice) transfer among different embodiments in a way that users can recognize the agent in different embodiments (Hassani & Lee, 2014). As an example, in LIREC project – a European open-source research project for designing social companions based on CMION architecture, the ultimate goal was to move beyond the novelty effect of the both social robots and ECAs towards the social companions to support long-term relationships between humans and synthetic companions (Segura et al. 2012; Aylett et al., 2013). Sarah is an example of a LIREC-based companions that can embodied within a robot, a large graphical screen, or a handheld device (Segura et al., 2012). In (Gomes et al., 2011), an artificial pet is proposed with two robotic and graphical embodiments in which the behavior is driven by some needs to maintain the coherence and motivate the user interactions.

A migrating agent relies on its cognitive processes managed by an underlying infrastructure known as a cognitive architecture (Langley et al., 2009). In literature, some commonly accepted architectures are introduced as follows.

1. **State, Operator, and Result (Soar):** Developed at the Carnegie Mellon University (Laird et al., 1987), it is an attempt to create an AGI agent based on the physical symbol system hypothesis. The underlying engine within the Soar architecture is symbol processing. It utilizes a set of production rules to decide its behaviors and a set of search algorithms within problem space for general problem solving tasks. Although the recent version of the Soar is equipped with reinforcement learning, yet it cannot efficiently support data-driven approaches (Laird, 2012).
2. **Clarion Architecture (Sun & Zhang, 2006):** Incorporates the distinction between implicit and explicit processes and focuses on capturing the interactions between them. It consists of four sub-systems including action-centered, non-action-centered, motivational, and meta-cognition subsystems each consisting of both implicit and explicit processes. The implicit layer of the subsystems

is made of neural networks whereas the explicit layer is made of associative rules. It also utilizes Q-learning for learning the implicit knowledge and rule-extraction-refinement algorithm (RER) for propagating the explicit knowledge.

3. **Adaptive Control of Thought-Rational (ACT-R) (Anderson et al., 2004):** Developed at Carnegie Mellon University, it is a cognitive framework that emphasizes human psychological verisimilitude. The ACT-R utilizes two main categories of knowledge representations including declarative and procedural. It consists of two types of modules including perceptual-motor modules and memory modules. The perceptual-motor type handles the interfacing with the real world. The memory modules can be either declarative memory consisting of facts or procedural memory consisting of productions.
4. **Belief-Desire-Intention (BDI) (Rao & Georgeff, 1995):** Developed based on the Dennett's theory of intentional systems. It triggers behaviors driven by conceptual models of intentions and goals in complex dynamic scenarios. BDI meets real-time constraints by reducing the time used in planning and reasoning. BBSOAA (Liu & Lu, 2006) is an extension of BDI architecture that enhances the knowledge representation and inference capabilities. BDI-inspired architectures such as IRMA (Bratman et al., 1988) support long term behaviors and are typically implemented as Procedural Reasoning System (PRS).
5. **ICARUS (Langley & Choi, 2006):** Model designed for embodied agents emphasizes perception and action over abstract problem solving. It aims to unify reactive execution with problem solving, combine symbolic structures with numeric utilities, and learn structures and utilities in a cumulative manner (Chong, Tan, & Ng, 2009).
6. **Other Systems:** SASE (Weng, 2004) is based on Markov decision processes and utilizes the concept of autonomous mental development. PRODIGY (VELOSO et al., 1995), DUAL (Kokinov, 1994) and Polyscheme (Cassimatis, 2002) are other examples of cognitive architectures.

There are various architectures proposed in literature addressing different aspects as follows.

1. Architectures that address object-oriented design such as CAA (Kim, 2005).
2. Data-driven architectures that exploit machine learning techniques such as reinforcement learning for behavioral animation (Conde et al., 2003) and FALCON-X (Kang & Tan, 2013). OML (Starzyk et al., 2013) is an agent architecture for virtual environments equipped with neural network-based learning mechanism. In this model, a sensory neuron represents an object, and a motor neuron represents an action. Neural-dynamic architecture (Sandamirskaya et

al., 2011) utilizes a dynamic neural field to describe and learn the behavioral state of the system, which in turn, enables the agent to select the appropriate action sequence regarding its environment.

3. Architectures that employ a middleware to integrate existing multi-agent systems such as 2APL (Dastani, 2008), GOAL (Hindriks, 2009), Jadex (Pokahr et al., 2005) and Jason (Bordini, Hübner, & Wooldridge, 2007) with existing game engines. As an example, CIGA (Oijen et al., 2012) is a middleware that amalgamates an arbitrary multi-agent system with a given game engine using a domain ontology.
4. Architectures that focus on the behavioral organization and action selection. SAIBA (Kopp et al., 2006) is a popular framework that defines a pipeline for abstract behavior generation. It consists of intent planner, behavior planner and behavior realizer. Thalamus (Ribeiro et al., 2012) is another framework that adds a perceptual loop to SAIBA to let the agent perform continuous interaction. AATP (Edward et al., 2009) is a coupled planning and execution architecture for action selection.
5. Layered (hybrid) architectures such as COGNITIVA (Spinola & Imbert, 2012) that concurrently perform deliberative and reactive operations. Hybrid architectures are widely utilized in space robotic systems (Hassani & Lee, 2013). Remote Agent (RA) (Muscettola et al., 1998) is a hybrid architecture designed to provide reliable autonomy for extended periods. IDEA (Gregory et al., 2002) is a multi-agent architecture that supports distributed autonomy by separating the layers of the architecture into independent agents. MDS (Horvath et al., 2006) is a hybrid software framework that emphasizes state estimation and control whereas TITAN (Horvath et al., 2006) emphasizes model-based programming. LAAS (Alami et al., 1998) and Claraty (Volpe et al., 2001) are other examples of hybrid architectures utilized in space missions.

UNIVERSAL MIGRATING COGNITIVE AGENT

The proposed UMCA is inspired by RA and IDEA architectures. It consists of two layers including cognitive and executive layers and utilizes a middleware as an interface between abstract agent and its embodied counterpart. The internal components of the UMCA are placed in different layers based on their operational frequency. The cognitive layer is responsible for providing cognitive functionalities such as decision making whereas the executive layer is responsible for executing the decisions and providing high-level feedbacks. The cognitive layer utilizes high-level knowledge representation, functions with low frequency, and plans for long temporal horizons.

On the other hand, the executive layer functions in high frequency and deals with the current situations in a reactive and soft real-time manner. The schematic of the architecture is shown in Figure 1.

Cognitive Layer

The Cognitive layer consists of three modules including Mission Manager (MM), Planning-Scheduling (PS) and Knowledgebase (KB). The MM module contains a goal network and a set of feasible actions. This module consists of three sub-modules including goal automaton, goal generator, and action database. The goal automaton is a Deterministic Finite Automaton (FDA) that keeps a network of predefined goals. It defines what the next goals are and when a goal is satisfied. It uses a transition function to decide the next goal based on the current state of the agent and whether the current goal is satisfied. The goal generator is the transition function of the goal network. In each time step, it evaluates the current goal and the received perceptions to determine whether the current goal is satisfied. If so, it transforms the goal state to a new goal and sends the new goal to the planner.

A residing body can only perform a restricted set of valid actions. These feasible actions are stored in the action database. An action is defined as an abstract activity that encapsulates a few low-level sub-actions. This abstraction scheme dramatically reduces the complexity of the planning and scheduling processes. Each action contains a set of preconditions and effects, estimated execution duration, a set of sub-actions, and their execution timeline. The preconditions of an action determine the constraints on state variables that must be satisfied so that the action can be

Figure 1. Proposed architecture for universal migrating cognitive agent (UMCA)

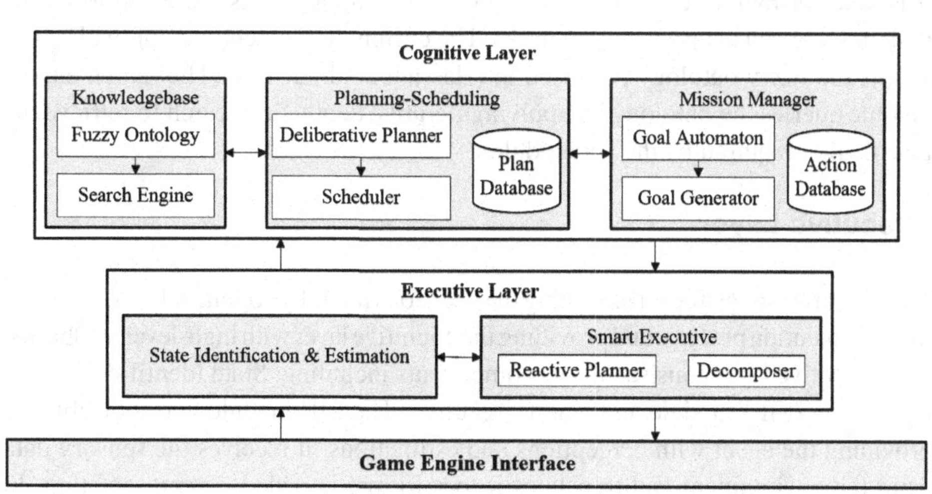

executed. The effects of an action determine how the execution of the action affects the state variables. The planner module relies on this information to determine the optimal sequence of actions.

The PS component consists of three sub-modules including deliberative planner, scheduler, and plan database. The deliberative planner decides the sequences of actions and the concurrency among them. These sequences are planned for long temporal horizons and executing them lets the agent to reach the mission objectives in an optimal trajectory. It is shown that different variations of hierarchical planners such as HTNs and STRIPS are expressive and can solve all domains solvable by a Turing machine (Lekavý & Návrat, 2007). In the current version of UMCA, planning is carried out using a backtracking algorithm with pruning strategy. The reason for exploiting the backtracking algorithm is its simplicity and efficiency in the toy world search spaces. The scheduler determines the start time of the planned sequences based on the estimated execution time of the actions within sequences. The planned and scheduled sequences are then inserted into the plan database. This database is a temporal database which constantly checks the start times of the sequences. In case the start time of a sequence matches the current time, it sends the sequence to the executive layer.

The knowledgebase module is essentially a profound memory that provides the agent with the knowledge acquired from the perception history. It consists of two sub-modules including a fuzzy ontology and a search engine. The fuzzy ontology represents the concepts, objects, features and their relations based on the agent's perceptional history. It utilizes a maintainer which updates the current knowledge. The update process relies on the knowledge extraction from perceptions and may insert new concepts, objects or relations to the ontology, update the current relations or attributes, or even prune the ontology to omit the redundancies or inconsistencies. It is noteworthy that the fuzzy extension of the ontology lets the agent to model both internal and external uncertainties. The current implementation of the UMCA adopts the fuzzy ontology proposed in (Hassani et al., 2013a). The search engine module queries the ontology by applying iterative depth-first search to retrieve the information required by the PS module.

Executive Layer

This layer is responsible for executing the decisions made by cognitive layer, monitoring the execution process, and providing the cognitive layer with high-level feedbacks. The executive layer consists of two components including: State Identification and Estimation (SIE) module and smart executive. The SIE module is responsible for providing the agent with perceptions and estimations. It receives the sensory data from the environment and translates it to a formal knowledge representation. In other words, it converts data acquired from sensors to the perceptions cognoscible by the agent. To carry out this task, it utilizes a Kalman filter and a fuzzifier for data assimilation and conceptualization, respectively. Moreover, it exploits variety of software packages to perform specialized data processing activities such as automatic speech recognition, image processing, etc. Hence, this module allows the agent to deal with a variety of sensory data acquired from different sensory channels.

The smart executive executes the sequences of planned actions and monitors the execution to prevent inconsistencies. It consists of two sub-modules including decomposer and reactive planner. The decomposer fetches the action sequences from the plan database and executes the actions using multi-threading to support the parallel plan execution. It starts an action thread according to its scheduled start time. The execution of an action is done by invoking the internal sub-actions following a predefined timeline. The execution of a sub-action is linked to an activity in the body and hence the action hierarchy maps the abstract decisions into embodied manipulations and actuations. The decomposer also employs a set of software libraries to provide the actions with a set of required facilities such as text-to-speech engine. Additionally, it prevents the inconsistencies by monitoring the current states and comparing them with the expected states predicted by the SIE module. In case of irregularities, it halts the inconsistent thread and sends a signal to the reactive planner so that it can take a proper action to eliminate the inconsistency.

The versatility of the UMCA stems from its capability to perform cognitive tasks regardless of the current feasible sets of actions and perceptions. One could easily tailor the actions and perceptions within the action database and the SIE module to adapt the UMCA to the associated body.

EVALUATION AND EXPERIMENTAL RESULTS

Three toy world scenarios are developed for evaluating the behavior of UMCA. In the first two scenarios, the agent must perform path planning and obstacle avoidance tasks while satisfying temporal and spatial constraints. In the third scenario, the UMCA is tailored to automatically generate 2D and 3D computer animation based on a natural language interface. In these scenarios, the agent is implemented in C#.Net programming language and the visualization is done via Microsoft XNA game engine.

Experiment I: World of Circles

The first scenario is a 2D environment called the world of circles in which three different colors (red, green and blue) are allowed for circles. The circles can pop into the world with a random color in a stochastic manner or enter the world in a predefined order from a fixed position called the origin. There are three non-stationary target zones in the world that are associated with the colors. A few non-stationary obstacles are also embedded within the world. The agent's goal is to pick up a circle from the origin and move it to one of the target zones based on the color similarity while avoiding the obstacles. Considering that the random circles can pop into the world among the predefined sequences of the circles and considering that both targets and obstacles are non-stationary, the world provides proper characteristics for evaluating the both reactive and deliberative behaviors. A sample scene of this world is shown in Figure 2.

The gray rectangles present the obstacles whereas the blue, purple and green rectangles present the target zones. In this simulation, user can determine the number of circles, their colors, and the randomness level (i.e. randomness of 0.1 means one out of every ten circles pops into the world randomly following a uniform distribution). The agent's knowledgebase contains the positions of the origin, the obstacles and the target zones. Its action database contains the following action:

Action Identifier: Move (Circle)

```
Preconditions:
 Status: Idle
```

Figure 2. A sample scene of the simulation of the world of circles

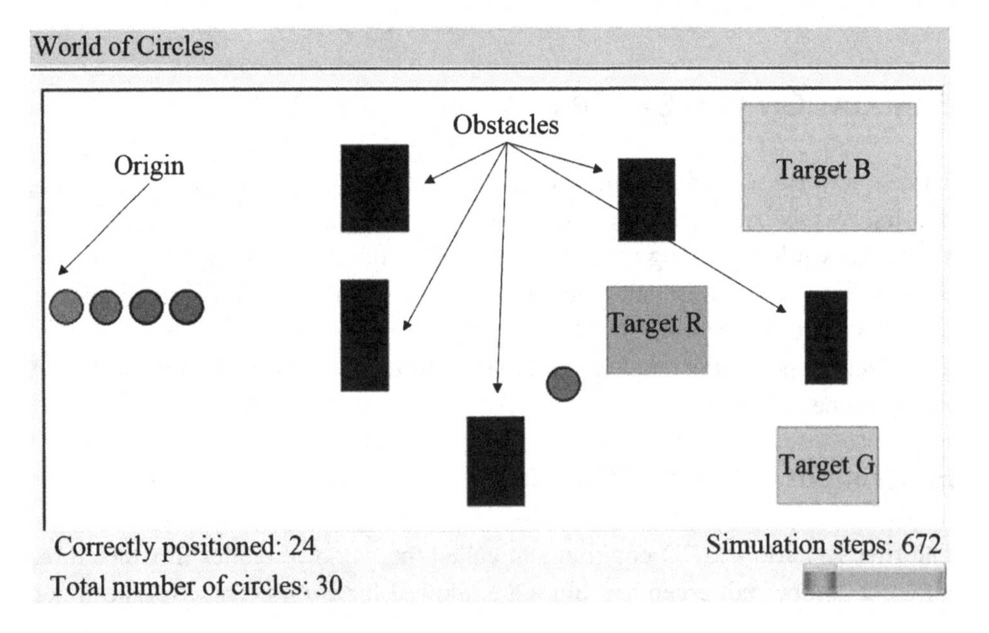

```
 #Circles>0
Effects:
 #Circles--
Estimated execution time: #Steps
Sub-Actions:
 Status: Moving
 Fetch(Circle);
 Loop: FindPath(Circle);
 Move(Circle);
 If(Failure)
 FindTarget(Circle);
 Update(KB);
 Goto Loop;
 Else
 Status: Idle
```

The preconditions state that to execute this action the agent must be idle and there must be new circles in the queue. The action first moves the agent to the origin and then finds an optimal path from the origin to the corresponding target using a simple A* search. It then moves the circle in front of the queue to its corresponding target zone. If the action fails, the agent finds a new target and repeats the path planning. As soon as the agent is provided with the information regarding the arrival of a new sequence, the deliberative planner plans a set of paths for upcoming circles. However, it is possible that a random circle arrives among the predetermined circles. In this case, the reactive planner plans the proper path for the random circle and asks the scheduler to reschedule the remaining actions.

In each simulation, 200 circles with uniform random colors are created. The length of an arrival ball sequence is a random variable following a uniform distribution in the range of [5,25]. The simulations are repeated for 1000 times and the randomness is increased by 0.1 after every 100 simulations. As shown by the simulation results in Figure 3, increasing the randomness decreases the activation frequency of the deliberative planner and increases this frequency for the reactive planner. The results also suggest that even when there is no randomness in the simulations, the reactive planner may activate when two consecutive sequences appear. In this case, the deliberative planner simply does not have enough time to plan for all the arrivals and hence it treats some of them as the unplanned circles and activates the reactive planner.

Figure 3. Average number of reactive and deliberative behaviors in response to changes in randomness of the simulations

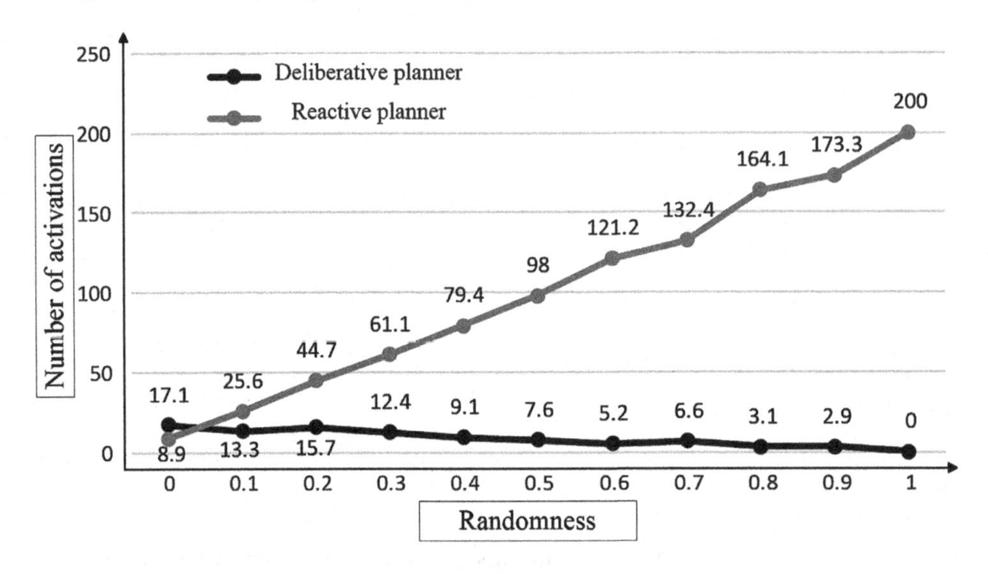

Experiment II: Interstellar World

The interstellar world consists of four 2D planets including Earth, Saturn, Neptune, and Jupiter, and four observatory satellites orbiting these planets with different velocities either in a circular or an elliptic orbit. It also contains two non-stationary asteroids moving in random and linear trajectories. A robot avatar is embedded within the world to serve as the embodiment of the UMCA. The robot starts from the origin and non-periodically receives sequences of mission goals. A mission sequence contains a set of goals with corresponding deadlines. A goal contains a start satellite that holds the payload to be picked up by the robot and a target satellite which the payload must be delivered to. The robot must transfer the payloads while satisfying the goal deadlines and avoiding the asteroids and planets. The robot plans and schedules a complete mission sequence in a deliberative manner and executes the plans for the whole sequence. However, due to the stochastic trajectories of the asteroids and the noise within the satellite orbits, it utilizes its reactive planner to tweak the plans in real-time. The robot sets a rendezvous with the start satellite to pick up the payload and then careens towards the target satellite and sets a rendezvous with that satellite to deliver the payload. Due to the noisy estimations of the satellite trajectories and the stochastic dynamics of the asteroids, the interstellar world provides a set of proper characteristics for investigating the both reactive and deliberative behaviors. A sample scene of the interstellar world is shown in Figure 4.

Figure 4. A sample scene of the simulation of the interplanetary world

As depicted in Figure 4, the mission goals are displayed on the left side and the movements of the satellites, asteroids and the robot are shown in the simulation panel. The agent's knowledgebase contains the positions of the planets, asteroids and the satellites. The action database contains the following action:

Action Identifier: SetRendezvous(Satellite)

```
Preconditions:
 Position: close to orbit
Effects:
 Change in payload
Estimated execution time: T_Rendezvous
Sub-Actions:
 Status: Moving
 Point←EstimatePoint(Satellite)
 MoveTo (Point)
 Stop ()
 Status: Fixed
```

Action Identifier: MovePayload(Satellite src, Satellite des)

```
Preconditions:
 Mission: Available
 Deadline not passed
Effects:
 Change in position
Estimated execution time: #Steps
Sub-Actions:
 Status: Moving
 Find Path (src,des)
 Move ()
 Status: Idle
```

Action Identifier: AvoidAsteroid()

```
Preconditions:
 Asteroid in sight
Effects:
 Change in direction
Estimated execution time: T_changeDirection
Sub-Actions:
 If (Collision is estimated)
 Status: Moving
 ChangeDirection()
 RefinePlans ()
 Status: Moving
```

A useful operational diagram of the systems with interacting components is the timing diagram which presents the activation sequence of the components in a casual manner. The timing diagram helps the designers to investigate the operational frequency of the components, their activation and deactivation sequences, and their interactions. In order to facilitate the evaluations, the timing diagrams of the simulations are used to investigate whether the internal modules of the UMCA interact and operate as expected. A sample timing diagram of the simulations is shown in Figure 5 in which vertical axis is the components (FO: fuzzy ontology, SE: search engine, PS: planner-scheduler, RP: reactive planner, MM: mission manager, SIE: state identification and estimation, DE: decomposer) and the horizontal axis represents the time flow.

Figure 5. A sample timing diagram of the simulations

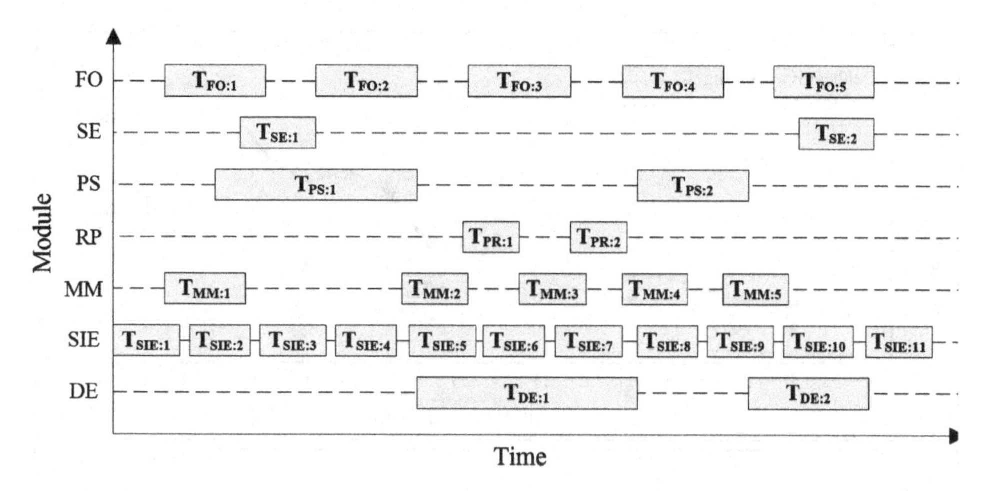

As expected, the internal modules of the UMCA function in different frequencies. Those modules that belong to cognitive layer have a lower frequency in comparison to those within the executive layer. As an example, the state identification and estimation modules is operating in a higher frequency than the deliberative planner.

The simulation setup of the interstellar world is identical to the setup of the world of circles. The simulation results are shown in Figure 6. As illustrated, increasing the length of the mission sequence increases the number of activations of the both deliberative and reactive planners. This is because it is more likely that the agent has to re-plan the already planned sequences with the longer sequences. It is also shown that as expected increasing the randomness results in increasing the activation of the reactive obstacle avoidance and the reactive rendezvous modification.

Experiment III: Automatic Animation Generation

The previous two scenarios investigate the consistency of the internal interactions and the deliberative and reactive behaviors of the UMCA architecture within random environments. The third scenario is designed to investigate the behavior of the UMCA in the presence of a continuous interaction with another agent and a tailored embodiment. For this purpose, the agent architecture is tailored in a way that it can process natural language commands and respond to those interactions in terms of generating animation. To carry out this task, the UMCA is coupled with a natural language processing agent which extracts the mission goals from the user's linguistic inputs. It is noteworthy that the focus of this chapter is not on the techniques

Figure 6. Reactive and deliberative behaviors of the agent

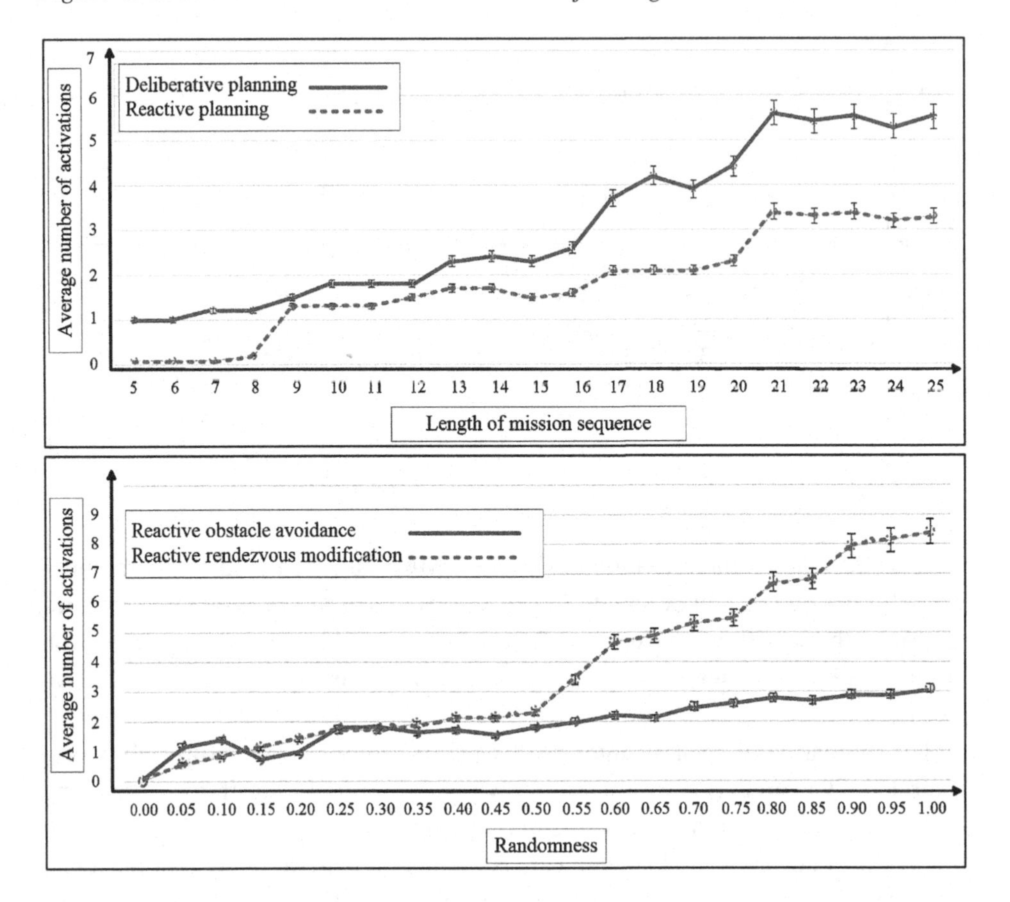

related to the natural language processing. Hence, the language processing agent is implemented as a simple agent which utilizes regular expressions for capturing shallow semantics and template-based language generation. This agent consists of two modules including natural language understanding and natural language generation.

The natural language understanding component is responsible for processing the user's textual inputs and capturing the semantics indicating user's intentions. To do so, it utilizes a lexicon, a set of patterns, a text processor and a knowledge mapping module. The lexicon is designed as a hierarchical ontology-like structure that contains nouns, verbs, and spatial prepositions. Each noun token in the lexicon is defined as a hextuple $<t,T,c,s,P,V>$ where t denotes the token, T denotes the part-of-speech tag (i.e. noun), c denotes the concept it belongs to, s denotes the sub-concept within c that it belongs to, P presents a set of its properties, and V is the corresponding values assigned to those properties. As an example, the noun "Earth" is represented as: <Earth, noun, object, planet, <model, texture, size, distance>, <V_model, V_tex-

ture, V_size, V_distance>>. Each verb is represented as a septuple <*t, T, c, s, a, P, V*> in which similar to the noun definition *t, T, c*, and *s* correspond to the token, part-of-speech tag, concept, and the sub-concept, respectively. *a* denotes the action or functionality that the verb represents. *P* is a set of action parameters that the verb represents, and *V* is the set of corresponding values of those parameters. As an instance, the verb "orbit" is represented as follows: <orbit, verb, move, rotate, rotate, <center, angular velocity, radius>, <V_center, V_angular velocity, V_radius>>. The spatial preposition is defined as a quadruple <*t,T,c,p*> where *t, T, c*, and *p* denote token, part-of-speech tag, concept and the number of parameters, respectively. As an example, for the preposition "between", *p* is set to two because it needs two parameters to correctly address this spatial relation. A semi-automatic approach is used for assigning the concepts and sub-concepts to the terms. In this approach, the terms are first grouped based on their part-of-speech tags and then are divided into sub-groups based on their occurrence frequency. The terms co-occurring with high frequencies are manually labeled as a same concept.

The patterns module is basically a rule-base that contains a set of general patterns for extracting the spatial relations, action parameters, and the user's intentions. These rules are represented as a predicates of the first order logic. The rule-base contains a set of rules related to the spatial relation (e.g. "X under Y", "X close to Y", "X is between Y and Z" etc.), a set of rules for capturing the semantics of the verbs and their corresponding parameters (e.g. "X orbit Y", "X move A→B" etc.), a set of manipulation rules for capturing the user manipulations (e.g. "increase speed X", "remove X" etc.). These rules are defined using the concepts defined in the lexicon in which synonyms are presented as a unique concept. These rules are crafted manually as a set of regular expressions.

The text processing module receives the user's textual inputs and pre-processes it. The pre-processing consists of normalizing the text, removing the function words, stemming and tagging the terms. In normalizing step, Unicode ambiguity is addressed, the lower and upper cases are mapped to a predefined format, and the punctuation marks are amputated. The normalized input is then tokenized by spaces. The function words except for the spatial prepositions are removed and the remaining terms are stemmed using Porter's stemming algorithm (Porter, 1980). The remaining terms are tagged using Apache OpenNLP library. The text processor traverses the tokenized terms and matches them against lexicon based on their concepts and part-of-speech tags. For each term matching an entry in the lexicon a hierarchical dataset is generated and the rest of the terms are ignored. Then the patterns are matched against the strings of the generated structures to capture the spatial and dynamic relations. Ultimately, knowledge mapping module maps the extracted relations to a formal knowledge representation that is understandable by the UMCA.

The natural language generating module is responsible for generating responses in natural language that inform the user about the outcome of the manipulation she has requested. It consists of two sub-modules including patterns and text generating module. The patterns contain a set of generic sentence structures that drive different sentences with same semantics. The patterns are defined as a set of stochastic finite state automata. This module is essentially an IF-THEN rule-base that detects the current context using the semantic information extracted by the natural language understanding module and the success or failure of performing the task, and activates the corresponding pattern automaton to generate the corresponding response in natural language. The Microsoft speech engine is utilized to convert the generated responses to the speech.

In order to evaluate the interactions between the language processing agent and the UMCA, two planetary scenarios are implemented. The first scenario is a 2D environment in which the user can add, remove, or resize the planets, insert or remove satellites, change their rotational velocities, and add meteors to the space and manipulate their speed. The second scenario is a 3D solar system generation scenario in which the user can add planets, change their angular velocities, sizes and relative distances, and manipulate the camera. A sample scene of the 2D and 3D scenarios are shown in Figures 7 and 8, respectively.

A sample conversation between the user and the system is shown in Table 1.

The results of the experiment suggest that the UMCA architecture can successfully communicate with other agents to collaboratively complete a set of given objectives. The results also show that the UMCA can be tailored to perform vast variety of actions such as animation generation.

Figure 7. A sample scene of 2D animation generated by UMCA

Figure 8. A sample scene of 3D animation generated by UMCA

Table 1. A sample conversation between the user and the system

Actor	Utterance
User	Add the sun to the center.
System	Alright, I inserted the sun to the simulation.
User	Good, now animate the outer planets within the solar system.
System	Done, Jupiter, Saturn, Uranus and Neptune are created.
User	Let's add the inner planets as well.
System	Mercury, Venus, Earth and Mars are animated.
User	Increase the size of Jupiter.
System	Its size is increased.
User	Ok, move the camera to the top.
System	I am moving the camera.
User	Increase the rotational velocity of Earth.
System	The angular velocity of planet Earth is increased.
User	Now set the camera to its default position and zoom in on the Sun.
System	The view is changed according to your request.
User	Zoom out and remove the Mars from system.
System	That is done. I deleted Mars from system.

FUTURE RESEARCH DIRECTIONS

The current focus of the UMCA architecture is to integrate cognitive-deliberative behaviors with operational-reactive behaviors in a consistent manner while supporting parallelism. The building blocks of the architecture are considered as black-boxes positioned within layers based on their activation frequency. In the current implementation, the building blocks mostly hold a set of simple algorithms that are suitable for addressing the toy-world problems. As an instance, the planner is based on a simple tree-pruning backtracking search which is neither efficient nor realistic for real-world applications. Thus, the first direction for the future research is to implement the internal blocks of the architecture based on the cutting-edge algorithms. This step will allow the developers to evaluate the UMCA in real-world scenarios by investigating the actual cross-body agent migration on robotic systems and virtual characters.

Currently, the UMCA architecture relies on a priori sets of perceptions-actions that define the perception and actions of a given body. These sets are manually defined by an expert following a specific knowledge representation scheme and programmed into the UMCA architecture. The second direction that can significantly enhance the versatility of the proposed architecture is to integrate reinforcement learning mechanism in a way that the agent can learn the possible actions by experiencing the interactions of the new body with the environment. This enhancement would require the agent to adapt to a new body like an infant learning her physical capabilities.

Another direction for further studies is to define a robust knowledge representation scheme. During the experimental results, authors used a customized integration of the first order logic and the belief networks as a knowledge representation scheme. However, this scheme had to been slightly manipulated for different evaluation scenarios. In order to have a universal agent, it is very crucial to develop a unified and robust knowledge representation scheme that can handle various scenarios without requiring altering the representation scheme.

Although the preliminary evaluations resulted in promising results, yet these results must be validated in real world scenarios. Therefore, an important research direction is to perform comprehensive experiments in stochastic, partially observable, dynamic, and continuous environments to emulate the situations in real world. For this purpose, the UMCA must be equipped with robust and adaptive algorithms within its internal modules. Also, to investigate the interactional performance of the UMCA architecture, comprehensive user studies are required.

CONCLUSION

In this chapter, universal migrating cognitive agent (UMCA) architecture is proposed and its requirements are discussed. This agent architecture integrates the autonomous characteristics of the onboard autonomy with the cognitive abilities of the intelligent agents to create an agent architecture that can freely move among different physical or virtual embodiments. To evaluate this architecture, three scenarios are implemented. The first two scenarios are toy worlds in which the agent concurrently performs path finding and obstacle avoidance tasks in both deliberative and reactive manners. In the third scenario, the agent automatically generates 3D animation via a natural language interface. In these scenarios agent receives some instructions in natural language and uses them to manipulate the environment.

The preliminary experimental results suggest that: (1) UMCA successfully integrates the cognitive-deliberative behaviors with the operational-reactive behaviors in a consistent manner while supporting parallelism, (2) It successfully communicates with other agents to collaboratively complete a set of given objectives, and (3) It can be tailored to perform vast variety of actions such as animation generation.

REFERENCES

Alami, R., Chatila, R., Fleury, S., Ghallab, M., & Ingrand, F. (1998). An Architecture for Autonomy. *The International Journal of Robotics Research*, *17*(4), 315–337. doi:10.1177/027836499801700402

Anderson, J. R., Bothell, D., Byrne, M. D., Douglass, S., Lebiere, C., & Qin, Y. (2004). An Integrated Theory of the Mind. *Psychological Review*, *111*(4), 1036–1060. doi:10.1037/0033-295X.111.4.1036

Arent, K., & Kreczmer, B. (2013). Identity of a companion, migrating between robots without common communication modalities: Initial results of VHRI study. In *2013 18th International Conference on Methods and Models in Automation and Robotics (MMAR)* (pp. 109–114). http://doi.org/ doi:<ALIGNMENT.qj></ALIGNMENT>10.1109/MMAR.2013.6669890

Aylett, R., Kriegel, M., Wallace, I., Marquez Segura, E., Mecurio, J., Nylander, S., & Vargas, P. (2013). *Do I remember you? Memory and identity in multiple embodiments*. IEEE RO-MAN; doi:10.1109/ROMAN.2013.6628435

Bordini, R. H., Hübner, J. F., & Wooldridge, M. (2007). *Programming Multi-Agent Systems in AgentSpeak using Jason*. Chichester, UK: Wiley-Interscience.

Bratman, M. E., Israel, D. J., & Pollack, M. E. (1988). Plans and resource-bounded practical reasoning. *Computational Intelligence*, *4*(3), 349–355. doi:10.1111/j.1467-8640.1988.tb00284.x

Cassimatis, N. L. (2002). *Polyscheme: A Cognitive Architecture for Integrating Multiple Representation and Inference Schemes*. Massachusetts Institute of Technology.

Chong, H.-Q., Tan, A.-H., & Ng, G.-W. (2009). Integrated cognitive architectures: A survey. *Artificial Intelligence Review*, *28*(2), 103–130. doi:10.1007/s10462-009-9094-9

Conde, T., Tambellini, W., & Thalmann, D. (2003). Behavioral Animation of Autonomous Virtual Agents Helped by Reinforcement Learning. In T. Rist, R. S. Aylett, D. Ballin, & J. Rickel (Eds.), *Intelligent Virtual Agents* (pp. 175–180). Springer Berlin Heidelberg. doi:10.1007/978-3-540-39396-2_28

Dastani, M. (2008). 2APL: A practical agent programming language. *Autonomous Agents and Multi-Agent Systems*, *16*(3), 214–248. doi:10.1007/s10458-008-9036-y

Edward, L., Lourdeaux, D., & Barthès, J.-P. (2009). An Action Selection Architecture for Autonomous Virtual Agents. In N. T. Nguyen, R. P. Katarzyniak, & A. Janiak (Eds.), *New Challenges in Computational Collective Intelligence* (pp. 269–280). Springer Berlin Heidelberg. doi:10.1007/978-3-642-03958-4_23

Fong, T., Nourbakhsh, I., & Dautenhahn, K. (2003). A survey of socially interactive robots. *Robotics and Autonomous Systems*, *42*(3–4), 143–166. doi:10.1016/S0921-8890(02)00372-X

Gomes, P. F., Segura, E. M., Cramer, H., Paiva, T., Paiva, A., & Holmquist, L. E. (2011). ViPleo and PhyPleo: Artificial Pet with Two Embodiments. In *Proceedings of the 8th International Conference on Advances in Computer Entertainment Technology* (pp. 3:1–3:8). New York, NY: ACM. http://doi.org/ doi:10.1145/2071423.2071427

Gregory, N. M., Dorais, G. A., Fry, C., Levinson, R., & Plaunt, C. (2002). IDEA: Planning at the Core of Autonomous Reactive Agents. In *Proceedings of the 3rd International NASA Workshop on Planning and Scheduling for Space*.

Hassani, K., & Lee, W.-S. (2013). A software-in-the-loop simulation of an intelligent microsatellite within a virtual environment. In *2013 IEEE International Conference on Computational Intelligence and Virtual Environments for Measurement Systems and Applications (CIVEMSA)* (pp. 31–36). http://doi.org/ doi:10.1109/CIVEMSA.2013.6617391

Hassani, K., & Lee, W. S. (2014). On Designing Migrating Agents: From Autonomous Virtual Agents to Intelligent Robotic Systems. In SIGGRAPH Asia 2014 Autonomous Virtual Humans and Social Robot for Telepresence (pp. 7:1–7:10). New York, NY: ACM. http://doi.org/ doi:<ALIGNMENT.qj></ALIGNMENT>10.1145/2668956.2668963

Hassani, K., & Lee, W.-S. (2015). An Intelligent Architecture for Autonomous Virtual Agents Inspired by Onboard Autonomy. In P. Angelov, K. T. Atanassov, L. Doukovska, M. Hadjiski, V. Jotsov, J. Kacprzyk, … S. Zadrożny (Eds.), *Intelligent Systems'2014* (pp. 391–402). Springer International Publishing. Retrieved from http://link.springer.com/chapter/10.1007/978-3-319-11313-5_35

Hassani, K., Nahvi, A., & Ahmadi, A. (2013a). Architectural design and implementation of intelligent embodied conversational agents using fuzzy knowledge base. *Journal of Intelligent and Fuzzy Systems*, *25*(3), 811–823. doi:10.3233/IFS-120687

Hassani, K., Nahvi, A., & Ahmadi, A. (2013b). Design and implementation of an intelligent virtual environment for improving speaking and listening skills. *Interactive Learning Environments*, *0*(0), 1–20. doi:10.1080/10494820.2013.846265

Hindriks, K. V. (2009). ProgrammingRationalAgents in GOAL. In A. E. F. Seghrouchni, J. Dix, M. Dastani, & R. H. Bordini (Eds.), *Multi-Agent Programming* (pp. 119–157). Springer, US. doi:10.1007/978-0-387-89299-3_4

Horvath, G., Ingham, M., Chung, S., Martin, O., & Williams, B. (2006). Practical application of model-based programming and state-based architecture to space missions. In *Second IEEE International Conference on Space Mission Challenges for Information Technology, 2006. SMC-IT 2006*. http://doi.org/ doi:<ALIGNMENT.qj></ALIGNMENT>10.1109/SMC-IT.2006.62

Kang, Y., & Tan, A.-H. (2013). Self-organizing Cognitive Models for Virtual Agents. In R. Aylett, B. Krenn, C. Pelachaud, & H. Shimodaira (Eds.), *Intelligent Virtual Agents* (pp. 29–43). Springer Berlin Heidelberg. doi:10.1007/978-3-642-40415-3_3

Kim, I.-C. (2005). CAA: A Context-Sensitive Agent Architecture for Dynamic Virtual Environments. In T. Panayiotopoulos, J. Gratch, R. Aylett, D. Ballin, P. Olivier, & T. Rist (Eds.), *Intelligent Virtual Agents* (pp. 146–151). Springer Berlin Heidelberg. doi:10.1007/11550617_13

Koay, K. L., Syrdal, D. S., Dautenhahn, K., Arent, K., Małek, Ł., & Kreczmer, B. (2011). Companion Migration – Initial Participants' Feedback from a Video-Based Prototyping Study. In X. Wang (Ed.), *Mixed Reality and Human-Robot Interaction* (pp. 133–151). Springer Netherlands. doi:10.1007/978-94-007-0582-1_8

Kokinov, B. N. (1994). The Context-Sensitive Cognitive Architecture DUAL. In *Proceedings of the Sixteenth Annual Conference of the Cognitive Science Society.*

Kopp, S., Krenn, B., Marsella, S., Marshall, A. N., Pelachaud, C., Pirker, H., … Vilhjálmsson, H. (2006). Towards a Common Framework for Multimodal Generation: The Behavior Markup Language. In *Proceedings of the 6th International Conference on Intelligent Virtual Agents* (pp. 205–217). Berlin: Springer-Verlag. http://doi.org/ doi:10.1007/11821830_17

Kriegel, M., Aylett, R., Cuba, P., Vala, M., & Paiva, A. (2011). Robots Meet IVAs: A Mind-Body Interface for Migrating Artificial Intelligent Agents. In H. H. Vilhjálmsson, S. Kopp, S. Marsella, & K. R. Thórisson (Eds.), *Intelligent Virtual Agents* (pp. 282–295). Springer Berlin Heidelberg. doi:10.1007/978-3-642-23974-8_31

Laird, J. E. (2012). *The Soar Cognitive Architecture*. Cambridge, MA: The MIT Press.

Laird, J. E., Newell, A., & Rosenbloom, P. S. (1987). SOAR: An Architecture for General Intelligence. *Artificial Intelligence*, *33*(1), 1–64. doi:10.1016/0004-3702(87)90050-6

Langley, P., & Choi, D. (2006). A Unified Cognitive Architecture for Physical Agents. In Proceedings of the 21st National Conference on Artificial Intelligence (Vol. 2, pp. 1469–1474). Boston, MA: AAAI Press. Retrieved from http://dl.acm.org/citation.cfm?id=1597348.1597422

Langley, P., Laird, J. E., & Rogers, S. (2009). Cognitive Architectures: Research Issues and Challenges. *Cognitive Systems Research*, *10*(2), 141–160. doi:10.1016/j.cogsys.2006.07.004

Lekavý, M., & Návrat, P. (2007). Expressivity of STRIPS-Like and HTN-Like Planning. In N. T. Nguyen, A. Grzech, R. J. Howlett, & L. C. Jain (Eds.), *Agent and Multi-Agent Systems: Technologies and Applications* (pp. 121–130). Springer Berlin Heidelberg. doi:10.1007/978-3-540-72830-6_13

Liu, J., & Lu, Y. (2006). Agent Architecture Suitable for Simulation of Virtual Human Intelligence. In *The Sixth World Congress on Intelligent Control and Automation, 2006. WCICA 2006* (*Vol. 1*, pp. 2521–2525). http://doi.org/ doi:<ALIGNMENT.qj></ALIGNMENT>10.1109/WCICA.2006.1712816

MacLennan, A. (1996). The artificial life route to artificial intelligence: Building embodied, situated agents. *Journal of the American Society for Information Science*, *47*(6), 482–483. doi:10.1002/(SICI)1097-4571(199606)47:6<482::AID-ASI14>3.0.CO;2-0

Mahner, M., & Bunge, M. (2001). Function and Functionalism: A Synthetic Perspective. *Philosophy of Science*, *68*(1), 75–94. doi:10.1086/392867

Muscettola, N., Nayak, P. P., Pell, B., & Williams, B. C. (1998). Remote Agent: To boldly go where no AI system has gone before. *Artificial Intelligence*, *103*(1–2), 5–47. doi:10.1016/S0004-3702(98)00068-X

Ogawa, K., & Ono, T. (2008). ITACO: Constructing an emotional relationship between human and robot. In *The 17th IEEE International Symposium on Robot and Human Interactive Communication, 2008. RO-MAN 2008* (pp. 35–40). doi:10.1109/ROMAN.2008.4600640

Pokahr, A., Braubach, L., & Lamersdorf, W. (2005). Jadex: A BDI Reasoning Engine. In R. H. Bordini, M. Dastani, J. Dix, & A. E. F. Seghrouchni (Eds.), *Multi-Agent Programming* (pp. 149–174). Springer, US. doi:10.1007/0-387-26350-0_6

Porter, M. F. (1980). An algorithm for suffix stripping. *Program*, *14*(3), 130–137. doi:10.1108/eb046814

Rao, A. S., & Georgeff, M. P. (1995). BDI Agents: From Theory to Practice. In Proceedings of the First International Conference on Multi-Agent Systems (pp. 312–319).

Ribeiro, T., Vala, M., & Paiva, A. (2012). Thalamus: Closing the Mind-Body Loop in Interactive Embodied Characters. In Y. Nakano, M. Neff, A. Paiva, & M. Walker (Eds.), *Intelligent Virtual Agents* (pp. 189–195). Springer Berlin Heidelberg. doi:10.1007/978-3-642-33197-8_19

Ribeiro, T., Vala, M., & Paiva, A. (2013). Censys: A Model for Distributed Embodied Cognition. In R. Aylett, B. Krenn, C. Pelachaud, & H. Shimodaira (Eds.), *Intelligent Virtual Agents* (pp. 58–67). Springer Berlin Heidelberg. doi:10.1007/978-3-642-40415-3_5

Russell, S., & Norvig, P. (2009). *Artificial Intelligence: A Modern Approach* (3rd ed.). Upper Saddle River, NJ: Prentice Hall.

Sandamirskaya, Y., Richter, M., & Schoner, G. (2011). A neural-dynamic architecture for behavioral organization of an embodied agent. In *2011 IEEE International Conference on Development and Learning (ICDL)* (*Vol. 2*, pp. 1–7). http://doi.org/doi:10.1109/DEVLRN.2011.6037353

Segura, E. M., Kriegel, M., Aylett, R., Deshmukh, A., & Cramer, H. (2012). How Do You Like Me in This: User Embodiment Preferences for Companion Agents. In Y. Nakano, M. Neff, A. Paiva, & M. Walker (Eds.), *Intelligent Virtual Agents* (pp. 112–125). Springer Berlin Heidelberg. doi:10.1007/978-3-642-33197-8_12

Siegwart, R., & Nourbakhsh, I. R. (2004). *Introduction to Autonomous Mobile Robots* (1st ed.). Cambridge, MA: The MIT Press.

Spinola, J., & Imbert, R. (2012). A Cognitive Social Agent Architecture for Cooperation in Social Simulations. In Y. Nakano, M. Neff, A. Paiva, & M. Walker (Eds.), *Intelligent Virtual Agents* (pp. 311–318). Springer Berlin Heidelberg. doi:10.1007/978-3-642-33197-8_32

Starzyk, J. A., Graham, J., & Puzio, L. (2013). Simulation of a Motivated Learning Agent. In H. Papadopoulos, A. S. Andreou, L. Iliadis, & I. Maglogiannis (Eds.), *Artificial Intelligence Applications and Innovations* (pp. 205–214). Springer Berlin Heidelberg. doi:10.1007/978-3-642-41142-7_21

Sun, R., & Zhang, X. (2006). Accounting for a variety of reasoning data within a cognitive architecture. *Journal of Experimental & Theoretical Artificial Intelligence*, *18*(2), 169–191. doi:10.1080/09528130600557713

Vala, M., Ribeiro, T., & Paiva, A. (2012). A Model for Embodied Cognition in Autonomous Agents. In Y. Nakano, M. Neff, A. Paiva, & M. Walker (Eds.), *Intelligent Virtual Agents* (pp. 505–507). Springer Berlin Heidelberg. doi:10.1007/978-3-642-33197-8_60

van Oijen, J., Vanhée, L., & Dignum, F. (2012). CIGA: A Middleware for Intelligent Agents in Virtual Environments. In M. Beer, C. Brom, F. Dignum, & V.-W. Soo (Eds.), *Agents for Educational Games and Simulations* (pp. 22–37). Springer Berlin Heidelberg. doi:10.1007/978-3-642-32326-3_2

Veloso, M., Carbonell, J., Pérez, A., Borrajo, D., Fink, E., & Blythe, J.VELOSO. (1995). Integrating planning and learning: The PRODIGY architecture. *Journal of Experimental & Theoretical Artificial Intelligence*, *7*(1), 81–120. doi:10.1080/09528139508953801

Volpe, R., Nesnas, I., Estlin, T., Mutz, D., Petras, R., & Das, H. (2001). The CLARAty architecture for robotic autonomy. In *Aerospace Conference, 2001, IEEE Proceedings.* (*Vol. 1*, pp. 1/121–1/132). http://doi.org/ doi:10.1109/AERO.2001.931701

Wang, P., & Goertzel, B. (2007). Introduction: Aspects of Artificial General Intelligence. In Proceedings of the 2007 Conference on Advances in Artificial General Intelligence: Concepts, Architectures and Algorithms: Proceedings of the AGI Workshop 2006 (pp. 1–16). Amsterdam, The Netherlands, The Netherlands: IOS Press. Retrieved from http://dl.acm.org/citation.cfm?id=1565455.1565457

Weng, J. (2004). Developmental robotics: Theory and experiments. *International Journal of Humanoid Robotics*, *01*(02), 199–236. doi:10.1142/S0219843604000149

Wooldridge, M. (2002). Intelligent Agents: The Key Concepts. In V. Mařík, O. Štěpánková, H. Krautwurmová, & M. Luck (Eds.), *Multi-Agent Systems and Applications II* (pp. 3–43). Springer Berlin Heidelberg. doi:10.1007/3-540-45982-0_1

KEY TERMS AND DEFINITIONS

Embodied Intelligent: Coupled mind-body loop in which high level deliberative processes working with symbolic representation of the world within the mind decide the behavior of the agent based on a collection of physical or virtual sensors and actuators within the body.

Migration: Ability of an abstract entity to morph from one embodiment into a different embodiment and control the new body, accordingly.

Onboard Autonomy: In space robotic systems refers to high-level behavioral controllers that adapt to failure at lower levels of the system.

Strong AI: Create intelligent machine that can successfully perform cognitive tasks that a human being can do.

Weak AI: Non-sentient machine intelligence that is domain-dependent and is developed for a specific task.

Chapter 10
Game AGI beyond Characters

Alexander Zook
Georgia Institute of Technology, USA

ABSTRACT

Artificial General Intelligence has traditionally used games as a testbed to develop domain-agnostic game playing techniques. Yet games are about more than winning. This chapter reviews recent efforts that have broadened the ways Artificial Intelligence (AI) is used in games, covering: modeling and managing player experiences, creating novel game structures based in interacting with AI, and enabling AI agents to make games. Many of the techniques used to address these challenges have been ad hoc approaches to solving specific problems. This chapter discusses open challenges in each of these areas and the potential for cognitive architectures to provide unified techniques that address these challenges.

INTRODUCTION

Artificial General Intelligence (AGI) in games has traditionally emphasized control over characters within virtual environments. Yet game experiences depend on the coupling of characters with their environments and players. Recent trends in games have begun to target a broader array of AI applications. *Left 4 Dead* and *Dark Spore* employ an AI director to manipulate the game environment in real time to adjust player experience. *Black & White* and *Prom Week* engage players directly in teaching an intelligent system or manipulating "social physics" as core gameplay mechan-

DOI: 10.4018/978-1-5225-0454-2.ch010

ics. And Procedural Content Generation (PCG)—the automated creation of game content like levels—has emerged as a core technique in popular games spanning *Spelunky* through *Diablo III*, making AI-driven content creation a growing interest across AAA and independent game creation.

These considerations have recently led many researchers to consider alternative applications of AGI in games beyond embodied agents within virtual environments. Rather than simulate the intelligence of non-player characters within a virtual environment, researchers are considering ways more general and autonomous agents can create gameplay experiences. Traditional character AI has been "weak AI" intended to provide adversaries designed for a single task, using techniques such as planning in *F.E.A.R.* (Orkin, 2005), behavior trees in *Halo* (Isla, 2005), or finite state machines in *Pacman* (Pittman, 2011). Research efforts on applying "strong AGI" to games have aimed to create intelligent adversaries in games like *Quake* (J. E. Laird, 2001) that engage in the full process of gathering visual information, processing that information, learning, and acting in the game environment. Most efforts, however, have aimed at highly controllable agents amenable to specific game designs with low computational demands. By approaching AGI from the perspective of agents that are core to gameplay there are new opportunities for AGI architectures that tackle the holistic problem of creating games for a variety of player experiences. Taking game AGI beyond characters has the potential to make AGI relevant to a broader range of game design concerns while enriching the knowledge and reasoning processes AGI is capable of.

This chapter reviews new avenues for AGI in games, highlighting emerging areas and discussing the potential for cognitive architectures and games to contribute to one another. Three primary topics are discussed, reviewing existing work as well as challenges and future directions for growing the areas and integrating AGI and cognitive architectures. *Player Experience Modeling and Management* highlights ways intelligent systems can model player experiences in a game to inform people making games about a game design and game environments that are automatically tailored to individual players. *AI-based Games* explores ways to make interaction with a general, autonomous intelligent system the core of gameplay, including learning techniques, social simulations, and strong AGI adversaries. *Game Creation* discusses ways to enable intelligent systems to augment and automate the process of creating part or all of games autonomously, simulating the intelligence of human creators to structure a game environment to create a game experience. By addressing AGI concerned with creating player experiences the chapter highlights opportunities for AGI to tackle the challenges of how designers and players reason about games as a whole.

The chapter first briefly reviews artificial general intelligence with an emphasis on existing cognitive architectures and their primary game application in game playing. Next is a discussion of three major avenues for AGI in games beyond characters: (1) systems for modeling and shaping player in-game experience; (2) approaches to making AGI the core of gameplay; and (3) applications of AGI to support and automate the creation of games. The chapter closes by discussing overarching challenges for game AGI beyond characters and encourages further work to model and augment the game design process. By reviewing these areas and their open challenges and directions this chapter provides an understanding of the state of AGI applications targeted toward creating gameplay and exposes opportunities to contribute to this growing area.

BACKGROUND

Researchers in AGI have developed cognitive architectures to model many forms of human knowledge, reasoning, and learning (Chong, Tan, & Ng, 2007; Duch, Oentaryo, & Pasquier, 2008). Cognitive architectures address the challenge of using multimodal information for complex, human tasks, ideally in a domain-agnostic fashion. Broadly speaking, cognitive architectures can be divided among three approaches: symbolic, subsymbolic, and hybrid models (Duch et al., 2008). Symbolic architectures emphasize the use of top-down, declarative representations of system knowledge in terms of atomic symbols. Well-known symbolic architectures include Soar (State, Operator, and Result) (J. Laird, 2012) and ICARUS (Langley & Choi, 2006). Subsymbolic architectures employ distributed representations that cannot be directly analyzed, such as neural networks, and aim to represent biological processes that underpin cognition. IBCA (Integrated Biologically-Inspired Cognitive Architecture) exemplifies this approach by modeling distributed information processing in the brain (O'Reilly & Munakata, 2000). Hybrid architectures combine symbolic and subsymbolic representations to model knowledge at multiple levels and transform knowledge between representations. The CLARION (Connectionist Learning Adaptive Rule Induction ON-line) (Sun, 2006) and ACT-R (Adaptive Components of Thought - Rational) (Anderson et al., 2004) architectures are hybrid approaches that use low-level distributed representations that are transformed into high-level symbols and back.

How should cognitive architectures be used in games? Existing work has primarily addressed ways AGI—and more rarely existing cognitive architectures—can be used to *play* games. Examples of existing cognitive architectures playing games include ICARUS learning football plays from video (Stracuzzi et al., 2011) and Soar playing the first-person shooter game *Quake* (J. E. Laird, 2001). Others have developed AGI agents in *Super Mario Bros.* with models of agent drives, speech processing and generation, and symbolic knowledge that can be learned through interacting with the environment (Ehrenfeld, Schrodt, & Butz, 2015). The ICARUS and Soar agents both demonstrate greater generality in game situations they can address compared to the narrow set of scripted tasks handled by the built-in agents for their corresponding games; the Mario agent illustrates the potential for more autonomous agents to take a human's position in a game with the capacity to interact via speech. Together these examples illustrate the potential for AGI to enhance the capabilities of agents *in* games.

Games have also been treated as a test domain for artificial general intelligence (J. Laird & VanLent, 2001), with competitions ranging from the General Game Playing competition that uses broad symbolic models (Genesereth, Love, & Pell, 2005; Love, Hinrichs, Haley, Schkufza, & Genesereth, 2008), to the General Video Game Playing competition that models a range of existing simple video games (Perez et al., 2015; Schaul, 2013), to directly playing in arcade game emulators using input controls and video (Bellemare, Naddaf, Veness, & Bowling, 2013; Murphy, 2013). These efforts illustrate the potential for AGI to reason on game structures to achieve game goals, applying a combination of methods for exploring the game domain and learning.

These efforts, however, have taken a limited view of how AGI should intersect with games, treating games as a means for testing agents with complexity that falls between the toy domains of the past and the real-world of embodied robotics (J. Laird & VanLent, 2001). AGI, however, can take an alternative role: using the capabilities of intelligent systems to model games and human cognition to extend the bounds of game experiences. Below three umbrella topics will be discussed where AGI can contribute to new game experiences: using AGI to model and improve player experiences, interacting with AGI systems as gameplay, and using AGI for (complete or partial) development of games. These topics see AGI as increasingly central to the game experience and demonstrate the importance of considering both game players and game designers.

PLAYER EXPERIENCE MODELING AND MANAGEMENT

Games are designed to create a desired experience (Fullerton, Swain, & Hoffman, 2008; Salen & Zimmerman, 2003). Yet designers face a fundamental challenge: they directly control the structure of a game's design, but seek to influence player behavior and cognition to create a desired experience (Hunicke, Leblanc, & Zubek, 2004). Current approaches to game design and development advocate an iterative process, where designers examine how players act in and experience a game to inform subsequent changes to the game's design. Techniques for understanding player experience include playtesting (Fullerton et al., 2008)—where people play a game and designers learn from subjective feedback and observation—games user research (Isbister & Schaffer, 2008)—studies of players, using techniques from psychometrics, biometrics, and human-computer interaction—and game analytics (Seif El-Nasr, Drachen, & Canossa, 2013)—aggregation of quantitative metrics on player activity through a game client. Studying game design has primarily been used to make hand-tuned adjustments to a game to craft a desired end experience. Applying AGI to player experiences requires addressing the problems of: (1) understanding how game design choices influence player behavior and experience; (2) developing tools to inform human designers of how design decisions shape possible player behavior in a game; and (3) developing techniques to employ cognitive architectures to make real-time changes to a game that support a designer's goals.

Narrow Player Modeling

Narrow player modeling work has emphasized models tailored to specific games or design scenarios. Studies of how game design choices influence behavior can be divided into approaches aiming primarily for strong model fit and approaches aiming for human-interpretable models. *Performant* models, which target strong fit, are generally subsymbolic and constructed with large feature spaces meant to provide good performance on a desired task. *Interpretable* models emphasize symbolic approaches with fewer, strongly correlated features. This distinction is not hard and many models fall between the extremes: the lens helps when understanding the intended model applications. Performant models are often used in real-time, online adaptation approaches where an intelligent system is responsible for manipulating a game environment outside the control of a human designer. Interpretable models are typically used during development for offline content modeling and consideration of design decisions.

Performant models have primarily used neural architectures to predict complex aspects of player subjective experience in specific game domains (Yannakakis & Togelius, 2011). Researchers have applied neuro-evolution to predict player ratings of frustration, boredom, or enjoyment in platformer games (run and jump games similar to *Super Mario Bros.*) (Pedersen, Togelius, & Yannakakis, 2009), player subjective responses in terms of preference between pieces of content including map designs (Liapis, Yannakakis, & Togelius, 2013), weapon particle effects (Hastings, Guha, & Stanley, 2009), and flower appearance (Risi, Lehman, D'Ambrosio, Hall, & Stanley, 2012). These efforts have been "narrow" AI efforts in the sense that the applications are limited to specific subsets of player experience and specific games, without generality to apply to other domains.

Interpretable models have studied player responses to parameterized changes in games, including variants in avatar colors (Domínguez & Roberts, 2014), how touch control pathfinding influences player behavior and perceived difficulty (Thompson & Vinciguerra, 2014), how design decisions influence challenge and engagement in games for education (Andersen et al., 2011; Y.-E. Liu, Mandel, Brunskill, & Popović, 2014; Lomas, Patel, Forlizzi, & Koedinger, 2013) and scientific data collection (Siu, Zook, & Riedl, 2014). As with performant models, research has been narrowly targeted at understanding a specific game or design problem, without efforts to provide general models applicable across design scenarios. Together performant and interpretable models represent a broad class of techniques for enabling intelligent agents to understand how to influence a game environment to shape player behavior. Performant and interpretable models have been used in two broad classes of tools: offline (before a game is played) and online (during gameplay) design tools.

Offline tools support designers through displaying ways players might play a game given a set of design choices. Offline content creation involves choosing among alternative designs for many types of game content. Existing approaches have supported level creation in a variety of domains, including 2D platformers (G. Smith, Whitehead, & Mateas, 2011; Tremblay, Borodovski, & Verbrugge, 2014), stealth games (Tremblay, Torres, & Verbrugge, 2014), and puzzle games (A. M. Smith, Andersen, Mateas, & Popović, 2012). Current approaches have focused on checking whether levels can be completed given the design and visualizations of the possible ways players can complete levels.

Online adaptation algorithms change a game in real time based on a model of desired player behavior. Adaptation algorithms act as surrogates for game designers, changing game content to tailor the game's design toward a designer-specified player experience. Online adaptation forgoes creating a single design for all players to provide player experiences sensitive to player differences in skill, preferences, or demographics. Existing approaches have adapted 2D platformer levels (Blom et al., 2014; Pedersen et al., 2009), space shooter weapons (Hastings et al., 2009), game stories (M. Riedl, Stern, Dini, & Alderman, 2008; M. O. Riedl & Bulitko, 2013; Robertson & Young, 2014), and game environments (Hartsook, Zook, Das, & Riedl, 2011). To date these techniques show promise for adapting content by optimizing a set of design features to expected player behavior or planning and switching discrete content to enforce guided sequences of player activity.

Toward General Player Modeling

A core challenge facing player experience modeling is creating reusable knowledge about player behavior that generalizes across games. Ideally player models should help inform a general *science of game design* grounded in models that predict how design features shape player behavior and experiences. How can player behavior be studied in game designs to scientifically model how game design influences player behavior? How can player models unify an understanding of per-player differences (e.g., reaction time or working memory capacity) across games and genres?

Existing cognitive architectures require extensive development of domain-specific knowledge to enable the architecture to play a game. The learning Mario agent requires a schema tailored to the Mario games, with the ICARUS football agent and Soar *Quake* bot facing similar limitations (Ehrenfeld, Schrodt, & Butz, 2015; J. E. Laird, 2001; Stracuzzi et al., 2011). In game development, however, creating this knowledge can be an expensive and time-consuming process, leading to contemporary techniques that use simple, shallow models of player behavior to address a small set of tasks in a game. As a result, current player models have only addressed single games. Developing unified models of player behavior in games can begin by capturing patterns common to genres of games, such as 2D platformer games or simple arcade games. Modeling how design features within these genres interact with player cognition to produce player behavior can lead to a new class of player behavior models that is game-agnostic and cognitively realistic. This poses many questions for future work using cognitive architectures: How should systems acquire this knowledge autonomously? Are there shared sets of knowledge that can be stored and reused across games within a genre? Across genres? What knowledge representations and learning processes are effective for transfer across games?

Developing cognitive architectures able to interact with a variety of games within a genre is poised to advance our understanding of how game design functions through generalized, reusable scientific knowledge of game design. Developing these architectures to model player differences (e.g., in working memory capacity, rate of learning, etc.) can enhance offline tools with the capability to display how a design may differentially impact players depending on their capabilities. Offline tools currently are limited to evaluating *what* people can potentially accomplish in a game. Cognitive architectures could augment this as "crash test dummies" for evaluating *how* people may be expected to play. As models become more general in their capabilities they can serve as tools across a wider range of games and a wider array of tasks within games. Rather than using agents in tools meant only to test reachability in a platformer, these agents could check learnability for when and what players might learn about how the game functions from playing a level. Going further, these same agents might eventually enable showing how players would adapt knowledge from playing one game to new games or genres. Online tools would similarly benefit from understanding differences in player cognition. Further, aspects of player cognition could be estimated using a cognitive model trained on observations of gameplay data. Providing game-agnostic player models stands to be a breakthrough for enabling game design and adaptation without extensive player data and may potentially fuel the creation of player models that evolve with players over time and across games.

Opportunities also exist for modeling other aspects of hidden subjective player state not directly observable through play behavior. Cognitive architectures have the potential to model player emotional responses or procedural and conceptual understanding over the course of a game. Extending existing tools, these models can inform offline design of how design decisions shape player experience at a level beyond direct gameplay. A new generation of tools might estimate player drives or interests in a game based on a cognitive model of how these drives lead to in-game behavior. Online adaptation can also benefit from the capability to consider what new content to present to maximize players learning game concepts—i.e., acting as dynamic tutorials for games or automated hinting systems. Overall, cognitive architectures have great potential to improve the process of designing games through deeper models of player in-game experiences. At the same time, these efforts at player modeling can explore open challenges for cognitive architectures in acquiring (perhaps through learning) knowledge for modeling player behavior and applying this knowledge in a domain-agnostic fashion when meeting design needs.

AI-BASED GAMES

Games create experiences for players through a set of rules and content for those rules. Game mechanics are these rules, providing players with verbs to take action in a game environment. Most contemporary games are 'physics-based'—emphasizing gameplay around interacting with a physical simulation of the world. Physics-based game mechanics center on actions like running, jumping, shooting, colliding, and so forth. Yet games need not be based solely in physical interactions. Could games be designed around interacting with AI as the core gameplay? What opportunities exist for gameplay around interacting with intelligent systems driven by AI techniques?

AI-based games have recently challenged the tradition of physics-based games by using AI technologies to provide the fundamental systems and actions of games (Treanor et al., 2015). Several approaches have emerged for using richer agent intelligence to create new types of player experiences. New designs include foregrounding agent teaching and learning as a game mechanic, managing social interactions and relations among agents as the core gameplay loop, or coping with autonomous agents with wide-ranging capabilities as opponents. These efforts highlight the value of more sophisticated agent architectures as the basis for gameplay while revealing design challenges associated with these approaches. This section reviews work in these efforts, highlight the core challenges of making AI-based games, and discusses future directions for AGI and cognitive architectures to expand this space.

AI as a Game

In-game non-player characters (NPCs) are traditionally short-lived opponents that have little need for learning or changing behavior. Building a game around *agent learning* can re-direct gameplay toward how agents learn and how players teach agents. *Black & White* (R. Evans, 2002) is a simulation game where the player trains a creature as an autonomous assistant to act in spatial regions where the player cannot directly take action. The agent learns behaviors through a reward signal based on agent needs and direct feedback from player actions (slapping or petting the creature after it takes actions). Players cannot directly control the actions of the avatar, but instead must use feedback to train the agent to perform actions that align with the player's desired strategy. *Black & White* converts traditional play around direct player control into play based on teaching an agent to have indirect control. The agents in *Black & White* illustrate the use of simple learning mechanisms to create interesting gameplay when an agent has a wide range of ways to interact with a game environment and is given a degree of autonomy that is core to play.

Beyond players training embodied avatars, learning can support gameplay based on players curating desired patterns from AI-based systems. In *NERO* (Stanley, Bryant, & Miikkulainen, 2005) players train a team of agents to compete against opponents. The agents use neuroevolution to acquire behaviors that succeed against training scenarios arranged by the player. *Galactic Arms Race* (Hastings et al., 2009) is a space shooter where players use different weapons to evolve an underlying neural network representation of weapon particle effects to produce new firing behavior. As players fire weapons to defeat enemies their weapon choices are used as a signal for preferences. Information on player preferences is used to update the neural network model to evolve toward player-preferred particle effects, creating new patterns and providing new weapons for players to pick up and test. *Petalz* (Risi et al., 2012) is a social game where players breed, trade, and buy and sell virtual flowers. Flowers evolve using similar underlying technology as in *NERO* and *GAR*—gameplay revolves around getting flowers with patterns the player enjoys. *NERO*, *GAR*, and *Petalz* highlight how learning architectures enable players to treat learning as a curation process, creating team behaviors, weapon patterns, or flower appearances through systems governed by preference learning algorithms. Compared to a tightly scripted narrow AI system, greater autonomy on the part of trained systems is a core part of the value of the game experience in these cases, showing a place for more autonomous AI in games.

Interactions with game avatars have primarily targeted adversarial combat behavior, rather than modeling the more complex, subtle interactions needed for *social behavior*. *Façade* (Mateas & Stern, 2002) addresses richer social models in a situational comedy setting where players interact with a couple nearing divorce. Social modeling in *Façade* use a specialized programming language—A Behavior Language (ABL)—for reactive planning to control how agents respond to player choices, including behaviors that coordinate multiple agents. ABL enables authors to specify knowledge about agent behaviors in individual interactions as well as joint behaviors that coordinate multiple agents to model appropriate social interaction. This technology allows agents to stage scenes dynamically in response to models of agent knowledge and behavior while following along a desired overall narrative trajectory.

Prom Week (McCoy et al., 2014) is set in the weeks leading up to a high school prom night and takes a systematic approach to AI-based social interactions. A full "social physics" model represents individual character traits, feelings, and relationships and enables characters to form intents, take actions, relate to a shared cultural space, and remember and refer to past events. Players direct the choices of a character, shaping the way relationships evolve over the days leading up to the prom. *Prom Week* developed the social modeling language Comme il Faut (CiF), allowing authors to give rules encoding appropriate behavior in a social setting. Authored rules control how agents respond to players while remaining consistent with character-level personal concerns and the broader social concerns of a situation. Unlike traditional narrow AI scripts for social behavior, CiF allows agent authors to create a rich space of social interactions. In *Prom Week* this leads to agents tracking a dynamic social state and being able to autonomously form their intentions that guide how they react to players, leading to gameplay built around how players learn to interact with this model of social interaction. The generality and autonomy of CiF agents creates rich gameplay around players learning the social rules of the world.

MKULTRA (Horswill, 2014) is set in an alternate future where mind control technologies are prevalent. Agents in *MKULTRA* use classical question answering technologies in gameplay where players talk with agents to complete missions and may directly manipulate agent beliefs through mind control mechanics. *MKULTRA* enables agents to retain a large base of declarative knowledge about the world and use AI models of belief revision to modify these beliefs based on interactions with players. This approach creates gameplay where players learn to explore and exploit the general capabilities of classical question answering and reasoning, refocusing gameplay on how AGI agents with general reasoning capabilities succeed (and fail) in the face of player interaction.

Versu (R. Evans & Short, 2014) is a storytelling language that provides a rich social model of small-group interactions, capturing abstract, context-specific and universal individual motivations; emotions; beliefs; social norms; and inter-personal relationships using a novel logical formalism. Players guide agents modeled in the language, with the social physics controlled through nuanced notions of actions and interactions among agents. *Versu*'s language allows authoring a variety of social scenarios where players can actively (or passively) engage with AGI agents that follow social norms while hewing toward their individual personality and individual motivations. The language allows players to fill any agent role in a scenario, providing the unique opportunity to try scenarios from the perspective of different parties involved. *Façade*, *Prom Week*, *MKULTRA* and *Versu* all developed underlying models of individual agents and social interactions. Rather than ground gameplay in physical interaction, a "social physics" governs these environments to create new kinds of experiences where players consider how agents navigate social environments.

Other efforts have explored how agents can create new kinds of gameplay experience as generalized opponents. *Spy Party* is a competitive multiplayer game that inverts the general trend of social agents of game designers strive for greater realism. In *Spy Party* one player must perform game-given actions at a party as a spy, while another player attempts to identify them from among a crowd of AI agents. The challenge of the game for the spy is imitating the actions of NPCs. This inversion forces both players to focus on understanding the realism or artificiality of different agent behaviors in the game. *Spy Party* illustrates how even limited agent models can be core gameplay systems, showing how AI-based games can gradually improve without perfectly replicating real-world behavior and phenomena.

Alien: Isolation (Creative Assembly Ltd., 2014) highlights ways to make an AI opponent core to gameplay. *Alien: Isolation* is a survival game where the player must avoid and escape from a single alien with general AI capabilities. The alien has capabilities for sensing the environment (light and sound), tracking and estimating the player's location, and learning from interactions with the player. Gameplay revolves around the player devising strategies to deceive the AI over long periods of time and adapt their strategies as the alien accommodates player tactics. Unlike traditionally short-lived and narrowly circumscribed AI opponents, the long-term interactions with the alien AI required more sophisticated and general capabilities to remain a realistic opponent to players. The alien AI highlights how enabling asymmetric AI opponents to have autonomy and general capabilities to interact with the game can yield engaging and novel gameplay. *Spy Party* and *Alien: Isolation* take advantage of controlled agents with a wide array of actions and pure autonomy from the player as a way to create a game where players must learn how agents work.

Toward General AI Games

Using AGI in general, and cognitive architectures in specific, as game mechanics introduces a number of design challenges. These systems face challenges in authoring expressivity, AI transparency to players, AI fragility, and enabling players to understand AI systems. The social interaction models reviewed above have all created specialized languages for expressing how people behave in a social domain and as a means for authors to create domain-specific instantiations of these systems for specific social interactions. Mateas (2001) champions this approach under the umbrella of *Expressive AI*: supporting authorship and providing "hooks" into an AI system that enable authors to create desired cultural products. Designing each of the above social models required careful consideration of the social theory being applied to the system (e.g., *Prom Week* adopting Goffman's dramaturgical view of social interactions or *Versu* drawing from Brandom's argument for foundational social practices) and the kinds of social interactions supported by those models. Thus, an open challenge for cognitive architectures that support socially-based games will be developing a rich set of models of social and cultural norms, interactions, and behaviors and sharing these social theories across architectures. Developing techniques for these architectures to acquire knowledge of social practices based on human player behavior stands as an open avenue for furthering the use of social interactions and knowledge as core to gameplay.

Horswill (2014) highlights three major challenges for making AI the core of gameplay: transparency of the AI to players, controlling player expectations about AI capabilities, and the fragility of the AI in unanticipated situations. *Transparency* of AI agent state through communicating cognitive models or states to the player is an ongoing challenge for game design, particularly for complex agents that are capable of learning, or for agents that model aspects of personality like emotion or belief. Some elements can be communicated naturally as people would, such as characters in *Versu* expressing distaste or surprise. But internal beliefs or learned knowledge are complex and are not easily represented to players. Researchers and designers currently lack an understanding of how to best signal when a system learns, how it learns, and what it has learned. Thus, an open challenge for using cognitive architectures as game systems is appropriately communicating how these systems function without spoiling gameplay that requires management of some amount of hidden information or uncertainty.

A lack of agent transparency often results in problems in *controlling player expectations*. Without effective ways to communicate agent states (and processes) to players it is impossible to convert these complex algorithms into systems that players indirectly control or directly interact with to play games. Managing player expectations requires ways to introduce players to how a complex system functions and guide them to learn how to interact with the system. Lacking this understanding, players either become overwhelmed by the perceived complexity of the system or disappointed by a perceived lack of flexibility for rich gameplay. A core challenge thus lies in deciding the appropriate level of detail in the underlying AI system to balance between providing rich behavior and an understandable system.

Fragility is the design challenge of addressing the limitations of any AI architecture when used in gameplay. Players will typically discover situations that the AI system (be it a cognitive architecture or simple finite state machine) is not equipped to handle. Classical AI techniques, such as rule-based systems, are often fragile in not gracefully degrading when encountering situations they were not coded for; the same critique applies to neural networks and subsymbolic techniques. Thus, a general challenge will be exploring how to expose AI-based game mechanics to players in a way that avoids making the AI fragility detrimental to player experience. *The Sims* addresses AI fragility by explicitly making the player's task one of protecting and nurturing (or willfully harming) a virtual character. Procedural content generation (PCG) for games has often addressed fragility through high-level authoring of large pieces of content that an AI system assembles in well-controlled ways. Alternatively, many games using PCG have gameplay that allows for the vagaries of the AI system—for example, in *Spelunky* the AI level generator may place tiles in a dungeon to make no exit paths possible; a core gameplay mechanic involves bombs that can destroy tiles, leaving players the option to clear a path even if the AI does not leave one. Other games make the randomness and harshness of AI generation part of gameplay itself: many "rogue-like" games are built around the experience of players mastering game systems to overcome potentially impossible odds created by the AI generator. Developing techniques for coping with AI fragility in games will be a key part of advancing the adoption of more sophisticated (but potentially brittle) AI models including AGI systems and cognitive architectures.

A number of technical challenges exist beyond these design challenges. Current approaches to AI as a game mechanic fall short of employing full cognitive architectures both in applying existing systems and in developing novel domain-specific architectures. How can cognitive architectures extend and improve the range of social models and "AI engines" available to be used in games? How can games designed around interacting with AI systems inform the design of better cognitive architectures? Broaching these questions will require developing holistic gameplay cognitive architectures that integrate many aspects of cognition and knowledge.

GAME CREATION

Going beyond shaping a player experience in a game or being the mechanics underlying a game, could an intelligent system be responsible for making the entire game? Automated game creation requires integrated reasoning on many aspects of games and modeling the game design and development process itself. Games integrate concepts, themes, stories, world, characters, mechanics, and a plethora of other types of content. How can an agent reason on the interrelations among these semantically distinct pieces of content? And how should an agent go about developing the game, from conceptualizing the design through creating the content and evaluating the final game structure or experience? This section addresses efforts to enable automated generation of aspects of game content, highlighting current work and future directions for this growing field.

Conceptualization

Conceptualization involves the process of creating ideas to build into a game and connecting these to game content. People often build games with a pre-conceived fictional setting (e.g., high fantasy) or a core message or theme (e.g., "murder is bad" or "about things under the ground"). Current approaches to conceptualization draw primarily from existing databases of conceptual linkages. Llano, Cook, Guckelsberger, Colton, and Hepworth (2014) generate fictional ideas to use in games by following links among concepts in ConceptNet (H. Liu & Singh, 2004). Other approaches use relationships from ConceptNet to select objects to populate predefined game structure templates (Treanor, Schweizer, Bogost, & Mateas, 2012; Nelson & Mateas, 2008). Cook and Colton (2014) circumvent ConceptNet by using *Google Milking* (Veale, 2012)—using the autocomplete functionality on the popular search engine to gather common beliefs about the world. The queries take a general form of "why do [NOUN]s [VERB]", where autocompletion provides a target object. Generated games use a predefined set of verb actions that substitute in appropriate player avatars (from nouns) and enemies (from objects). These approaches highlight how existing knowledge for other purposes can be used to extend the bounds of AGI agents to new task domains.

Conceptualization is currently limited by the knowledge available to AI game designers: a broad challenge for future work will be developing new techniques to gather conceptual knowledge for game designs. One important source of conceptual knowledge is how things are meaningful to people in terms of the real world: ConceptNet and other knowledge bases have begun to compile such commonsense knowledge. Yet this is limited—across game design and the arts people draw from a cornucopia of folklore sources, common media tropes, and personal experiences. How can systems tap into this broader array of semantic knowledge for richer concepts for games? Further, how can systems tap into the ways people create metaphors and use non-realistic settings (e.g., science fiction writing or surrealist art) as vehicles to convey meaning? Could AI game designers create games that convey meaning through their structure and the ways concepts behind a game are manifest in game content? Realizing these goals will require access to a much broader set of human knowledge and experience than is currently used in cognitive architectures.

Applying such knowledge to games will further require exploration of how games convey meaning to enable systems to mimic those techniques or even discover novel methods previously unknown to human creators.

Level Design

Games typically revolve around a series of challenges in the form of goals or end states for the player to reach. Level design is concerned with the structure of individual challenges, spanning single boards in a puzzle game to stages in a platformer game to dungeons and worlds in a role-playing game. Designing levels involves a number of challenges: ensuring levels can be completed, understanding how players might complete a level, and synthesizing considerations of level structure with other game features including theme and story.

Level design methods include using hand-authored templates, generate and test approaches evaluated by quality criteria, and declarative techniques to ensure content constraints. Several level generation methods have focused on design patterns for levels that ensure playability: *Launchpad* uses a hand-authored grammar (G. Smith, Whitehead, Mateas, Treanor, et al., 2011), while others have mined design knowledge from existing level design examples (Dahlskog & Togelius, 2014; Guzdial & Riedl, 2015; Snodgrass & Ontañón, 2014). Level design can also be cast as aiming to optimize abstract fitness criteria (Togelius, Yannakakis, Stanley, & Browne, 2011). Examples include optimizing dungeon and cave layouts and optimizing 2D platformer or adventure game levels for challenge models while obeying design constraints (Sorenson, Pasquier, & DiPaola, 2011).

Design constraints are often central to levels, most importantly ensuring playability. Several approaches have ensured playability through declarative programming techniques including: constraint solving to ensure fraction puzzles are solvable (A. M. Smith et al., 2012; A. M. Smith, Butler, & Popović, 2013) and controlling level progressions to introduce concepts sequentially (Butler, Andersen, Smith, Gulwani, & Popović, 2015; Butler, Smith, Liu, & Popović, 2013); ensuring adventure puzzle solvability by checking for action chains (Dart & Nelson, 2012); and ensuring 3D worlds embed semantic constraints on content (van der Linden, Lopes, & Bidarra, 2013; Lopes, Tutenel, & Bidarra, 2012).

Level design does not stand on its own, but is often integrated with other design features at the level of quest structure, story, or aesthetic theme. Dormans (2010) integrates action role-playing game quest structure and level layout through a pipeline of shape grammars. Hartsook et al. (2011) transform hand-annotated story event locations into an abstract world layout that is then grounded in a 2D world.

Cook, Colton, and Pease (2012) parse news articles and uses sentiment analysis to generate titles, music, commentary, and art assets to generate 2D platformer games.

Level generation is typically a self-contained process that cannot be influenced by the player except through the selection of parameters before the generation starts. In *Endless Web* (G. Smith, Othenin-Girard, Whitehead, & Wardrip-Fruin, 2012) the player controls aspects of a procedural level generator by playing the game in a particular way and making decisions about how to explore. Player choices of level selection alter the generation of the subsequent levels to emphasize a characteristic of the levels. For example, if the player chooses an exit that is surrounded by dangerous enemies, the next level might have a higher number of enemies in it for the player to get past.

These approaches have begun to create levels for a variety of content domains, but have emphasized techniques that depend heavily on domain knowledge in the form of grammars, evaluation functions for optimization, or domain-specific specifications of level playing and solution criteria. Could level generators acquire more knowledge from online corpora? What would a cross-genre or domain-agnostic level generator look like? For example, 2D platformers span a variety of level designs, including everything from *Mario Bros.* reflex-based designs to *Metroid*'s more exploratory structures. Could a system generate levels that create novel genres or innovate in design for an existing genre?

To date, integrating multiple design features has worked by using pipelines that do not reconsider prior choices or model the influence of design decisions on other features (e.g., aesthetics). What level design approaches exist outside content assembly pipelines? If content creation systems are to be treated as cognitive architectures modeling the *process* of game development they will require new methods to learn from how the content they create works for making revisions. Non-pipeline models will need to account for how designers go about revising a design, revising their goals for a design, or using information from how players interact with levels of a design to assign credit and blame to their goals for level design versus the broader concept of the game.

Player-controlled PCG is still relatively unexplored by games research and development, leaving many opportunities for future work exploring these ideas (G. Smith, 2014). Currently video streaming of people playing games has become a popular pastime. How might level generators function when attempting to create content that is not only adapted to a player, but also suited to entertain an audience watching that player? Could level generation be shaped by audience reactions? Developing these technologies will require approaches that integrate models of theory of mind with existing approaches to consider not only the experience of the player, but also observers of the game.

Mechanics

Game mechanic generation requires creating rules for how a game operates: the way game systems function and/or the actions available to players. One broad class of rules concerns how game events resolve: research has primarily focused on collision rules and movement patterns in 2D arcade-like environments. Togelius and Schmidhuber (2008) evolve game rules as collision tables describing the outcome of collisions between entities in two broad classes in a simple arcade game resembling *Pacman*. Rule sets are evolved to optimize for a learning curve based on a separate learning agent that learns to play the game. Lim and Harrell (2014) similarly evolve rules from the puzzle game grammar, *PuzzleScript* (Lavelle, 2013), evaluating rules by whether games can be completed. A. M. Smith and Mateas (2010) use constraint solving to create rules and environments in the same domain, requiring checks on level solvability rather than optimizing for agent learning patterns. Togelius (2011) takes an alternative, player-driven approach that learns rules in this domain by logging instances of player actions that dictate collision resolution. Past patterns of how players have chosen to manipulate the environment following agent collisions are generalized to rules for all future resolution of those collisions.

A second broad class of rules concerns the actions players may take in a game. Cook, Colton, Raad, and Gow (2013) manipulate code for player actions in a 2D platformer domain using reflection on the source game code. This enables the system to alter the functionality of existing mechanics and create new mechanics based on patterned manipulations of how code alters game parameters. Mechanics are evaluated by simulating agent play to solve levels. Zook and Riedl (2014) generate player actions in the class of fully observable, deterministic, discrete, turn-based games using a combination of constraint solving and planning. Actions are represented as modified planning operators and are generated from a declarative representation of game state and pre-existing game levels. Mechanics are evaluated by using a planner to determine whether pre-authored win states can be reached with mechanics without entering pre-authored loss states while optimizing to designer criteria for mechanic quality.

Mechanic generation efforts are still largely domain-specific and lack a shared understanding of where, when, and how mechanics may be generated. Existing approaches are primarily offline techniques and vary widely in sources of knowledge: from depending heavily on high-level declarative domain descriptions to low-level code manipulation. How can mechanic generation learn from prior sets of generated and tested mechanics to improve? Where does mechanic generation fit in the full generation process? The question of what a mechanic may represent or mean, both to the player and in the context of the theme and message of a game, remains under-studied, particularly when generating from low-level sources of data. When a character touches some food and the food disappears, players build and use mental models of meaning to interpret what has happened (in this case, perhaps eating). Building agents that can predict and understand these associations ahead of time, so that mechanics can be built with specific goals, can contribute greatly to game design practice. Beyond game meaning, there are similar opportunities for cognitive architectures to provide a view into how players learn game rules and how rules influence player capabilities (including learning or serving as a creative medium—for example, user-generated content for *Minecraft*).

Full Games

Full game generation involves assembling all game content from conceptualization through level design and mechanics. To date, few systems have achieved this goal, though many have addressed large subsets of the problem. *ANGELINA*$_5$ (Cook, Colton, & Gow, 2014) generates full 3D maze games starting from a single word or phrase concept. *ANGELINA*$_5$ uses a pipeline starting from a word or phrase theme: it collects associated words or phrases from online sources; uses these to define a game title, font, sound effects, and textures; and uses a set of evolutionary processes to evolve a level design, arrangement of visual and auditory assets in levels, positioning of the player and exit, and set of rules for the movement behaviors and scoring/failure effects of entities. *ANGELINA*$_5$ generates a full 3D game, though it is limited to working on games that involve walking through 3D environments from a start to end position and depends on designer knowledge of valid entity behaviors.

GameForge (Hartsook et al., 2011) generates 2D role-playing games and personalizes the game story and world to player preferences. *GameForge* starts from an input designer-authored story and adapts this story based on player indications of desirable or undesirable story events using automated planning on the story plot. Game world layouts then evolve to fit player preferences for exploration or linear paths and fit designer-authored aesthetics for world environments. The 2D world tile placement, environment decoration, and NPC behavior scripting are then assigned using story and world layout information. *GameForge* generates full 2D role-playing games, though it is limited to 2D role-playing game worlds and depends heavily on input designer knowledge for NPC scripting, story assets, and world aesthetics.

Other efforts have addressed broad subsets of the game generation process. Previously mentioned, A. M. Smith and Mateas (2010) generate game rules and environments for those games using constraint solving. Zook and Riedl (2014) generate mechanics and use their mechanic and game domain representation to generate levels, level progressions, and game control assignments. An earlier version of *ANGELINA* also generates 2D platformer games based on interpretations of popular news media, though addressing a smaller set of aesthetic criteria.

Full game generation represents a grand challenge for intelligent agents in games. Future work will need to address the general lack of sources of knowledge for these games and explore techniques that increase the control systems have over content choices. How can game generation enable systems to invent new genres or game systems? How will cognitive architectures assemble the knowledge of individual content needed for creating full games and integrate these pieces of content to create a full coherent product? What kinds of design and development processes remain to be modeled to capture the rich set of ways people design games? Cognitive architectures that create games can ultimately explore new kinds of games not conceived of by people, act as intelligent mixed-initiative tools to help amateur creators, perform design services, and augment expert design practice.

Future Directions

Summarizing the discussion above, several key challenges emerge for AGI and cognitive architectures in game creation: having relevant knowledge, modeling player experience, and modeling design processes. Agents for game creation to date have suffered from the *knowledge bottleneck*: the inability of authors to create enough content about a domain for a system to have sufficient expertise. Consequently, most existing approaches for creating specific game content have been framed as tools rather than cognitive architectures: these systems lack the deep and broad knowledge of a domain necessary to enable automated expertise, instead serving to support humans that provide that knowledge. Current work has shown the potential to repurpose existing knowledge corpora, but the range of knowledge sources remains largely untapped. Further, little work has gone into methods that gather new knowledge relevant to a task, addressing the general challenge of exploratory learning in ill-defined domains like game design. Proposals to address this challenge include mining design corpora (Snodgrass & Ontañón, 2014), interfacing with existing game development communities (Grace & Maher, 2014), and using crowdsourcing (hiring untrained groups of individuals) to gather task-specific knowledge (Hodhod, Huet, & Riedl, 2014). Extending cognitive architectures to directly address the knowledge bottleneck as part of their functionality will be crucial to creating the next generation of intelligent game creation agents. Developing design formalisms that can be shared across systems (on analogy to systems like ConceptNet) can accelerate progress in these efforts and enable cumulative efforts at understanding the variety of formalisms possible for design.

Player models for game generation have been limited to simple checks for game playability or optimization for fairly rudimentary models of player experience. Using cognitive architectures as a way to provide more in-depth understanding of nuanced player experience can improve how game generation systems evaluate content. Developing these as architectures that can be reused by others stands to further accelerate progress in this domain by providing parameterizable (by player cognitive architecture parameters such as working memory capacity) benchmark models. Employing richer models of cognition can lead to systems that intelligently design games not only for player "fun," but also to enhance player learning of concepts (e.g., in educational games) or better meet player capabilities (through tailoring content to player knowledge or skills). Modeling player declarative and procedural knowledge based on player behavior holds great potential to enhance the design of games and enable new models for automated game creation. Doing so, however, will require dedicated work on general models of how people learn in games in architectures with a wide range of game playing capabilities.

How can cognitive architectures model the *process* of game design itself? Approaches to automated game creation have emphasized pipelines or creation of individual pieces of content, falling far from cognitively plausible models of how people perform game design. Developing cognitive architectures that perform game design stands to revolutionize this area by changing this model to one that captures aspects of the game design process spanning game conceptualization through testing games with players and refining game design goals. Creating these cognitive architectures will require technical advances in how agents create games (e.g., using means-ends analysis as in Soar) as well as studies of the many ways people go about game design to inform the creation of these architectures. Intelligent design process models have the potential to rigorously explore the space of game design in ways that yield deeper, more systemic of the possibilities for game design than currently considered by people.

CONCLUSION

Using AGI and cognitive architectures in games outside virtual characters is a rich field with many open challenges. In player modeling and experience management, cognitive architectures stand to provide more detailed, reusable models of how people play games and are influenced by that play in ways that inform offline content authoring and online content adaptation. As game mechanics, cognitive architectures can unlock a new class of AI-based games that make interaction with an intelligent agent the core to gameplay. Acting as game creators, cognitive architectures have the potential to create more systemic models of game design and open new avenues for gathering and integrating knowledge about game design and development. These application domains all show great promise for future development while benefiting research in cognitive architectures and AGI with systems that cope with the highly knowledge-intensive, complex technical and artistic tasks involved in games.

REFERENCES

Andersen, E., Liu, Y.-E., Snider, R., Szeto, R., Cooper, S., & Popović, Z. (2011). On the harmfulness of secondary game objectives. In *Proceedings of the 6th International Conference on the Foundations of Digital Games*. doi:10.1145/2159365.2159370

Anderson, J. R., Bothell, D., Byrne, M. D., Douglass, S., Lebiere, C., & Qin, Y. (2004). An integrated theory of the mind. *Psychological Review*, *111*(4), 1036–1060. doi:10.1037/0033-295X.111.4.1036 PMID:15482072

Bellemare, M. G., Naddaf, Y., Veness, J., & Bowling, M. (2013). The Arcade Learning Environment: An evaluation platform for general agents. *Journal of Artificial Intelligence Research*, *47*, 253–279.

Blom, P. M., Bakkes, S., Tan, C. T., Whiteson, S., Roijers, D., Valenti, R., & Gevers, T. (2014). Towards personalised gaming via facial expression recognition. In *Proceedings of the 10th AAAI Conference on Artificial Intelligence and Interactive Digital Entertainment*.

Butler, E., Andersen, E., Smith, A. M., Gulwani, S., & Popović, Z. (2015). Automatic game progression design through analysis of solution features. In *Proceedings of the ACM SIGCHI conference on Human Factors in Computing*. doi:10.1145/2702123.2702330

Butler, E., Smith, A. M., Liu, Y.-E., & Popović, Z. (2013). A mixed-initiative tool for designing level progressions in games. In *Proceedings of the ACM Symposium on User Interface Software and Technology*. doi:10.1145/2501988.2502011

Chong, H.-Q., Tan, A.-H., & Ng, G.-W. (2007). Integrated cognitive architectures: A survey. *Artificial Intelligence Review*, *28*(2), 103–130. doi:10.1007/s10462-009-9094-9

Cook, M., & Colton, S. (2014). A Rogue Dream: Automatically generating meaningful content for games. In *Proceedings of the Experimental AI in Games Workshop*.

Cook, M., Colton, S., & Gow, J. (2012). Initial results from co-operative co-evolution for automated platformer design. In EvoGAMES. doi:10.1007/978-3-642-29178-4_20

Cook, M., Colton, S., & Gow, J. (2014). Automating game design in three dimensions. In *AISB Symposium on AI and Games*.

Cook, M., Colton, S., & Pease, A. (2012). Aesthetic considerations for automated platformer design. In *Proceedings of the 8th AAAI Conference on Artificial Intelligence and Interactive Digital Entertainment*.

Cook, M., Colton, S., Raad, A., & Gow, J. (2013). Mechanic Miner: Reflection-driven game mechanic discovery and level design. In EvoGAMES.

Creative Assembly Ltd. (2014). *Alien: Isolation*. Tokyo, Japan: Sega Games Co., Ltd.

Dahlskog, S., & Togelius, J. (2014). Procedural content generation using patterns as objectives. In *Applications of evolutionary computation*. Springer Berlin Heidelberg. doi:10.1007/978-3-662-45523-4_27

Dart, I., & Nelson, M. J. (2012). Smart terrain causality chains for adventure-game puzzle generation. In *Proceedings of the IEEE Conference on Computational Intelligence and Games*. doi:10.1109/CIG.2012.6374173

Domínguez, I. X., & Roberts, D. L. (2014). Asymmetric virtual environments: Exploring the effects of avatar colors on performance. In *Proceedings of the Experimental AI in Games Workshop*.

Dormans, J. (2010). Adventures in level design: Generating missions and spaces for action adventure games. In *Proceedings of the Workshop on Procedural Content Generation in Games*. doi:10.1145/1814256.1814257

Duch, W., Oentaryo, R. J., & Pasquier, M. (2008). Cognitive architectures: Where do we go from here? In *Proceedings of the Artificial General Intelligence Conference*.

Ehrenfeld, S., Schrodt, F., & Butz, M. V. (2015). *Mario Lives! An Adaptive Learning AI Approach for Generating a Living and Conversing Mario Agent* [Video file]. Retrieved from http://www.aaaivideos.org/2015/14_mario_lives/

Evans, R. (2002). AI game programming wisdom. In S. Rabin (Ed.), *chap. Varieties of Learning). Charles River Media.*

Evans, R., & Short, E. (2014). Versu—a simulationist storytelling system. *IEEE Transactions on Computational Intelligence and AI in Games*, *6*(2), 113–130. doi:10.1109/TCIAIG.2013.2287297

Fullerton, T., Swain, C., & Hoffman, S. (2008). *Game design workshop: A playcentric approach to creating innovative games*. Morgan Kaufmann.

Genesereth, M., Love, N., & Pell, B. (2005). General game playing: Overview of the AAAI competition. *AI Magazine*, *26*(2), 62.

Grace, K., & Maher, M. L. (2014). Towards computational co-creation in modding communities. In *Proceedings of the Experimental AI in Games Workshop*.

Guzdial, M., & Riedl, M. O. (2015). Toward Game Level Generation from Gameplay Videos. In *Proceedings of the 2015 Workshop on Procedural Content Generation*.

Hartsook, K., Zook, A., Das, S., & Riedl, M. (2011). Toward supporting storytellers with procedurally generated game worlds. In *Proceedings of the IEEE Conference on Computational Intelligence and Games*.

Hastings, E., Guha, R. K., & Stanley, K. (2009). Automatic content generation in the galactic arms race video game. *IEEE Transactions on Computational Intelligence and AI in Games*, *1*(4), 245–263. doi:10.1109/TCIAIG.2009.2038365

Hodhod, R., Huet, M., & Riedl, M. (2014). Toward generating 3D games with the help of commonsense knowledge and the crowd. In *Proceedings of the Experimental AI in Games Workshop*.

Horswill, I. D. (2014). Game design for classical AI. In *Proceedings of the Experimental AI in Games Workshop*.

Hunicke, R., Leblanc, M., & Zubek, R. (2004). MDA: A formal approach to game design and game research. In *Proceedings of the Workshop on Challenges in Game AI*.

Isbister, K., & Schaffer, N. (2008). *Game usability: Advancing the player experience*. Morgan Kaufmann.

Isla, D. (2005). *Dude: Where's My Warthog? From Pathfinding to General Spatial Competence* [Powerpoint slides]. Retrieved from http://www.aaai.org/Papers/AIIDE/2005/AIIDE05-040.ppt

Laird, J. (2012). *The Soar cognitive architecture*. MIT Press.

Laird, J., & VanLent, M. (2001). Human-level AI's killer application: Interactive computer games. *AI Magazine*, *22*(2), 15.

Laird, J. E. (2001). It knows what you're going to do: adding anticipation to a Quakebot. In *Proceedings of the International Conference on Autonomous Agents*. doi:10.1145/375735.376343

Langley, P., & Choi, D. (2006). A unified cognitive architecture for physical agents. In *Proceedings of the National Conference on Artificial Intelligence*.

Lavelle, S. (2013). *PuzzleScript*. Retrieved from http://www.puzzlescript.net/

Liapis, A., Yannakakis, G. N., & Togelius, J. (2013). Sentient sketchbook: Computer-aided game level authoring. In *Proceedings of the 8th International Conference on the Foundations of Digital Games*.

Lim, C.-U., & Harrell, D. F. (2014). An approach to general videogame evaluation and automatic generation using a description language. In *Proceedings of the IEEE Conference on Computational Intelligence and Games*.

Liu, H., & Singh, P. (2004). ConceptNet—a practical commonsense reasoning tool-kit. *BT Technology Journal*, *22*(4), 211–226. doi:10.1023/B:BTTJ.0000047600.45421.6d

Liu, Y.-E., Mandel, T., Brunskill, E., & Popović, Z. (2014). Trading off scientific knowledge and user learning with multi-armed bandits. In *Proceedings of the International Conference on Educational Data Mining*.

Llano, M. T., Cook, M., Guckelsberger, C., Colton, S., & Hepworth, R. (2014). Towards the automatic generation of fictional ideas for games. In *Proceedings of the Experimental AI in Games Workshop*.

Lomas, D., Patel, K., Forlizzi, J. L., & Koedinger, K. R. (2013). Optimizing challenge in an educational game using large-scale design experiments. In *Proceedings of the ACM SIGCHI Conference on Human Factors in Computing Systems*. doi:10.1145/2470654.2470668

Lopes, R., Tutenel, T., & Bidarra, R. (2012). Using gameplay semantics to procedurally generate player-matching game worlds. In *Proceedings of the Workshop on Procedural Content in Games*. doi:10.1145/2538528.2538531

Love, N., Hinrichs, T., Haley, D., Schkufza, E., & Genesereth, M. (2008). *General game playing: Game description language specification (Tech. Rep.)*. Stanford University.

Mateas, M. (2001). Expressive AI: A hybrid art and science practice. *Leonardo*, *34*(2), 142–153. doi:10.1162/002409401750184717

Mateas, M., & Stern, A. (2002). *Architecture, authorial idioms and early observations of the interactive drama Façade (Tech. Rep.)*. Carnegie Mellon University.

McCoy, J., Treanor, M., Samuel, B., Reed, A., Mateas, M., & Wardrip-Fruin, N. (2014). Social story worlds with Comme il Faut. *IEEE Transactions on Computational Intelligence and AI in Games*, *6*(2), 97–112. doi:10.1109/TCIAIG.2014.2304692

Murphy, T. (2013). *The first level of super mario bros. is easy with lexicographic ordering and time travel ...after that it gets a little tricky*. SIGBOVIK.

Nelson, M. J., & Mateas, M. (2008). An interactive game-design assistant. In *Proceedings of the 13th International Conference on Intelligent User Interfaces*. doi:10.1145/1378773.1378786

O'Reilly, R. C., & Munakata, Y. (2000). *Computational explorations in cognitive neuroscience: Understanding the mind by simulating the brain*. MIT Press.

Orkin, J. (2005). Agent architecture considerations for real-time planning in games. In *Proceedings of the 1st AAAI Conference on Artificial Intelligence and Interactive Digital Entertainment*.

Pedersen, C., Togelius, J., & Yannakakis, G. N. (2009). Modeling player experience in Super Mario Bros. In *Proceedings of the IEEE Symposium on Computational Intelligence and Games*. doi:10.1109/CIG.2009.5286482

Perez, D., Spyridon, Togelius, J., Schaul, T., Lucas, S. M., Couëtoux, A., . . . Thompson, T. (2015). The 2014 general video game playing competition. *IEEE Transactions on Computational Intelligence and AI in Games*.

Pittman, J. (2011, June 16). *The Pac-Man Dossier*. Retrieved from http://home.comcast.net/~jpittman2/pacman/pacmandossier.html

Riedl, M., Stern, A., Dini, D., & Alderman, J. (2008). Dynamic experience management in virtual worlds for entertainment, education, and training. *International Transactions on Systems Science and Applications*, *4*(2), 23–42.

Riedl, M. O., & Bulitko, V. (2013). Interactive narrative: An intelligent systems approach. *AI Magazine*, *34*(1), 67–77.

Risi, S., Lehman, J., D'Ambrosio, D. B., Hall, R., & Stanley, K. O. (2012). Combining search-based procedural content generation and social gaming in the petalz video game. In *Proceedings of the 8th AAAI Conference on Artificial Intelligence and Interactive Digital Entertainment*.

Robertson, J., & Young, R. M. (2014). Gameplay as on-line mediation search. In *Proceedings of the Experimental AI in Games Workshop*.

Salen, K., & Zimmerman, E. (2003). *Rules of play: Game design fundamentals*. Cambridge, MA: MIT Press.

Schaul, T. (2013). A video game description language for model-based or interactive learning. In *Proceedings of the IEEE Conference on Computational Intelligence and Games*. doi:10.1109/CIG.2013.6633610

Seif El-Nasr, M., Drachen, A., & Canossa, A. (Eds.). (2013). *Game analytics*. Springer London. doi:10.1007/978-1-4471-4769-5

Siu, K., Zook, A., & Riedl, M. O. (2014). Collaboration versus competition: Design and evaluation of mechanics for games with a purpose. In *Proceedings of the 9th International Conference on the Foundations of Digital Games*.

Smith, A., Lewis, C., Hullett, K., Smith, G., & Sullivan, A. (2011). An inclusive view of player modeling. In *Proceedings of the 6th International Conference on the Foundations of Digital Games*.

Smith, A. M., Andersen, E., Mateas, M., & Popović, Z. (2012). A case study of expressively constrainable level design automation tools for a puzzle game. In *Proceedings of the 7th International Conference on the Foundations of Digital Games*. doi:10.1145/2282338.2282370

Smith, A. M., Butler, E., & Popović, Z. (2013). Quantifying over play: Constraining undesirable solutions in puzzle design. In *Proceedings of the 8th International Conference on the Foundations of Digital Games*.

Smith, A. M., & Mateas, M. (2010). Variations Forever: Flexibly generating rulesets from a sculptable design space of mini-games. In *Proceedings of the IEEE Conference on Computational Intelligence and Games*. doi:10.1109/ITW.2010.5593343

Smith, G. (2014). The future of procedural content generation in games. In *Proceedings of the Experimental AI in Games Workshop*.

Smith, G., Othenin-Girard, A., Whitehead, J., & Wardrip-Fruin, N. (2012). PCG-based game design: creating Endless Web. In *Proceedings of the 7th International Conference on the Foundations of Digital Games*.

Smith, G., Whitehead, J., & Mateas, M. (2011). Tanagra: Reactive planning and constraint solving for mixed-initiative level design. *IEEE Transactions on Computational Intelligence and AI in Games*, *3*(3), 201–215. doi:10.1109/TCIAIG.2011.2159716

Smith, G., Whitehead, J., Mateas, M., Treanor, M., March, J., & Cha, M. (2011). Launchpad: A rhythm-based level generator for 2-D platformers. *IEEE Transactions on Computational Intelligence and AI in Games*, *3*(1), 1–16. doi:10.1109/TCIAIG.2010.2095855

Snodgrass, S., & Ontañón, S. (2014). A hierarchical approach to generating maps using markov chains. In *Proceedings of the 10th AAAI Conference on Artificial Intelligence and Interactive Digital Entertainment*.

Sorenson, N., Pasquier, P., & DiPaola, S. (2011). A generic approach to challenge modeling for the procedural creation of video game levels. *IEEE Transactions on Computational Intelligence and AI in Games*, *3*(3), 229–244. doi:10.1109/TCIAIG.2011.2161310

Stanley, K. O., Bryant, B. D., & Miikkulainen, R. (2005). Real-time neuroevolution in the NERO video game. *IEEE Transactions on Evolutionary Computation*, *9*(6), 653–668. doi:10.1109/TEVC.2005.856210

Stracuzzi, D. J., Fern, A., Ali, K., Hess, R., Pinto, J., Li, N., & Shapiro, D. G. et al. (2011). An application of transfer to american football: From observation of raw video to control in a simulated environment. *AI Magazine*, *32*(2), 107–125.

Sun, R. (2006). *Cognition and multi-agent interaction: from cognitive modeling to social simulation*. Cambridge Univ Press.

Thompson, T., & Vinciguerra, M. (2014). Integrated pathfinding and player analysis for touch-driven games. In *Proceedings of the Experimental AI in Games Workshop*.

Togelius, J. (2011). A procedural critique of deontological reasoning. In *Proceedings of the Digital Games Research Association Conference*.

Togelius, J., & Schmidhuber, J. (2008). An experiment in automatic game design. In *Proceedings of the IEEE Symposium on Computational Intelligence and Games*.

Togelius, J., Yannakakis, G., Stanley, K., & Browne, C. (2011). Search-based procedural content generation: A taxonomy and survey. *IEEE Transactions on Computational Intelligence and AI in Games*, *3*(3), 172–186. doi:10.1109/TCIAIG.2011.2148116

Treanor, M., Schweizer, B., Bogost, I., & Mateas, M. (2012). The micro-rhetorics of Game-O-Matic. In *Proceedings of the 7th International Conference on the Foundations of Digital Games*.

Treanor, M., Zook, A., Eladhari, M. P., Togelius, J., Smith, G., Cook, M., & Smith, A. M. et al. (2015). AI-Based Game Design Patterns. In *Proceedings of the 10th International Conference on the Foundations of Digital Games*.

Tremblay, J., Borodovski, A., & Verbrugge, C. (2014). I can jump! exploring search algorithms for simulating platformer players. In *Proceedings of the Experimental AI in Games Workshop*.

Tremblay, J., Torres, P. A., & Verbrugge, C. (2014). An algorithmic approach to analyzing combat and stealth games. In *Proceedings of the IEEE Conference on Computational Intelligence and Games*. doi:10.1109/CIG.2014.6932898

van der Linden, R., Lopes, R., & Bidarra, R. (2013). Procedural generation of dungeons. *IEEE Transactions on Computational Intelligence and AI in Games*, *6*(1), 78–89. doi:10.1109/TCIAIG.2013.2290371

Veale, T. (2012). From conceptual "mash-ups" to "bad-ass" blends: A robust computational model of conceptual blending. In *Proceedings of the International Conference on Computational Creativity*.

Yannakakis, G., & Togelius, J. (2011). Experience-driven procedural content generation. *IEEE Transactions on Affective Computing*, *2*(3), 147–161. doi:10.1109/T-AFFC.2011.6

Zook, A., & Riedl, M. O. (2014). Automatic game design via mechanic generation. In *Proceedings of the 28th AAAI Conference on Artificial Intelligence*.

KEY TERMS AND DEFINITIONS

AI-Based Game: A game in which interaction with an artificial intelligence is the core of gameplay.

Experience Management: The real-time, automated modification of a virtual environment to create a desired experience for a user within that environment.

Game Design: The process of constructing a set of game assets and rules to create an interactive artifact (game).

Game Mechanic: A rule in a game describing conditions for transitioning from one game state to another. May be an action based on player input or fixed rule of the game simulation.

Knowledge Bottleneck: The challenge of providing sufficient knowledge to an artificial intelligence to enable intelligent behavior when performing a desired task.

Player Model: A predictive computational model that takes input features describing a game design or setting and outputs features describing expected player objective behavior or subjective experience.

Procedural Content Generation: The automatic creation of game assets or data by software processes, either during the running of the game or during its development.

Conclusion

OVERVIEW

Interfacing cognitive architectures with virtual agents is still in its infancy and remains an open teleological question. However, we hope that this book encourages readers to seriously consider the particular role of cognitive architectures for state-of-the-art virtual agents. The chapters in this book consider the advantages of specific cognitive architectures over others depending on the nature of the virtual agent implementation. Deciding to use a cognitive architecture in the first place is a meaningful decision, and the book's chapters have discussed the utility of doing so.

Cognitive architectures can be used to support arbitrary complex agents for contexts ranging from videogames to training simulations. Currently, advanced cognitive abilities are not necessary for videogames, but their progression may allow novel forms to emerge. We believe that as Artificial General Intelligences improve, they will become more accepted, and cognitive architectures will begin to assume a more prominent and explicit role with the deliberative management of intelligent behaviors.

COGNITION WITHOUT ARCHITECTURES: THE SEMANTIC AMBIGUITY BETWEEN COGNITIVE MODELS AND ARCHITECTURES

Reflecting on the book's theme as a whole, most of the chapter authors focused on cognition that would be represented and embodied within virtual characters. However, the thematic scope was not limited to an analysis of virtual characters. Cognition was also represented in terms of human behavioral templates and character authoring systems.

Virtual characters are sometimes implemented in immersive virtual reality (VR) or mixed (augmented) reality domains. In Chapter 1, Borovikov et al. argued that "the assumption of an *imperfect* human participant is critical to the development of a robust VC cognition system." Essentially, they suggest that the cognition of virtual characters could be expedited if there were human-avatar templates from which to mimic behavioral characteristics, which assists its goal-oriented learning model. Further, a realistic human-template representing a virtual agent would allow humans themselves to project their own expectations for realistic human-like cognition from themselves to the virtual agent they are interacting with. This initial chapter prefers instead to maintain a focus on bypassing the uncanny valley for externally perceived poses and behaviors (e.g. facial, head and eye movements, Borovikov et al., Ch. 1) in order to provide a placebo for cognition rather than to focus on those actual internal states underlying cognitive deliberation. Eventually, these authors should address those cognitive aspects revealed behind the virtual guise of a humanoid form.

These authors have also suggested that virtual agents will one day need to use a human-like connectionist approach involving deep learning in the form of convolutional neural nets (CNNs, Borovikov et al.). Using CNNs, virtual agents can use a large data-set (involving raw pixels and other visual features) to learn, recognize, and store semantically salient and spatially situated visual imagery (e.g. foreground and background) in their short-term and long-term memories. Ultimately, a human-like virtual agent does not need a humanoid avatar form, per se, to employ "similarity measure learning" as a means to learn about feature representations from new domains. Unfortunately, training any deep learning network is a slow and cumbersome process (Borovikov et al.). This neural training latency puts virtual humans at a cognitive disadvantage compared with biological humans who can intuitively process multiple stochastic scenarios in real-time. Consequently, Borovikov et al. feel that priorities should be made towards reducing the set of feature parameters that a virtual human should have to learn and process (Borovikov et al.). This will not be a trivial challenge - even if these parameters have associative and contextual semantic links. The reason why this challenge will remain non-trivial for a while is because these authors also proposed to have the virtual agent account for learning additional sensorial parameters beyond vision to also include the processing of haptic, olfactory, and auditory experiences (Borovikov et al.).

The third chapter by Corradini & Mehta discusses the development of a cognition authoring system – showcasing a behavior recommendation engine – within the popular virtual world, Second Life (ca. 2003-present) called "Second Mind". Second Life is a social virtual world that relies on user-generated content for creation. Corradini & Mehta selected Second Life so that cognition designers can constantly iterate and test the capabilities of their virtual agents in real-time. Using Second Mind, creative avatar users – with little or no programming experience - can manually customize the cognition of virtual agents (aka. NPCs, bots). This customization process allows human controlled avatars to create, represent and model a virtual agent's particular idiosyncratic traits by cognitively associating between "real-world [hardware and software] experiences" and their own episodic experiences (Corradina & Mehta). Ultimately, to extend such an authoring system to that of an architecture – capable of cognitive generality – would require that the human designers possess relevant knowledge as well as the capability to appropriately model both an interactor's experience and their own design processes (Zook, Ch. 10). The video-game domain, for example, has reached a "knowledge bottleneck" where cognition engineers and designers are still struggling to properly "[…] create enough content about a domain for a system to have sufficient expertise" (Zook).

These authors learned from this particular implementation experience with Second Mind that cognition designers must simplify and limit processes that "[...] link behavior constraints to lower level percepts" even when behavioral expressivity might be compromised (Corradini & Mehta). Furthermore, these authors dissuade character cognition designers from deploying "deep hierarchical structures" (Corradini & Mehta). Instead they endorse a sequential approach to authoring cognition (Ibid.). One reason why Second Life in particular, was chosen as the preferred virtual world for designing cognitively enhanced characters was so these designers could continuously observe and resolve any implementation problems that might arise (Corradini & Mehta). Specifically, Second Mind's authoring system -and their reliance on event-driven "triggers" and story-like "timelines" - does not seem to diverge in any significant way from the affordances of Second Life's finite-state architecture already in place for hand-scripted bot authoring. Basically this particular authoring system has been designed with accessibility in mind. Second Mind's purpose is to reduce technical implementation obstacles for non-programmers so that they can design compelling cognitively enhanced characters without having to manually code finite-state scripts.

A third of the authors in this book focused exclusively on issues relating to a virtual character's cognitive believability. Proportionally, this furthers the thematic notion that it is more of a current priority to enhance the externally perceivable "illusion of life" over the underlying functionalist engineering principles that conceptually organizes a virtual agent's internal cognitive states and flow. In other words, action(s) that suspend a human interactor's belief as to whether the virtual agent seems intelligent still takes a precedence over deliberation(s) (Forgette & Katchabaw, p. 1). Ultimately, cognition for now, is easier and more sensible to implement virtually than in actuality. Having said this, virtual cognition still "[…] requires autonomous and flexible behavior that is not defined a-priori" (Forgette & Katchabaw) and this is why alternatives – outside of "narrow problem domains" - should be made to pre-scripting cognitive knowledge. Forgette & Katchabaw suggest that reinforcement learning implementations provide the technical bridge between virtual and actual cognition. Further, these authors believe that in fact, reinforcement learning is a prerequisite for any sufficiently complex cognitive architecture. Under the hood, these authors also suggest that varying decay functions will cement idiosyncratic traits for each unique virtual agent (Forgette & Katchabaw). Idiosyncratic virtual character traits can also be reinforced for voluntarily producing appropriate behaviors - represented as drives - that match that particular character's motivational structure. In fact, these authors go as far as to claim that intrinsic and extrinsic motivations directly influence a character's "[...] perception, cognition, emotion, and behavior" (Forgette & Katchabaw) - even if the behavior is performed as an *"end motive"* for its own teleological sake rather than for strictly instrumental ends (e.g. intrinsic behavior).

This quasi-behaviorist approach brought up by reinforcement learning techniques also applies to the continuing endorsements of affective finite-state machines. Finite-state cognitive architectures – rather than more elaborate architectural representations – remain as the conventional state-of-the art for virtual character implementations. For non-verbal communication, virtual agents are still designed with the goal of expressing archetypal personalities relating to events that are categorized according to modulatory and continuous affective states. These finite-states are mapped to low-level parameters (e.g. movement speed, gesture repetition, Saberi, Ch. 5). Saberi and her colleagues have demonstrated a test-case scenario between a human-user and a humanoid avatar body (inhabited by the virtual agent) that was developed using the proprietary SmartBody software. Smartbody can allow the agent's scripted avatar body to track and respond to locomotion, gaze, facial feature configurations and other physical representations of proxemically situated nonverbal behavior. This scripting is enabled via Behavior Markup Language (BML). Saberi's demonstration allows the human user - wearing a data-glove and tracked via a Microsoft Kinect device - to physically interact with a life-size virtual humanoid agent by playing the childhood game, "rock-paper-scissors" (Saberi). After performing 10 rounds of this game, Saberi's research team claims that the virtual agent will be able to properly respond to nonverbal behavior by predicting the appropriate state transitions based on interactions with the human-user.

Aside from behavioral approaches, another issue that was brought up related to the idea of the degree to which a virtual agent can retain isomorphic cognitive capabilities when migrating across different environments, platforms and domains. For example, Hassani & Lee's chapter discusses a universal architecture for migrating cognitive agents where their aim is to conserve and transfer cognitive features representing "sets of goals, emotional reaction rules, action tendencies, emotional thresholds, decay rates of the emotions" (Hassani & Lee) For Hassani & Lee, virtual agents can act as world-agnostic "migrating companions" that require "[...] a sufficient level of abstraction, modularity, [and a] flexible definition of identity [identities]" (Hassani & Lee). Specifically, these authors concretize this abstraction by discussing three implementation modules - subject to universal migration - that form a virtual agent's cognitive layer. These modules are comprised of a "Mission Manager" (MM), a "Planning-Scheduling" (PS) component, and a "Knowledge Base" (KB). The first module is basically a "goal network" that orders "a set of feasible actions" (Hassani & Lee). Generally speaking, the mission manager module represents the cognition of each virtual character as a deterministic finite-goal machine (rather than simply, a finite-state machine, Hassani & Lee). In addition to a transition function that bridges one goal to the next, the universal migrating virtual agent also evaluates the teleology of each goal. Once a goal has been evaluated and is determined to be salient enough for prioritization, this agent then passes this goal to the next component for planning and scheduling. This module plans goals in a hierarchical manner (i.e. similar to STRIPS AND HTNs) while using a backtracking algorithm and a tree-pruning heuristic (Hassani & Lee). From these plans, the agent can then select valid actions from an action database. Finally, the knowledge base module includes a fuzzy ontology and a search engine that semantically orders and prunes every event from its perception history for retrieval and execution.

There are emerging assumptions that virtual agents can migrate universally across various worlds. If true, then perhaps authors within this book are also anticipating a future time where AGI (artificial general intelligence) can enable these virtual agents with universal intelligence capabilities as well (as Zook discussed in Chapter 10). At present, there does exist cognitive dilutions of universal intelligence (see DeepMind's Q-Learning and MC-AIXI video-game NPC implementations). Some authors have even proposed a non-modular cognitive architecture that claims to address the full capabilities of "realistic" behaviors produced by an Artificial General Intelligence - ordered by probabilities and factor graphs (see Ustun & Rosenbloom, Ch. 8). Sigma, for example, evolved out of Rosenbloom's own innovative development with helping to design SOAR.

Sigma might be based historically on issues raised with designing the SOAR architecture. Sigma, however, is a unique offshoot of SOAR. Unlike a conventional instantiation of SOAR, the Sigma cognitive architecture orders symbolic, discrete and continuous predicates as well as conditionals (relating to rule-systems and probabilistic networks) according to factor graphs and sub-graphs (e.g. a sub-graph reserved for working memory). These graphs have the capability for long-term memory storage even though they are not having each cognitive category sorted into self-contained memory modules, as such. Further, Sigma deals with what are known as "condacts". Condacts are "predicate patterns that support the bidirectional processing that is key to probabilistic reasoning, partial matching, constraint satisfaction and signal processing" (Ustun & Rosenbloom). Sigma's virtual agents are similar to SOAR-agents in that they utilize a decision cycle. In the case of Sigma, however, these agents base their decision cycle on predicates that are addressed a "summary product algorithm" (hence the architecture's title, Ustun & Rosenbloom). This decision-selection algorithm on input messages (perhaps represented as a continuous percept stream). Sigma's agents focus on the relationship between perception and action primarily on a sub-symbolic level. As far as these authors are concerned, the level of higher-cognition - ranging from reactivity through deliberation to full-blown reflection - is contingent on the number of cognitive cycles being processed (Ustun & Rosenbloom).

Overall, the near-future possibility of interacting directly with AGI characters imbued with genuine universal intelligence remains an open question – at least as far as these authors are concerned. Beyond characters themselves, AGI has an additional extra-architectural role by extending "[...] the capabilities of intelligent systems to model games and human cognition to extend the bounds of game experiences" (Zook).

ASSESSING THE UTILITY OF COGNITIVE ARCHITECTURES

Cognitive architectures realize, organize and structure the various cognitive processes that can be used for the implementation of the intelligent part a virtual agent's "mind". Chapter authors in the present book have ultimately made the case for and against the teleology of state-of-the-art cognitive architectures. However, there is still no guarantee that cognitive architectures will always reliably serve as a robust alternative to the mere implementation of narrow algorithms and heuristics for externally perceivable outcomes. Consequently, some authors have pondered the overall utility towards realizing cognitive architectures for virtual characters in the near and far future.

The fundamental characteristic of cognitive architectures is their domain generality (Sun, 2004, p. 341). However, this fundamental characteristic is not always viewed as an asset when applied to virtual character implementations. Zook i.e. criticizes cognitive architectures for lacking in-depth knowledge of a particular untapped domain of expertise (Zook). In a similar vein, Carstensdóttir and colleagues, for example, claimed that established cognitive architectures are too generic and over-simplified to properly implement a temporal synchronization of believable virtual agent action sequences (E. Carstensdóttir, personal correspondence, January, 2016). In fact, they say that "[...] most cognitive architectures (SOAR, CLARION, ACT-R, etc.) have focused more on emulating the internal cognitive processes, [consequently] leaving out details on manifestation at the behavior level" (Ibid.). It is clear that not all virtual agents should require the partial or full implementation of cognitive architectures in every context that requires complex deliberations and actions. For more rudimentary cognitive implementations, most virtual agents function appropriately without the need to be more than mere finite-state machines augmented by hierarchical planners. Corroborating this view, in domains related to consumer entertainment, there is a compelling case to be made that "[...] most existing approaches for creating specific game content have been framed as tools rather than cognitive architectures" (Zook). Cognitive architectures might simply augment rather than supplant preexisting industry-approved heuristics for intelligent behavior. For example, Zook's chapter concludes by suggesting that "cognitive architectures can unlock a new class of AI-based games that make interaction with an intelligent agent the core to gameplay" (Zook). On the other extreme where cerebrally complex cognitive interactions are essential, it is even possible that some diluted forms of Universal AGI will render many – if not all – cognitive architectures obsolete. In the meantime, some progress has been made to show that AGI architectures are beginning to emerge out of symbolic/sub-symbolic hybrids which increasingly seem to be an "[...] important subset of the capabilities that virtual humans should support" (Ustun & Rosenbloom).

CHALLENGES WITH INTEGRATING ESTABLISHED ARCHITECTURES

Arguably, some cognitive architectures have proven to be quite useful and even seem uniquely suited for domain specific implementations. As an example, the most canonical cognitive architecture, ACT-R, has been optimized the cognitive implementation of first-person shooter video game non-player characters, (Smart et al., Ch. 2).

On a technical level, ACT-R is probably the most optimized architecture for 3D video game environments despite containing some implementation challenges when being interfaced with commercially available engines such as Unity or Unreal (Smart et al., Ch. 2). Despite this implementation advantage, ACT-R still consumes a large amount of computational resources for any multi-agent system to properly function in a real-time gaming environment (Smart et al., Ch. 2).

These authors address some key integration challenges as well as propose solutions to these challenges faced when interfacing ACT-R and the Unity platform. For example, they needed to address the issue of dealing with two large system based on entirely different programming languages; C++/C# for Unity and the classical AI language, Lisp for ACT-R. Naturally, this posed a fundamental implementation hurdle when trying to embed ACT-R within the Unity platform.

Another issue that had to be addressed related to the available computational resources and processing power: To date, the graphics and physics rendering capabilities of contemporary computer are not yet optimal for real-time cognitive processing - especially for properly handling multi-agent simulations (Smart et al., Ch. 2).

To be fair, ACT-R – an academic architecture – was historically never intended for integration with real-time gaming engines of the kind we take for granted today. However, these authors see this integration challenge as an opportunity rather than as an insurmountable technical hurdle. One proposed solution to this computational bottleneck was to have ACT-R and Unity run semi-autonomously while communicating using standard network protocols and messaging systems across different machines (Smart et al.). Technical hurdles aside, both ACT-R and SOAR can generate relatively quick and robust behavioral responses to cognitive deliberations and can alleviate the agent's taxing computational resources when performing animated movements (Smart et al., Ch. 2; Turner, Ch. 7). This is an important ability since Intelligent virtual agents are usually embodied in stochastic virtual environments and therefore, require "high-level behavioral controllers to be able to adapt to the failures at lower levels of the system" (Hassani & Lee).

Virtual agents in entertainment domains often employ a quick decision cycle for real-time implementations of cognitively-informed action sequences. ACT-R was one of the first architectures to consider a decision cycle but SOAR made this decision cycle its conceptual focus. In some domains, it might be more preferable for a virtual agent to spend some time and computational resources logically deliberating over an action before acting upon it (if an explicit action is even required at all). SOAR's trademark decision cycle feature and its capacity to choose and prefer operators does suggest an attempt at careful logical deliberation but does this make SOAR the most logically expressive architecture? With its communication between explicit and implicit processes, CLARION has been championed as the most logically expressive of all the canonical cognitive architectures discussed in this book (Bringsjord et al., Ch. 6). For virtual agents deployed in non-ludic virtual environments, CLARION seems "[…] better suited to emergent narrative approaches in which agents are given initial personalities and behavioral suites and are thereafter allowed to interact with other agents (and the player) without supervision from a 'drama manager' or similar component" (Lynch et al., 2011, p. 460).

Though CLARION might be more amenable to virtual environments with emergent narratives but Bringsjord et al. identify some technical limitations within CLARION itself where those chunks compared by CLARION need to be more cognitively distinguished than they are at present. A focus on Cognitively Distinguished Chunks (CDCs), neurobiologically based on adult human reasoning, might require further iterated versions of CLARION and/or the development of a brand new cognitive architecture devoted to non-ludic virtual agents (Bringsjord et al.). Generally speaking, there is still an engineering opportunity for cognitive architectures to transcend the decision-trees generated during interaction by instead, capturing a higher-order logic that can adequately "[…] carry out inferences or produce new structures, if and only if the formal semantics of the logic allows them" (Bringsjord et al.).

META AND HYBRID COGNITIVE ARCHITECTURES

Ideally, virtual agents are not limited by computational and time resources. Under such unconstrained conditions, virtual agents will eventually be able to use a meta-architecture – such as GOLEM (Goertzel, 2014) – to select from a corpus of A(G)I heuristics and algorithms as well as distinctive cognitive features common to various established architectures for remixing and customization. In this way, virtual agents can migrate across cognitive architectures in addition to world-domains (Hassani & Lee). For the present time, it is likely more practical that canonical architectures such as SOAR focus on enhancing its cognitive capabilities to include ontologies, connectionist representations and additional memory banks etc. (Turner, Ch. 7). If anything, cognitive architects should work towards a unified universal architecture (and not just a meta-architecture such as GOLEM) that handles implicit and explicit as well as symbolic and sub-symbolic processing.

From the perspective of personality modeling, some authors have endorsed the idea that future cognitive architectures might wish to more explicitly represent numerically timed affective finite-states – custom tailored to each virtual agent's character traits. However, this particular advocacy seems to reflect more of the state-of-the-art implementations that deal with virtual characters as embodied finite-state machines than as agents with a deeper sense of cerebral deliberation or mathematical rigour (as Bringsjord et al. argue in Chapter 6). To advocate that cognitive architectures merely reflect the explicit engineering of timed finite-states is hardly a speculative endeavor.

Overall, many of the authors in this book seem content with finite-state machine implementations and rudimentary cognitive models for virtual agents. In those cases where cognitive architectures are endorsed at all, this book's majority consensus appears to be in favour of advocating the deployment of more universal hybrid architectures – similar to CLARION – that combines implicit/explicit processes with classical (symbolic logic centered) and connectionist approaches in the services of embedded and emergent interoperable (i.e. domain and platform agnostic) virtual environments.

Jeremy Owen Turner
Simon Fraser University, Canada

Michael Nixon
Simon Fraser University, Canada

Ulysses Bernardet
Simon Fraser University, Canada

Steve DiPaola
Simon Fraser University, Canada

REFERENCES

Anderson, J. R. (1996). ACT: A simple theory of complex cognition. *The American Psychologist*, *51*(4), 355–365. doi:10.1037/0003-066X.51.4.355

Bernardet, U., & Verschure, P. F. M. J. (2010). iqr: A tool for the construction of multi-level simulations of brain and behaviour. *Neuroinformatics*, *8*(2), 113–134. doi:10.1007/s12021-010-9069-7 PMID:20502987

Carstensdóttir,E., Nguyen, T., El-Nasr, M. "Believable Personality Model: Non-verbal Behavior for Warmth and Competence". Proposal draft, submitted July 11, 2015.

Goertzel, B. (2014). GOLEM: Towards an AGI meta-architecture enabling both goal preservation and radical self-improvement. *Journal of Experimental & Theoretical Artificial Intelligence*, *26*(3), 391–403. doi:10.1080/0952813X.2014.895107

Goertzel, B., & Wang, P. (2007). Advances in Artificial General Intelligence: Concepts, Architectures and Algorithms: *Proceedings of the AGI Workshop 2006* (Vol. 157). IOS Press.

Laird, J. (2012). *The Soar cognitive architecture*. MIT Press.

Laird, J. E. (2008). Extending the Soar cognitive architecture. *Frontiers in Artificial Intelligence and Applications*, *171*, 224.

Laird, J. E., Newell, A., & Rosenbloom, P. S. (1987). Soar: An architecture for general intelligence. *Artificial Intelligence*, *33*(1), 1–64. doi:10.1016/0004-3702(87)90050-6

Lynch, M. F., Sun, R., & Wilson, N. (2011, January). CLARION as a cognitive framework for intelligent virtual agents. In *Intelligent Virtual Agents* (pp. 460–461). Springer Berlin Heidelberg. doi:10.1007/978-3-642-23974-8_62

Magerko, B., Laird, J., Assanie, M., Kerfoot, A., & Stokes, D. (2004). AI characters and directors for interactive computer games. In Ann Arbor, 1001(48), 109-2110.

Nareyek, A. (2004). AI in Computer Games. In Queue 1, 10 (February 2004), 58-65. DOI= doi:10.1145/971564.971593

Nguyen, T. H. D., Carstensdottir, E., Ngo, N., El-Nasr, M. S., Gray, M., Isaacowitz, D., & Desteno, D. (2015, August). Modeling Warmth and Competence in Virtual Characters. In Intelligent Virtual Agents (pp. 167-180). Springer International Publishing. doi:10.1007/978-3-319-21996-7_18

Ranathunga, S., Cranefield, S., & Purvis, M. 2011. Interfacing a Cognitive Agent Platform with a Virtual World: A Case Study using Second Life (Extended Abstract). In Proceedings of the 10th International Conference on Autonomous Agents and Multiagent Systems (AAMAS 2011). Turner, Yolum, Sonenberg and Stone (eds.). May 2-6, 2011, Taipei, Taiwan, pp. 1181-1182.

Russell, S., & Norvig, P. (2010). Artificial Intelligence: A Modern Approach. Third Edition. Chps: Intro, 3, 12, 7-17. Prentice Hall.

Sun, R. (2004). Desiderata for cognitive architectures. *Philosophical Psychology*, *17*(3), 341–373. doi:10.1080/0951508042000286721

Sun, R. (2006). The CLARION cognitive architecture: Extending cognitive modeling to social simulation. *Cognition and multi-agent interaction*, 79-99.

Compilation of References

Acton, G. (2009). *Playing the Role: Towards an Action Selection Architecture for Believable Behaviour in Non Player Characters and Interactive Agents.* (Masters Thesis). Department of Computer Science, The University of Western Ontario.

Alami, R., Chatila, R., Fleury, S., Ghallab, M., & Ingrand, F. (1998). An Architecture for Autonomy. *The International Journal of Robotics Research, 17*(4), 315–337. doi:10.1177/027836499801700402

Albus, J. S. (2000). 4-D/RCS reference model architecture for unmanned ground vehicles. In *Proceedings of the 2000 IEEE International Conference on Robotics and Automation.* San Francisco, CA: IEEE. doi:10.1109/ROBOT.2000.845165

Andersen, E., Liu, Y.-E., Snider, R., Szeto, R., Cooper, S., & Popović, Z. (2011). On the harmfulness of secondary game objectives. In *Proceedings of the 6th International Conference on the Foundations of Digital Games.* doi:10.1145/2159365.2159370

Anderson, J. (1976). Formal semantics of ACT representations. In Language, Memory, and Thought (pp. 220-251). Hillsdale, NJ: Psychology Press.

Anderson, J. R. (1993). *Rules of mind.* Hillsdale, NJ: Lawrence Erlbaum Associates.

Anderson, J. R. (1996). ACT: A simple theory of complex cognition. *The American Psychologist, 51*(4), 355–365. doi:10.1037/0003-066X.51.4.355

Anderson, J. R. (2007). *How Can the Human Mind Occur in the Physical Universe?* Oxford, UK: Oxford University Press. doi:10.1093/acprof:oso/9780195324259.001.0001

Anderson, J. R., Albert, M. V., & Fincham, J. M. (2005). Tracing problem solving in real time: fMRI analysis of the subject-paced Tower of Hanoi. *Journal of Cognitive Neuroscience, 17*(8), 1261–1274. doi:10.1162/0898929055002427 PMID:16197682

Anderson, J. R., Bothell, D., Byrne, M. D., Douglass, S., Lebiere, C., & Qin, Y. (2004). An integrated theory of the mind. *Psychological Review, 111*(4), 1036–1060. doi:10.1037/0033-295X.111.4.1036 PMID:15482072

Anderson, J. R., & Lebiere, C. (1998). *The atomic components of thought.* Mahwah, NJ: Lawrence Erlbaum Associates.

Anderson, J. R., Qin, Y., Jung, K.-J., & Carter, C. S. (2007). Information-processing modules and their relative modality specificity. *Cognitive Psychology*, *54*(3), 185–217. doi:10.1016/j.cogpsych.2006.06.003 PMID:16919255

Anderson, J., & Lebiere, C. (2003). The Newell Test for a theory of cognition. *Behavioral and Brain Sciences*, *26*(05), 587–640. doi:10.1017/S0140525X0300013X PMID:15179936

André, E., Klesen, M., Gebhard, P., Allen, S., & Rist, T. (2000). Integrating models of personality and emotions into lifelike characters. In *Affective interactions* (pp. 150–165). Springer Berlin Heidelberg. doi:10.1007/10720296_11

Andrews, P. (2002). *An introduction to mathematical logic and type theory: To truth through proof*. New York, NY: Springer. doi:10.1007/978-94-015-9934-4

Anonymous. (n.d.). *CLARION Cognitive Architecture*. Retrieved February 27, 2015, from https://sites.google.com/site/clarioncognitivearchitecture

Anonymous. (n.d.). *Kinect for Windows*. Retrieved March 15, 2015, from http://www.microsoft.com/en-us/kinectforwindows/

Anonymous. (n.d.). *LSL access portal*. Retrieved March 16, 2015, from http://wiki.secondlife.com/wiki/LSL_Portal

Anonymous. (n.d.). *Second Life. LSL access portal*. Retrieved March 16, 2015, from http://secondlife.com

Anonymous. (n.d.). *Soar Home Page*. Retrieved February 27, 2015, from http://Soar.eecs.umich.edu/

Anonymous. (n.d.). *SoarTech Corporate Brochure*. Retrieved February 28, 2015, from http://ai.eecs.umich.edu/Soar/sitemaker/docs/misc/SoarRBSComparison.pdf

Anonymous. (n.d.). *SoarTech* webpage. Retrieved March 15, 2015, from http://www.Soartech.com/about/

Arent, K., & Kreczmer, B. (2013). Identity of a companion, migrating between robots without common communication modalities: Initial results of VHRI study. In *2013 18th International Conference on Methods and Models in Automation and Robotics (MMAR)* (pp. 109–114). http://doi.org/ doi:<ALIGNMENT.qj></ALIGNMENT>10.1109/MMAR.2013.6669890

Aristotle, , Thomson, J. A. K., & Tredennick, H. (1976). *The Ethics of Aristotle: The Nicomachean Ethics*. Penguin.

Arrabales, R., Ledezma, A., & Sanchis, A. (2009). *Towards conscious-like behavior in computer game characters.IEEE Symposium on Computational Intelligence and Games*, Milan, Italy. doi:10.1109/CIG.2009.5286473

Aylett, R., Kriegel, M., Wallace, I., Marquez Segura, E., Mecurio, J., Nylander, S., & Vargas, P. (2013). *Do I remember you? Memory and identity in multiple embodiments*. IEEE RO-MAN; doi:10.1109/ROMAN.2013.6628435

Bailey, C., & Katchabaw, M. (2008, November). An emergent framework for realistic psychosocial behaviour in non player characters. In *Proceedings of the 2008 Conference on Future Play: Research, Play, Share.* doi:10.1145/1496984.1496988

Bailey, C., You, J., Acton, G., Rankin, A., & Katchabaw, M. (2012, December). Believability through psychosocial behaviour: Creating bots that are more engaging and entertaining. In P. Hingston (Ed.), *Believable Bots: Can Computers Play Like People?* Springer.

Barrera, A., & Weitzenfeld, A. (2007). *Bio-inspired model of robot spatial cognition: Topological place recognition and target learning.International Symposium on Computational Intelligence in Robotics and Automation*, Jacksonville, FL. doi:10.1109/CIRA.2007.382839

Barto, A., Singh, S., & Chentanez, N. (2004, October). Intrinsically motivated learning of hierarchical collections of skills. In *Proceedings of the 3rd International Conference on Developmental Learning.*

Barwise, J., & Etchemendy, J. (1999). *Language, proof, and logic.* New York, NY: Seven Bridges.

Becker-Asano, C., & Wachsmuth, I. (2009). Affective computing with primary and secondary emotions in a virtual human. *Autonomous Agents and Multi-Agent Systems, 20*(1), 32–49. doi:10.1007/s10458-009-9094-9

Bellemare, M. G., Naddaf, Y., Veness, J., & Bowling, M. (2013). The Arcade Learning Environment: An evaluation platform for general agents. *Journal of Artificial Intelligence Research, 47*, 253–279.

Bernardet, U., & Verschure, P. F. M. J. (2010). iqr: A tool for the construction of multi-level simulations of brain and behaviour. *Neuroinformatics, 8*(2), 113–134. doi:10.1007/s12021-010-9069-7 PMID:20502987

Best, B. J., & Lebiere, C. (2006). Cognitive agents interacting in real and virtual worlds. In R. Sun (Ed.), Cognition and Multi-Agent Interaction: From Cognitive Modeling to Social Interaction. New York: Cambridge University Press, 2006.

Best, B. J., & Lebiere, C. (2006). Cognitive Agents Interacting in Real and Virtual Worlds. In R. Sun (Ed.), *Cognition and Multi-Agent Interaction: From Cognitive Modeling to Social Interaction.* New York: Cambridge University Press.

Bevacqua, E., de Sevin, E., Pelachaud, C., McRorie, M., & Sneddon, I. (2010, March). Building credible agents: Behaviour influenced by personality and emotional traits. In *Proceedings of International Conference on Kansei Engineering and Emotion Research.*

Blackburn, P., & Bos, J. (2005). *Representation and inference for natural language: A first course in computational semantics.* Stanford, CA: CSLI.

Blom, P. M., Bakkes, S., Tan, C. T., Whiteson, S., Roijers, D., Valenti, R., & Gevers, T. (2014). Towards personalised gaming via facial expression recognition. In *Proceedings of the 10th AAAI Conference on Artificial Intelligence and Interactive Digital Entertainment.*

Bordini, R. H., Hübner, J. F., & Wooldridge, M. (2007). *Programming Multi-Agent Systems in AgentSpeak using Jason*. Chichester, UK: Wiley-Interscience.

Borkenau, P., & Liebler, A. (1992). Trait inferences: Sources of validity at zero acquaintance. *Journal of Personality and Social Psychology*, *62*(4), 645–657. doi:10.1037/0022-3514.62.4.645

Borkenau, P., Mauer, N., Riemann, R., Spinath, F. M., & Angleitner, A. (2004). Thin Slices of Behaviour as Cues of Personality and Intelligence. *Journal of Personality and Social Psychology*, *86*(4), 599–614. doi:10.1037/0022-3514.86.4.599 PMID:15053708

Bourg, D. M., & Seaman, G. (2004). *AI for Game Developers*. O'Reilly Media.

Bratman, M. E., Israel, D. J., & Pollack, M. E. (1988). Plans and resource-bounded practical reasoning. *Computational Intelligence*, *4*(3), 349–355. doi:10.1111/j.1467-8640.1988.tb00284.x

Bringsjord, S. (2001). Is it possible to build dramatically compelling interactive digital entertainment (in the form, e.g., of computer games)? *Game Studies, 1*(1). http://www.gamestudies.org

Bringsjord, S. (1992). CINEWRITE: An algorithm-sketch for writing novels cinematically and two mysteries therein. In M. Sharples (Ed.), *Computers and writing: State of the art. Kluwer*. Dordrecht, The Netherlands: Springer. doi:10.1007/978-94-011-2674-8_12

Bringsjord, S. (1995a). Could, how could we tell if, and why should—androids have inner lives? In K. Ford, C. Glymour, & P. Hayes (Eds.), *Android Epistemology* (pp. 93–122). Cambridge, MA: MIT Press.

Bringsjord, S. (1995b). Pourquoi Hendrik Ibsen Est-II une menace pour la littérature générée par ordinateur? In A. Vuillemin (Ed.), *Littérature et informatique la littérature générée par orinateur*. Arras, France: Artois Presses Universite.

Bringsjord, S. (2008). Declarative/logic-based cognitive modeling. In R. Sun (Ed.), *The handbook of computational psychology* (pp. 127–169). Cambridge, UK: Cambridge University Press. doi:10.1017/CBO9780511816772.008

Bringsjord, S., & Bringsjord, A. (2009). Synthetic worlds and characters and the future of creative writing. In C. A. P. Smith, K. Kisiel, & J. Morrison (Eds.), *Working through synthetic worlds* (pp. 235–255). Surrey, UK: Ashgate.

Bringsjord, S., & Ferrucci, D. (2000). *Artificial intelligence and literary creativity: Inside the mind of Brutus, a storytelling machine*. Mahwah, NJ: Lawrence Erlbaum Associates.

Bringsjord, S., Govindarajulu, N., Ellis, S., McCarty, E., & Licato, J. (2014). Nuclear deterrence and the logic of deliberative mindreading. *Cognitive Systems Research*, *28*, 20–43. doi:10.1016/j.cogsys.2013.08.001

Bringsjord, S., Khemlani, S., Arkoudas, K., McEvoy, C., Destefano, M., & Daigle, M. (2005) *Advanced synthetic characters, evil, and E.6th Annual European Game-On Conference*, Leicester, UK.

Bringsjord, S., Khemlani, S., Arkoudas, K., McEvoy, C., Destefano, M., & Daigle, M. (2005). Advanced synthetic characters, evil, and E. *Game-On, 6*, 31–39.

W. P. Brinkman, J. Broekens, & D. Heylen (Eds.). (2015). Intelligent Virtual Agents. In *15th International Conference, IVA 2015*, (Vol. 9238). Springer.

Brooks, R. (1991). Intelligence Without Representation. *Artificial Intelligence, 47*(1-3), 139–159. doi:10.1016/0004-3702(91)90053-M

Brooks, R. A. (1986). A robust layered control system for a mobile robot. *Robotics and Automation. IEEE Journal of, 2*(1), 14–23.

Butler, E., Andersen, E., Smith, A. M., Gulwani, S., & Popović, Z. (2015). Automatic game progression design through analysis of solution features. In *Proceedings of the ACM SIGCHI conference on Human Factors in Computing*. doi:10.1145/2702123.2702330

Butler, E., Smith, A. M., Liu, Y.-E., & Popović, Z. (2013). A mixed-initiative tool for designing level progressions in games. In *Proceedings of the ACM Symposium on User Interface Software and Technology*. doi:10.1145/2501988.2502011

Campbell, A., Rushton, J. (1978). Bodily communication and personality. *The British Journal of Social and Clinical Psychology, 17*(1), 31-36.

Campbell, J. (2011). Developing INOTS to Support Interpersonal Skills Practice. In *Proceedings of the Thirty-second Annual IEEE Aerospace Conference (IEEEAC)*. Washington, DC: IEEE doi:10.1109/AERO.2011.5747535

Carney, D. R., Colvin, C. R., & Hall, J. A. (2007). A thin slice perspective on the accuracy of first impressions. *Journal of Research in Personality, 41*(5), 1054–1072. doi:10.1016/j.jrp.2007.01.004

Carstensdóttir,E., Nguyen, T., El-Nasr, M. "Believable Personality Model: Non-verbal Behavior for Warmth and Competence". Proposal draft, submitted July 11, 2015.

Carver, C. S., & White, T. L. (1994). Behavioural inhibition, behavioural acti-vation, and affective responses to impending reward and punishment: The BIS/BAS scales. *Journal of Personality and Social Psychology, 67*(2), 319–333. doi:10.1037/0022-3514.67.2.319

Cassimatis, N. (2010). An architecture for adaptive algorithmic hybrids. *Systems, Man, and Cybernetics, Part B: Cybernetics. IEEE Transactions on, 40*(3), 903–914.

Cassimatis, N. L. (2002). *Polyscheme: A Cognitive Architecture for Integrating Multiple Representation and Inference Schemes*. Massachusetts Institute of Technology.

Charniak, E., & McDermott, D. (1985). *Introduction to artificial intelligence*. Reading, MA: Addison- Wesley.

Chen, J., Demski, A., Han, T., Morency, L. P., Pynadath, D. V., Rafidi, N., & Rosenbloom, P. S. (2011). Fusing Symbolic and Decision-Theoretic Problem Solving+ Perception in a Graphical Cognitive Architecture. In Biologically Inspired Cognitive Architectures.

Chentanez, N., Barto, A., & Singh, S. (2004, December). Intrinsically motivated reinforcement learning. In *Proceedings of the Eighteenth Annual Conference on Neural Information Processing Systems.*

Chi, M. T. H., Glaser, R., & Farr, M. J. (Eds.). (1988). *The Nature of Expertise.* Hillsdale, NJ: Erlbaum.

Chong, H.-Q., Tan, A.-H., & Ng, G.-W. (2009). Integrated cognitive architectures: A survey. *Artificial Intelligence Review, 28*(2), 103–130. doi:10.1007/s10462-009-9094-9

Christel, M. G., Stevens, S. M., Maher, B. S., Brice, S., Champer, M., Jayapalan, L., & Bastida, N. et al. (2012) *RumbleBlocks: Teaching science concepts to young children through a Unity game.17th International Conference on Computer Games*, Louisville, KY. doi:10.1109/CGames.2012.6314570

Clark, A. (2008). *Supersizing the Mind: Embodiment, Action, and Cognitive Extension.* New York: Oxford University Press. doi:10.1093/acprof:oso/9780195333213.001.0001

Cochran, R., Lee, F., & Chown, E. (2006). Modeling emotion: Arousal's impact on memory. *Proceedings of the 28th Annual Conference of the Cognitive Science Society* (pp. 1133–1138).

Cohen, L. H. (2014). Suddenly that summer. *The New York Times Sunday Book Review,* p. BR15

Conde, T., Tambellini, W., & Thalmann, D. (2003). Behavioral Animation of Autonomous Virtual Agents Helped by Reinforcement Learning. In T. Rist, R. S. Aylett, D. Ballin, & J. Rickel (Eds.), *Intelligent Virtual Agents* (pp. 175–180). Springer Berlin Heidelberg. doi:10.1007/978-3-540-39396-2_28

Cook, M., Colton, S., & Gow, J. (2012). Initial results from co-operative co-evolution for automated platformer design. In EvoGAMES. doi:10.1007/978-3-642-29178-4_20

Cook, M., Colton, S., Raad, A., & Gow, J. (2013). Mechanic Miner: Reflection-driven game mechanic discovery and level design. In EvoGAMES.

Cook, M., & Colton, S. (2014). A Rogue Dream: Automatically generating meaningful content for games. In *Proceedings of the Experimental AI in Games Workshop.*

Cook, M., Colton, S., & Gow, J. (2014). Automating game design in three dimensions. In *AISB Symposium on AI and Games.*

Cook, M., Colton, S., & Pease, A. (2012). Aesthetic considerations for automated platformer design. In *Proceedings of the 8th AAAI Conference on Artificial Intelligence and Interactive Digital Entertainment.*

Cook, M., & Smith, J. M. (1975). The role of gaze in impression formation. *The British Journal of Social and Clinical Psychology, 14*(1), 19–25. doi:10.1111/j.2044-8260.1975.tb00144.x PMID:1122344

Costa, P. T. Jr, & McCrae, R. R. (1992). Four ways why five factors are basic. *Personality and Individual Differences*, *13*(6), 653–665. doi:10.1016/0191-8869(92)90236-I

Cox, M. T., & Raja, A. (2011). Metareasoning: An Introduction. In M. T. Cox & A. Rajam (Eds.), *Metareasoning: Thinking About Thinking* (pp. 3–14). Cambridge, MA: MIT Press. doi:10.7551/mitpress/9780262014809.003.0001

Crawford, C. (2003). *The Art of Interactive Design: A euphonius and illuminating guide to building successful software* (pp. 77–90). San Francisco: No Starch Press.

Creative Assembly Ltd. (2014). *Alien: Isolation*. Tokyo, Japan: Sega Games Co., Ltd.

Dahlskog, S., & Togelius, J. (2014). Procedural content generation using patterns as objectives. In *Applications of evolutionary computation*. Springer Berlin Heidelberg. doi:10.1007/978-3-662-45523-4_27

Damasio, A. (1994). *Descartes' error: Emotion, reason, and the human brain*. New York: Putnam.

Danks, D. (2014). *Unifying the mind: Cognitive representations as graphical models*. MIT Press.

Dart, I., & Nelson, M. J. (2012). Smart terrain causality chains for adventure-game puzzle generation. In *Proceedings of the IEEE Conference on Computational Intelligence and Games*. doi:10.1109/CIG.2012.6374173

Dastani, M. (2008). 2APL: A practical agent programming language. *Autonomous Agents and Multi-Agent Systems*, *16*(3), 214–248. doi:10.1007/s10458-008-9036-y

DeVault, D. (2014). SimSensei Kiosk: A virtual human interviewer for healthcare decision support. In *Proceedings of the 2014 International Conference on Autonomous Agents and Multi-agent systems*.

Dias, J., Mascarenhas, S., & Paiva, A. (2011). FAtiMA Modular: Towards an agent architecture with a generic appraisal framework.*Proceedings of the International Workshop on Standards for Emotion Modeling*.

Doce, T., Dias, J., Prada, R., & Paiva, A. (2010, September). Creating individual agents through personality traits. In *Proceedings of the 10th International Conference on Intelligent Virtual Agents*. doi:10.1007/978-3-642-15892-6_27

Doirado, E., & Martinho, C. (2010, May). I mean it! Detecting user intentions to create believable behaviour for virtual agents in games. In *Proceedings of the 9th International Conference on Autonomous Agents and Multiagent Systems*.

Domínguez, I. X., & Roberts, D. L. (2014). Asymmetric virtual environments: Exploring the effects of avatar colors on performance. In *Proceedings of the Experimental AI in Games Workshop*.

Dormans, J. (2010). Adventures in level design: Generating missions and spaces for action adventure games. In *Proceedings of the Workshop on Procedural Content Generation in Games*. doi:10.1145/1814256.1814257

Droney, J. M., & Brooks, C. I. (1993). Attributions of Self-Esteem as a Function of Duration of Eye Contact. *The Journal of Social Psychology*, *133*(5), 715–722. doi:10.1080/00224545.1993.9713927 PMID:8283864

Duch, W., Oentaryo, R. J., & Pasquier, M. (2008). Cognitive architectures: Where do we go from here? In *Proceedings of the Artificial General Intelligence Conference.*

Eaves, L. J., Eysenck, H. J., & Martin, N. G. (1989). *Genes, culture, and personality: An empirical approach.* New York: Academic Press.

Ebbinghaus, H. D., Flum, J., & Thomas, W. (1984). *Mathematical logic.* New York, NY: Springer- Verlag.

Edward, L., Lourdeaux, D., & Barthès, J.-P. (2009). An Action Selection Architecture for Autonomous Virtual Agents. In N. T. Nguyen, R. P. Katarzyniak, & A. Janiak (Eds.), *New Challenges in Computational Collective Intelligence* (pp. 269–280). Springer Berlin Heidelberg. doi:10.1007/978-3-642-03958-4_23

Ehrenfeld, S., Schrodt, F., & Butz, M. V. (2015). *Mario Lives! An Adaptive Learning AI Approach for Generating a Living and Conversing Mario Agent* [Video file]. Retrieved from http://www.aaaivideos.org/2015/14_mario_lives/

Ehrenfeld, S., Schrodt, F., & Butz, M. (2015, January). Mario lives! An adaptive learning AI approach for generating a living and conversing Mario agent. In *Video Proceedings of the 29th Conference of the Association for the Advancement of Artificial Intelligence (AAAI 2015).*

Ekman, P. (1977). *Biological and cultural contributions to body and facial movement.* Academic Press.

El-Nasr, M. S., Yen, J., & Ioerger, T. R. (2000). Flame—fuzzy logic adaptive model of emotions. *Autonomous Agents and Multi-Agent Systems*, *3*(3), 219–257. doi:10.1023/A:1010030809960

Evans, R. (2002). AI game programming wisdom. In S. Rabin (Ed.), *chap. Varieties of Learning). Charles River Media.*

Evans, R., & Short, E. (2014). Versu—a simulationist storytelling system. *IEEE Transactions on Computational Intelligence and AI in Games*, *6*(2), 113–130. doi:10.1109/TCIAIG.2013.2287297

Exline, W. (1965). Affect Relations and Mutual Gaze in Dyads. In S. Tomkins & C. Izard (Eds.), *Affect, Cognition and Personality.* Springer.

Eysenck, H.J., (1967). *The Biological Basis of Personality.* Thomas.

Fagin, R., Halpern, J., Moses, Y., & Vardi, M. (2004). *Reasoning about knowledge.* Cambridge, MA: MIT Press.

Farabee, D., Nelson, R., & Spence, R. (1993). Psychosocial Profiles of Crim-inal Justice- and Noncriminal Justice-Referred Substance Abusers in Treatment. *Criminal Justice and Behavior*, *20*(4), 336–346. doi:10.1177/0093854893020004002

Fong, T., Nourbakhsh, I., & Dautenhahn, K. (2003). A survey of socially interactive robots. *Robotics and Autonomous Systems*, *42*(3–4), 143–166. doi:10.1016/S0921-8890(02)00372-X

Forgette, J. (2013). *Reinforcement Learning With Motivations For Realistic Agents*. (Masters Thesis). Department of Computer Science, The University of Western Ontario.

Fukayama, A., Ohno, T., Mukawa, N., Sawaki, M., & Hagita, N. (2002). Messages embedded in gaze of interface agents — impression management with agent's gaze.*Proceedings of the SIGCHI Conference on Human Factors in Compu-ting Systems, CHI '02* (pp. 41–48). New York: ACM. doi:10.1145/503376.503385

Fullerton, T., Swain, C., & Hoffman, S. (2008). *Game design workshop: A playcentric approach to creating innovative games*. Morgan Kaufmann.

Fum, D., & Stocco, A. (2004). Memory, emotion, and rationality: An ACT-R interpretation for gambling task results.*Proceedings of the Sixth International Conference on Cognitive Modeling* (pp. 106–111).

Funder, D. C. (2001). Personality. *Annual Review of Psychology*, *52*(1), 197–221. doi:10.1146/annurev.psych.52.1.197 PMID:11148304

Funder, D. C., & Sneed, C. D. (1993). Behavioural manifestations of person-ality: An ecological approach to judgmental accuracy. *Journal of Personality and Social Psychology*, *64*(3), 479–490. doi:10.1037/0022-3514.64.3.479 PMID:8468673

Gall, D., & Frühwirth, T. (2014). A formal semantics for the cognitive architecture ACT-R. In M. Proietti & H. Seki (Eds.), Lecture notes in computer science: Vol. 8981. Logic-based program synthesis and transformation (pp. 74–91). Cham, Switzerland: Springer. Retrieved from http://www.informatik.uniulm.de/pm/fileadmin/pm/home/fruehwirth/drafts/act_r_semantics.pdf

Gebhard, P. (2005, July). ALMA: a layered model of affect. In *Proceedings of the fourth international joint conference on Autonomous agents and multiagent systems* (pp. 29-36). ACM. doi:10.1145/1082473.1082478

Genesereth, M., Love, N., & Pell, B. (2005). General game playing: Overview of the AAAI competition. *AI Magazine*, *26*(2), 62.

Gill & Oberlander, J. (2002). Taking Care of the Linguistic Features of Ex-traversion. *Proc. 24th Ann. Conf. Cognitive Science Soc.*, 363-368.

Goble, L. (Ed.). (2001). *The Blackwell guide to philosophical logic*. Oxford, UK: Blackwell Publishing. doi:10.1111/b.9780631206934.2001.x

Goertzel, B. (2014). Artificial General Intelligence: Concept, state of the art, and future prospects. *Journal of Artificial General Intelligence.*, *5*, 1–46.

Goertzel, B. (2014). GOLEM: Towards an AGI meta-architecture enabling both goal preservation and radical self-improvement. *Journal of Experimental & Theoretical Artificial Intelligence*, *26*(3), 391–403. doi:10.1080/0952813X.2014.895107

Goertzel, B., & Wang, P. (2007). Advances in Artificial General Intelligence: Concepts, Architectures and Algorithms: *Proceedings of the AGI Workshop 2006* (Vol. 157). IOS Press.

Gomes, P. F., Segura, E. M., Cramer, H., Paiva, T., Paiva, A., & Holmquist, L. E. (2011). ViPleo and PhyPleo: Artificial Pet with Two Embodiments. In *Proceedings of the 8th International Conference on Advances in Computer Entertainment Technology* (pp. 3:1–3:8). New York, NY: ACM. http://doi.org/ doi:10.1145/2071423.2071427

Grace, K., & Maher, M. L. (2014). Towards computational co-creation in modding communities. In *Proceedings of the Experimental AI in Games Workshop.*

Gray, J. A. (1987). The neuropsychology of emotion and personality. In S. M. Stahl, S. D. Iversen & E. C. Goodman (Eds.), Cognitive neurochemistry (pp. 171-190). Oxford, UK: Oxford University Press.

Gregory, N. M., Dorais, G. A., Fry, C., Levinson, R., & Plaunt, C. (2002). IDEA: Planning at the Core of Autonomous Reactive Agents. In *Proceedings of the 3rd International NASA Workshop on Planning and Scheduling for Space.*

Grisetti, G., Kummerle, R., Stachniss, C., & Burgard, W. (2010). A tutorial on graph-based SLAM. *Intelligent Transportation Systems Magazine, IEEE, 2*(4), 31–43. doi:10.1109/MITS.2010.939925

Guye-Vuilleme, A., & Thalmann, D. (2000). A high-level architecture for believable social agents. *Virtual Reality (Waltham Cross), 5*(2), 95–106. doi:10.1007/BF01424340

Guzdial, M., & Riedl, M. O. (2015). Toward Game Level Generation from Gameplay Videos. In *Proceedings of the 2015 Workshop on Procedural Content Generation.*

Harmon, M., & Harmon, S. (1996). *Reinforcement Learning: A Tutorial.* Retrieved March 16, 2015, from http://citeseerx.ist.psu.edu/viewdoc/summary?doi=10.1.1.33.2480

Hartsook, K., Zook, A., Das, S., & Riedl, M. (2011). Toward supporting storytellers with procedurally generated game worlds. In *Proceedings of the IEEE Conference on Computational Intelligence and Games.*

Hassani, K., & Lee, W.-S. (2013). A software-in-the-loop simulation of an intelligent microsatellite within a virtual environment. In *2013 IEEE International Conference on Computational Intelligence and Virtual Environments for Measurement Systems and Applications (CIVEMSA)* (pp. 31–36). http://doi.org/ doi:10.1109/CIVEMSA.2013.6617391

Hassani, K., & Lee, W.-S. (2014). On Designing Migrating Agents: From Autonomous Virtual Agents to Intelligent Robotic Systems. In SIGGRAPH Asia 2014 Autonomous Virtual Humans and Social Robot for Telepresence (pp. 7:1–7:10). New York, NY: ACM. http://doi.org/ doi:<ALIGNMENT.qj></ALIGNMENT>10.1145/2668956.2668963

Hassani, K., & Lee, W.-S. (2015). An Intelligent Architecture for Autonomous Virtual Agents Inspired by Onboard Autonomy. In P. Angelov, K. T. Atanassov, L. Doukovska, M. Hadjiski, V. Jotsov, J. Kacprzyk, … S. Zadrożny (Eds.), *Intelligent Systems'2014* (pp. 391–402). Springer International Publishing. Retrieved from http://link.springer.com/chapter/10.1007/978-3-319-11313-5_35

Hassani, K., Nahvi, A., & Ahmadi, A. (2013a). Architectural design and implementation of intelligent embodied conversational agents using fuzzy knowledge base. *Journal of Intelligent and Fuzzy Systems*, *25*(3), 811–823. doi:10.3233/IFS-120687

Hassani, K., Nahvi, A., & Ahmadi, A. (2013b). Design and implementation of an intelligent virtual environment for improving speaking and listening skills. *Interactive Learning Environments*, *0*(0), 1–20. doi:10.1080/10494820.2013.846265

Hastings, E., Guha, R. K., & Stanley, K. (2009). Automatic content generation in the galactic arms race video game. *IEEE Transactions on Computational Intelligence and AI in Games*, *1*(4), 245–263. doi:10.1109/TCIAIG.2009.2038365

Hindriks, K. V. (2009). ProgrammingRationalAgents in GOAL. In A. E. F. Seghrouchni, J. Dix, M. Dastani, & R. H. Bordini (Eds.), *Multi-Agent Programming* (pp. 119–157). Springer, US. doi:10.1007/978-0-387-89299-3_4

Hodhod, R., Huet, M., & Riedl, M. (2014). Toward generating 3D games with the help of commonsense knowledge and the crowd. In *Proceedings of the Experimental AI in Games Workshop.*

Holyoak, K. J., & Hummel, J. E. (2000). The proper treatment of symbols in a connectionist architecture. In E. Deitrich & A. Markman (Eds.), *Cognitive dynamics: Conceptual change in humans and machines*. Cambridge, MA: MIT Press.

Hope, R. M., Schoelles, M. J., & Gray, W. D. (2014). Simplifying the interaction between cognitive models and task environments with the JSON Network Interface. *Behavior Research Methods*, *46*(4), 1007–1012. doi:10.3758/s13428-013-0425-z PMID:24338626

Horswill, I. D. (2014). Game design for classical AI. In *Proceedings of the Experimental AI in Games Workshop*.

Horvath, G., Ingham, M., Chung, S., Martin, O., & Williams, B. (2006). Practical application of model-based programming and state-based architecture to space missions. In *Second IEEE International Conference on Space Mission Challenges for Information Technology, 2006. SMC-IT 2006*. http://doi.org/ doi:<ALIGNMENT.qj></ALIGNMENT>10.1109/SMC-IT.2006.62

Hull, C. L. (1943). *Principles of Behavior*. New York: Appleton-Century-Crofts.

Hummel, J. E., & Holyoak, K. J. (1997). Distributed representations of structure: A theory of analogical access and mapping. *Psychological Review*, *104*(3), 427–466. doi:10.1037/0033-295X.104.3.427

Hummel, J. E., & Holyoak, K. J. (2003). Relational reasoning in a neurally-plausible cognitive architecture: An overview of the LISA project. *Cognitive Studies: Bulletin of the Japanese Cognitive Science Society*, *10*, 58–75.

Hunicke, R., Leblanc, M., & Zubek, R. (2004). MDA: A formal approach to game design and game research. In *Proceedings of the Workshop on Challenges in Game AI.*

Isbister, K. (2006). *Better Game Characters By Design: A Psychological Approach.* San Francisco: Elsevier.

Isbister, K., & Schaffer, N. (2008). *Game usability: Advancing the player experience.* Morgan Kaufmann.

Isla, D. (2005). *Dude: Where's My Warthog? From Pathfinding to General Spatial Competence* [Powerpoint slides]. Retrieved from http://www.aaai.org/Papers/AIIDE/2005/AIIDE05-040.ppt

Jacob, M., Coisne, G., Gupta, A., Sysoev, I., Verma, G., & Magerko, B. (2013). Viewpoints AI.*Proceedings of the Ninth Annual AAAI Conference on Artificial Intelligence and Interactive Digital Entertainment (AIIDE).*

Jaskowski, S. (1934). On the rules of suppositions in formal logic. *Studia Logica, 1*, 5–32.

Ji, Q., Gray, W. D., Guhe, M., & Schoelles, M. J. (2004). Towards an integrated cognitive architecture for modeling and recognizing user affect. In *AAAI Spring Symposium on Architectures for Modeling emotion: cross-disciplinary foundations.*

Joshi, H., Rosenbloom, P. S., & Ustun, V. (2014). Isolated word recognition in the Sigma cognitive architecture. In Biologically Inspired Cognitive Architectures. doi:10.1016/j.bica.2014.11.001

Kang, Y., & Tan, A.-H. (2013). Self-organizing Cognitive Models for Virtual Agents. In R. Aylett, B. Krenn, C. Pelachaud, & H. Shimodaira (Eds.), *Intelligent Virtual Agents* (pp. 29–43). Springer Berlin Heidelberg. doi:10.1007/978-3-642-40415-3_3

Keisler, H. J. (1986). *Elementary calculus: An infinitesimal approach.* Amsterdam, The Netherlands: Prindle, Weber and Schmidt.

Kim, I.-C. (2005). CAA: A Context-Sensitive Agent Architecture for Dynamic Virtual Environments. In T. Panayiotopoulos, J. Gratch, R. Aylett, D. Ballin, P. Olivier, & T. Rist (Eds.), *Intelligent Virtual Agents* (pp. 146–151). Springer Berlin Heidelberg. doi:10.1007/11550617_13

Kleinsmith, L. J., & Kaplan, S. (1963). Paired-associate learning as a function of arousal and interpolated interval. *Journal of Experimental Psychology, 65*(2), 190–193. doi:10.1037/h0040288 PMID:14033436

Koay, K. L., Syrdal, D. S., Dautenhahn, K., Arent, K., Małek, Ł., & Kreczmer, B. (2011). Companion Migration – Initial Participants' Feedback from a Video-Based Prototyping Study. In X. Wang (Ed.), *Mixed Reality and Human-Robot Interaction* (pp. 133–151). Springer Netherlands. doi:10.1007/978-94-007-0582-1_8

Kokinov, B. N. (1994). The Context-Sensitive Cognitive Architecture DUAL. In *Proceedings of the Sixteenth Annual Conference of the Cognitive Science Society.*

Koller, D., & Friedman, N. (2009). *Probabilistic Graphical Models: Principles and Techniques.* MIT Press.

Kope, A., Rose, C., & Katchabaw, M. (2013, October). Modeling autobiographical memory for believable agents. *Proceedings of the 9th AAAI Conference on Artificial Intelligence and Interactive Digital Entertainment (AIIDE'13).*

Kopp, S., Krenn, B., Marsella, S., Marshall, A. N., Pelachaud, C., Pirker, H., & Vilhjálmsson, H. (2006). Towards a common framework for multimodal generation: The behavior markup language. In Intelligent Virtual Agents.

Kopp, S., Krenn, B., Marsella, S., Marshall, A. N., Pelachaud, C., Pirker, H., ... Vilhjálmsson, H. (2006). Towards a Common Framework for Multimodal Generation: The Behavior Markup Language. In *Proceedings of the 6th International Conference on Intelligent Virtual Agents* (pp. 205–217). Berlin: Springer-Verlag. http://doi.org/ doi:10.1007/11821830_17

Kriegel, M., Aylett, R., Cuba, P., Vala, M., & Paiva, A. (2011). Robots Meet IVAs: A Mind-Body Interface for Migrating Artificial Intelligent Agents. In H. H. Vilhjálmsson, S. Kopp, S. Marsella, & K. R. Thórisson (Eds.), *Intelligent Virtual Agents* (pp. 282–295). Springer Berlin Heidelberg. doi:10.1007/978-3-642-23974-8_31

Kschischang, F. R., Frey, B. J., & Loeliger, H. A. (2001). Factor graphs and the sum-product algorithm. *IEEE Transactions on Information Theory*, *47*(2), 498–519. doi:10.1109/18.910572

Kshirsagar, S. (2002, June). A multilayer personality model. In *Proceedings of the 2nd international symposium on Smart graphics* (pp. 107-115). ACM. doi:10.1145/569005.569021

Kurup, U., & Lebiere, C. (2012). What can cognitive architectures do for robotics? *Biologically Inspired Cognitive Architectures*, *2*, 88–99. doi:10.1016/j.bica.2012.07.004

La France, B., Heisel, A., & Beatty, M. (2004). Is There Empirical Evidence for a Nonverbal Profile of Extraversion? A Meta-Analysis and Critique of the Literature. *Communication Monographs*, *71*(1), 28–48. doi:10.1080/03634520410001693148

Laird, J. E., & Rosenbloom, P. S. (1993). Integrating execution, planning, and learning in Soar for external environments. Academic Press.

Laird, J. E., Assanie, M., Bachelor, B., Benninghoff, N., Enam, S., Jones, B., . . . Wallace, S. (2002). A Test Bed for Developing Intelligent Synthetic Characters. *AAAI 2002 Spring Symposium Series: Artificial Intelligence and Interactive Entertainment.*

Laird, J. (2012). *The Soar Cognitive Architecture.* Cambridge, MA: MIT Press.

Laird, J. (2012). *The Soar cognitive architecture.* MIT Press.

Laird, J. E. (2001). It knows what you're going to do: adding anticipation to a Quakebot. In *Proceedings of the International Conference on Autonomous Agents.* doi:10.1145/375735.376343

Laird, J. E. (2008). Extending the Soar cognitive architecture. *Frontiers in Artificial Intelligence and Applications*, *171*, 224.

Laird, J. E. (2012). *The SOAR Cognitive Architecture.* Cambridge, MA: MIT Press.

Laird, J. E., Kinkade, K. R., Mohan, S., & Xu, J. Z. (2012). Cognitive robotics using the Soar cognitive architecture. *Cognitive Robotics AAAI Technical Report, WS-12*, 6.

Laird, J. E., Newell, A., & Rosenbloom, P. S. (1987). SOAR: An architecture for general intelligence. *Artificial Intelligence, 33*(1), 1–64. doi:10.1016/0004-3702(87)90050-6

Laird, J., Hucka, M., Huffman, S., & Rosenbloom, P. (1991). An analysis of Soar as an integrated architecture. *ACM SIGART Bulletin, 2*(4), 98–103. doi:10.1145/122344.122364

Laird, J., & van Lent, M. (2001). Human-level AI's killer application: Interactive computer games. *AI Magazine, 22*(2), 15.

Langley, P., & Choi, D. (2006). A Unified Cognitive Architecture for Physical Agents. In Proceedings of the 21st National Conference on Artificial Intelligence (Vol. 2, pp. 1469–1474). Boston, MA: AAAI Press. Retrieved from http://dl.acm.org/citation.cfm?id=1597348.1597422

Langley, P., & Choi, D. (2006). A unified cognitive architecture for physical agents. In *Proceedings of the National Conference on Artificial Intelligence.*

Langley, P., Laird, J. E., & Rogers, S. (2009). Cognitive architectures: Research issues and challenges. *Cognitive Systems Research, 10*(2), 141–160. doi:10.1016/j.cogsys.2006.07.004

Larsen, R. J., & Ketelaar, T. (1991). Personality and susceptibility to posi-tive and negative emotional states. *Journal of Personality and Social Psychology, 61*(1), 132–140. doi:10.1037/0022-3514.61.1.132 PMID:1890584

Larsen, R. J., & Shackelford, T. K. (1996). Gaze avoidance: Personality and social judgments of people who avoid direct face-to-face contact. *Personality and Individual Differences, 21*(6), 907–917. doi:10.1016/S0191-8869(96)00148-1

Larson, R., & Edwards, B. (2014). *Calculus*. Boston, MA: Brooks and Cole.

Lathrop, S. D., & Laird, J. E. (2009, March). Extending cognitive architectures with mental imagery.*Proceedings of the second conference on artificial general intelligence.*

Lavelle, S. (2013). *PuzzleScript*. Retrieved from http://www.puzzlescript.net/

Lawson, G., & Burnett, G. (2015). Simulation and Digital Human Modelling. In J. R. Wilson & S. Sharples (Eds.), *Evaluation of Human Work* (4th ed.). Boca Raton, FL: CRC Press.

Lazarus, R. S., & Folkman, S. (1984). Stress. *Appraisal, and Coping*, 456.

Lee-Urban, S., Trewhitt, E., Bieder, I., Odom, J., Boone, T., & Whitaker, E. (2015). CORA: A Flexible Hybrid Approach to Building Cognitive Systems. In *Proceedings of the Third Annual Conference on Advances in Cognitive Systems Poster Collection* (p. 28).

Lehman, J. F., Laird, J. E., & Rosenbloom, P. S. (1996). A gentle introduction to Soar, an architecture for human cognition. *Invitation to Cognitive Science, 4,* 212-249.

Lekavý, M., & Návrat, P. (2007). Expressivity of STRIPS-Like and HTN-Like Planning. In N. T. Nguyen, A. Grzech, R. J. Howlett, & L. C. Jain (Eds.), *Agent and Multi-Agent Systems: Technologies and Applications* (pp. 121–130). Springer Berlin Heidelberg. doi:10.1007/978-3-540-72830-6_13

Leonard, A. (1997). *Bots: The Origin of New Species*. San Francisco: HardWired.

Liapis, A., Yannakakis, G. N., & Togelius, J. (2013). Sentient sketchbook: Computer-aided game level authoring. In *Proceedings of the 8th International Conference on the Foundations of Digital Games.*

Licato, J. (2015). *Analogical constructivism: The emergence of reasoning through analogy and action schemas.* (PhD thesis). Retrieved from Rensselaer Polytechnic Institute, Troy, NY.

Licato, J., Sun, R., & Bringsjord, S. (2014a). Structural representation and reasoning in a hybrid cognitive architecture. In *Proceedings of the 2014 International Joint Conference on Neural Networks (IJCNN)*. doi:10.1109/IJCNN.2014.6889895

Licato, J., Sun, R., & Bringsjord, S. (2014b). Using a hybrid cognitive architecture to model children's errors in an analogy task. In *Proceedings of CogSci 2014.*

Licato, J., Sun, R., & Bringsjord, S. (2014c). Using meta-cognition for regulating explanatory quality through a cognitive architecture. In *Proceedings of the 2nd International Workshop on Artificial Intelligence and Cognition.*

Li, J., & Laird, J. E. (2013). The Computational Problem of Prospective Memory Retrieval. *Proceedings of the 17th International Conference on Cognitive Modeling.*

Lim, C.-U., & Harrell, D. F. (2014). An approach to general videogame evaluation and automatic generation using a description language. In *Proceedings of the IEEE Conference on Computational Intelligence and Games.*

Lim, M. Y., Dias, J., Aylett, R., & Paiva, A. (2012). Creating adaptive af-fective autonomous NPCs. *Autonomous Agents and Multi-Agent Systems, 24*(2), 287–311. doi:10.1007/s10458-010-9161-2

Liu, J., & Lu, Y. (2006). Agent Architecture Suitable for Simulation of Virtual Human Intelligence. In *The Sixth World Congress on Intelligent Control and Automation, 2006. WCICA 2006* (*Vol. 1*, pp. 2521–2525). http://doi.org/ doi:<ALIGNMENT.qj></ALIGNMENT>10.1109/WCICA.2006.1712816

Liu, H., & Singh, P. (2004). ConceptNet—a practical commonsense reasoning tool-kit. *BT Technology Journal, 22*(4), 211–226. doi:10.1023/B:BTTJ.0000047600.45421.6d

Liu, Y.-E., Mandel, T., Brunskill, E., & Popović, Z. (2014). Trading off scientific knowledge and user learning with multi-armed bandits. In *Proceedings of the International Conference on Educational Data Mining.*

Llano, M. T., Cook, M., Guckelsberger, C., Colton, S., & Hepworth, R. (2014). Towards the automatic generation of fictional ideas for games. In *Proceedings of the Experimental AI in Games Workshop*.

Lomas, D., Patel, K., Forlizzi, J. L., & Koedinger, K. R. (2013). Optimizing challenge in an educational game using large-scale design experiments. In *Proceedings of the ACM SIGCHI Conference on Human Factors in Computing Systems*. doi:10.1145/2470654.2470668

Lopes, R., Tutenel, T., & Bidarra, R. (2012). Using gameplay semantics to procedurally generate player-matching game worlds. In *Proceedings of the Workshop on Procedural Content in Games*. doi:10.1145/2538528.2538531

Love, N., Hinrichs, T., Haley, D., Schkufza, E., & Genesereth, M. (2008). *General game playing: Game description language specification (Tech. Rep.)*. Stanford University.

Loyall, A. B. (1997). *Believable Agents: Building Interactive Personalities*. (PhD Thesis). Stanford University.

Lucas, S. M., & Kendall, G. (2006). Evolutionary Computation and Games. *IEEE Computational Intelligence Magazine*, *1*(1), 10–19. doi:10.1109/MCI.2006.1597057

Lynch, M. F., Sun, R., & Wilson, N. (2011, January). CLARION as a cognitive framework for intelligent virtual agents. In *Intelligent Virtual Agents* (pp. 460–461). Springer Berlin Heidelberg. doi:10.1007/978-3-642-23974-8_62

MacLennan, A. (1996). The artificial life route to artificial intelligence: Building embodied, situated agents. *Journal of the American Society for Information Science*, *47*(6), 482–483. doi:10.1002/(SICI)1097-4571(199606)47:6<482::AID-ASI14>3.0.CO;2-0

Maffre, E., Tisseau, J., & Parenthoen, M. (2001). Virtual Agent's Self-Perception in Story Telling. In *Virtual Storytelling: Using Virtual Reality Technologies for Storytelling*. Berlin: Springer-Verlag.

Magerko, B., Laird, J., Assanie, M., Kerfoot, A., & Stokes, D. (2004). AI characters and directors for interactive computer games. Ann Arbor, 1001(48), 109-2110.

Magerko, B., Laird, J., Assanie, M., Kerfoot, A., & Stokes, D. (2004). AI characters and directors for interactive computer games. In Ann Arbor, 1001(48), 109-2110.

Magerko, B., Laird, J., Assanie, M., Kerfoot, A., & Stokes, D. (2004). AI characters and directors for interactive computer games. In *IAAI'04 Proceedings of the 16th conference on Innovative applications of artifical intelligence*.

Mahner, M., & Bunge, M. (2001). Function and Functionalism: A Synthetic Perspective. *Philosophy of Science*, *68*(1), 75–94. doi:10.1086/392867

Marinier, R. III, & Laird, J. (2008, July). Emotion-driven reinforcement learning. *Proceedings of the 30th Annual Meeting of the Cognitive Science Society*.

Marinier, R. III, Laird, J., & Lewis, R. (2009). A computational unification of cognitive behaviour and emotion. *Cognitive Systems Research, 10*(1), 48–69. doi:10.1016/j.cogsys.2008.03.004

Marsella, S. C., & Gratch, J. (2009). EMA: A process model of appraisal dynamics. *Cognitive Systems Research, 10*(1), 70–90. doi:10.1016/j.cogsys.2008.03.005

Mateas, M. (1997). *Computational Subjectivity in Virtual World Avatars.* AAAI Technical Report FS-97-02, 1-6.

Mateas, M. (1997). *Computational subjectivity in virtual world avatars.* Technical Report FS-97-02. AAAI. Retrieved from http://www.aaai.org/Papers/Symposia/Fall/1997/FS-97-02/FS97-02-021.pdf

Mateas, M. (1999). *An Oz-centric review of interactive drama and believable agents.* Springer Berlin Heidelberg. doi:10.1007/3-540-48317-9_12

Mateas, M. (2001). Expressive AI: A hybrid art and science practice. *Leonardo: Journal of the International Society for Arts, Sciences, and Technology, 34*(2), 147–153. doi:10.1162/002409401750184717

Mateas, M., & Stern, A. (2002). *Architecture, authorial idioms and early observations of the interactive drama Façade (Tech. Rep.).* Carnegie Mellon University.

Mattingly, W. A., Chang, D.-j., Paris, R., Smith, N., Blevins, J., & Ouyang, M. (2012) *Robot design using Unity for computer games and robotic simulations.17th International Conference on Computer Games,* Louisville, KY. doi:10.1109/CGames.2012.6314552

McCoy, J., Treanor, M., Samuel, B., Reed, A., Mateas, M., & Wardrip-Fruin, N. (2014). Social story worlds with Comme il Faut. *IEEE Transactions on Computational Intelligence and AI in Games, 6*(2), 97–112. doi:10.1109/TCIAIG.2014.2304692

McCrae, R. R., & Costa, P. T. (1986). Personality, coping, and coping ef-fectiveness in an adult sample. *Journal of Personality, 54*(2), 385–404. doi:10.1111/j.1467-6494.1986.tb00401.x

McCrae, R. R., & John, O. P. (1992). An Introduction to the Five-Factor Model and Its Applications. *Journal of Personality, 60*(2), 175–215. doi:10.1111/j.1467-6494.1992.tb00970.x PMID:1635039

McRorie, M., Sneddon, I., McKeown, G., Bevacqua, E., de Sevin, E., & Pe-lachaud, C. (2012). Evaluation of four designed virtual agent personalities. *Affective Computing. IEEE Transactions on, 3*(3), 311–322.

Merrick, K. & Shah, K. (2011, February). Achievement, affiliation, and power: Motive profiles for artificial agents. *Adaptive Behavior: Animals, Animats, Software Agents, Robots, Adaptive Systems, 19*(l).

Merrick, K. (2008). Modeling motivation for adaptive nonplayer characters in dynamic computer game worlds. Computers in Entertainment, 5(4), 5:l-5:32. doi:10.1145/1324198.1324203

Merrick, K. (2011). A computational model of achievement motivation for artificial agents. In *Proceedings of the 10th International Conference on Autonomous Agents and Multiagent Systems*.

Merrick, K., & Maher, M. (2006, May). Motivated reinforcement learning for non-player characters in persistent computer game worlds. In *Proc. of the 2006 ACM SIGCHI International Conference on Advances in Computer Entertainment Technology*. doi:10.1145/1178823.1178828

Merrick, K., & Maher, M. (2009). *Motivated Reinforcement Learning: Curious Characters for Multiuser Games* (1st ed.). Springer Publishing Company, Incorporated. doi:10.1007/978-3-540-89187-1

Mikolov, T., Joulin, A., & Baroni, M. (2015). *A roadmap towards machine intelligence*. Retrieved from http://arxiv.org/abs/1511.08130

Minsky, M. (1975). A Framework for Representing Knowledge. In The Psychology of Computer Vision (pp. 211-277). New York: McGraw Hill.

Mitchell, T. M. (1997). *Machine Learning*. McGraw-Hill.

Mnih, V., Kavukcuoglu, K., Silver, D., Rusu, A. A., Veness, J., Bellemare, M. G., & Hassabis, D. et al. (2015). Human-level control through deep reinforcement learning. *Nature*, *518*(7540), 529–533. doi:10.1038/nature14236 PMID:25719670

Moffat, D. (1997). Personality Parameters and Programs. In R. Trappl & P. Petta (Eds.), *Creating Personalities for Synthetic Actors* (pp. 120–165). New York: Springer Verlag. doi:10.1007/BFb0030575

Moon, J., & Anderson, J. R. (2012) *Modeling Millisecond Time Interval Estimation in Space Fortress Game.34th Annual Conference of the Cognitive Science Society*, Sapporo, Japan.

Morrison, J. E. (2003) A Review of Computer-Based Human Behavior Representations and Their Relation to Military Simulations. Institute for Defense Analyses. doi:10.1037/e427382005-001

Murphy, T. (2013). *The first level of super mario bros. is easy with lexicographic ordering and time travel ...after that it gets a little tricky*. SIGBOVIK.

Muscettola, N., Nayak, P. P., Pell, B., & Williams, B. C. (1998). Remote Agent: To boldly go where no AI system has gone before. *Artificial Intelligence*, *103*(1–2), 5–47. doi:10.1016/S0004-3702(98)00068-X

Myers, I., Briggs, McCaulley, M., Quenk, N., & Hammer, A. (1998). MBTI Manual: A Guide To The Development And Use Of The Myers Briggs Type Indicator. Consulting Psychologists Press.

Napieralski, L. P., Brooks, C. I., & Droney, J. M. (1995). The effect of du-ration of eye contact on American college students' attributions of state, trait, and test anxiety. *The Journal of Social Psychology*, *135*(3), 273–280. doi:10.1080/00224545.1995.9713957 PMID:7650931

Nareyek, A. (2004). AI in Computer Games. In Queue 1, 10 (February 2004), 58-65. DOI= doi:10.1145/971564.971593

Neff, M., Toothman, N., Bowmani, R., Tree, J. E. F., & Walker, M. A. (2011, January). Don't scratch! self-adaptors reflect emotional stability. In *Intelligent Virtual Agents* (pp. 398–411). Springer Berlin Heidelberg. doi:10.1007/978-3-642-23974-8_43

Nelson, M. J., & Mateas, M. (2008). An interactive game-design assistant. In *Proceedings of the 13th International Conference on Intelligent User Interfaces.* doi:10.1145/1378773.1378786

Newell, A. (1981). The knowledge level. *Artificial Intelligence, 18*, 81–132.

Newell, A. (1990). *Unified theories of cognition.* Cambridge, MA: Harvard University Press.

Nguyen, T. H. D., Carstensdottir, E., Ngo, N., El-Nasr, M. S., Gray, M., Isaacowitz, D., & Desteno, D. (2015, August). Modeling Warmth and Competence in Virtual Characters. In Intelligent Virtual Agents (pp. 167-180). Springer International Publishing. doi:10.1007/978-3-319-21996-7_18

Niehaus, J. (2013) *Mobile, Virtual Enhancements for Rabilitation.* Charles River Analytics.

Norris, C. J., Larsen, J. T., & Cacioppo, J. T. (2007). Neuroticism is associated with larger and more prolonged electrodermal responses to emotionally evocative pictures. *Psychophysiology, 44*(5), 823–826. doi:10.1111/j.1469-8986.2007.00551.x PMID:17596178

Northey, M. (2008). *Impact: a guide to business communication.* Toronto: Pearson Prentice Hall.

Noy, N. F., & McGuinness, D. L. (2001). *Ontology Development 101: A Guide to Creating Your First Ontology.* Stanford Knowledge Systems Laboratory Technical Report KSL-01-05 and Stanford Medical Informatics Technical Report SMI-2001-0880, March 2001.

O'Hair, D., Friedrich, G. W., & Dixon, L. D. (2011). *Strategic communica-tion in business and the professions.* Boston: Allyn and Bacon.

O'Reilly, R. C., & Munakata, Y. (2000). *Computational explorations in cognitive neuroscience: Understanding the mind by simulating the brain.* MIT Press.

Ogawa, K., & Ono, T. (2008). ITACO: Constructing an emotional relationship between human and robot. In *The 17th IEEE International Symposium on Robot and Human Interactive Communication, 2008. RO-MAN 2008* (pp. 35–40). doi:10.1109/ROMAN.2008.4600640

Orkin, J. (2005). Agent architecture considerations for real-time planning in games. In *Proceedings of the 1st AAAI Conference on Artificial Intelligence and Interactive Digital Entertainment.*

Ortony, A. (2002). On making believable emotional agents believable. Academic Press.

Ortony, A., Norman, D. A., & Revelle, W. (2005). Effective functioning: A three level model of affect, motivation, cognition, and behaviour. In J. Fellous & M. Arbib (Eds.), *Who needs emotions? The brain meets the machine* (pp. 173–202). New York: Oxford Univeristy Press. doi:10.1093/acprof:oso/9780195166194.003.0007

Ostendorf, & Angleitner. (1994). A comparison of different in-struments proposed to measure the Big Five. *European Review of Applied Psychology, 44*(1), 45-55.

Partee, B., Meulen, A., & Wall, R. (1990). *Mathematical methods in linguistics*. Dordrecht, The Netherlands: Kluwer.

Patrick, A., Gittens, C., & Katchabaw, M. (2015, October). The virtual little Albert experiment: Creating conditioned emotion response in virtual agents. In *Proceedings of 2015 IEEE Games, Entertainment, and Media Conference*. doi:10.1109/GEM.2015.7377228

Pedersen, C., Togelius, J., & Yannakakis, G. N. (2009). Modeling player experience in Super Mario Bros. In *Proceedings of the IEEE Symposium on Computational Intelligence and Games*. doi:10.1109/CIG.2009.5286482

Perez, D., Spyridon, Togelius, J., Schaul, T., Lucas, S. M., Couëtoux, A., . . . Thompson, T. (2015). The 2014 general video game playing competition. *IEEE Transactions on Computational Intelligence and AI in Games*.

Pervin, L. A., & Cervone, D. (2005). *Personality: Theory and Research*. Academic Press.

Pittman, J. (2011, June 16). *The Pac-Man Dossier*. Retrieved from http://home.comcast.net/~jpittman2/pacman/pacmandossier.html

Pokahr, A., Braubach, L., & Lamersdorf, W. (2005). Jadex: A BDI Reasoning Engine. In R. H. Bordini, M. Dastani, J. Dix, & A. E. F. Seghrouchni (Eds.), *Multi-Agent Programming* (pp. 149–174). Springer, US. doi:10.1007/0-387-26350-0_6

Porter, M. F. (1980). An algorithm for suffix stripping. *Program*, *14*(3), 130–137. doi:10.1108/eb046814

Poznanski, M., & Thagard, P. (2005). Changing personalities: Towards real-istic virtual characters. *Journal of Experimental & Theoretical Artificial Intelligence*, *17*(3), 221–241. doi:10.1080/09528130500112478

Prada, R., Raimundo, G., Dimas, J., Martinho, C., Pena, J. F., Baptista, M., & Ribeiro, L. L. et al. (2012, June). The role of social identity, rationality and anticipation in believable agents. In *Proceedings of the 11th International Conference on Autonomous Agents and Multiagent Systems*.

Pynadath, D. V., & Marsella, S. C. (2005). PsychSim: Modeling theory of mind with decision-theoretic agents. IJCAI.

Pynadath, D. V., Rosenbloom, P. S., & Marsella, S. C. (2014). Reinforcement Learning for Adaptive Theory of Mind in the Sigma Cognitive Architecture. In Artificial General Intelligence.

Pynadath, D. V., Rosenbloom, P. S., Marsella, S. C., & Li, L. (2013). Modeling two-player games in the Sigma graphical cognitive architecture. In Artificial General Intelligence. doi:10.1007/978-3-642-39521-5_11

Rabin, S. (2010). *Introduction to Game Development*. Game Development Series. Course Technology.

Ranathunga, S., Cranefield, S., & Purvis, M. 2011. Interfacing a Cognitive Agent Platform with a Virtual World: A Case Study using Second Life (Extended Abstract). In Proceedings of the 10th International Conference on Autonomous Agents and Multiagent Systems (AAMAS 2011). Turner, Yolum, Sonenberg and Stone (eds.). May 2-6, 2011, Taipei, Taiwan, pp. 1181-1182.

Rao, A. S., & Georgeff, M. P. (1995). BDI Agents: From Theory to Practice. In Proceedings of the First International Conference on Multi-Agent Systems (pp. 312–319).

Rauthmann, J. F., Seubert, C. T., Sachse, P., & Furtner, M. R. (2012). Eyes as windows to the soul: Gazing behaviour is related to personality. *Journal of Research in Personality*, *46*(2), 147–156. doi:10.1016/j.jrp.2011.12.010

Read, S., Monroe, B., Brownstein, A., Yang, Y., Chopra, G., & Miller, L. (2010). A neural network model of the structure and dynamics of human personality. *Psychological Review*, *117*(1), 61–92. doi:10.1037/a0018131 PMID:20063964

Reidl, M. O., & Stern, A. (2006). Failing Believably: Toward Drama Management with Autonomous Actors in Interactive Narratives. In S. Göbel, R. Malkewitz, & I. Iurgel (Eds.), *TIDSE 2006, LNCS 4326* (pp. 195–206). Berlin: Springer-Verlag. doi:10.1007/11944577_21

Reiss, S. (2000). Why people turn to religion: A motivational analysis. *Journal for the Scientific Study of Religion*, *39*(1), 47–52. doi:10.1111/0021-8294.00004

Reiss, S. (2004). Multifaceted nature of intrinsic motivation: The theory of 16 basic desires. *Review of General Psychology*, *8*(3), 179–193. doi:10.1037/1089-2680.8.3.179

Reiss, S., & Havercamp, S. (1998). Toward a comprehensive assessment of fundamental motivation: Factor structure of the Reiss Profiles. *Psychological Assessment*, *10*(2), 97–106. doi:10.1037/1040-3590.10.2.97

Reiss, S., Wiltz, J., & Sherman, M. (2001). Trait motivational correlates of athleticism. *Personality and Individual Differences*, *30*(7), 1139–1145. doi:10.1016/S0191-8869(00)00098-2

Reitter, D., & Lebiere, C. (2012) *Social Cognition: Memory Decay and Adaptive Information Filtering for Robust Information Maintenance.26th AAAI Conference on Artificial Intelligence*, Toronto, Canada.

Ribeiro, T., Vala, M., & Paiva, A. (2012). Thalamus: Closing the Mind-Body Loop in Interactive Embodied Characters. In Y. Nakano, M. Neff, A. Paiva, & M. Walker (Eds.), *Intelligent Virtual Agents* (pp. 189–195). Springer Berlin Heidelberg. doi:10.1007/978-3-642-33197-8_19

Ribeiro, T., Vala, M., & Paiva, A. (2013). Censys: A Model for Distributed Embodied Cognition. In R. Aylett, B. Krenn, C. Pelachaud, & H. Shimodaira (Eds.), *Intelligent Virtual Agents* (pp. 58–67). Springer Berlin Heidelberg. doi:10.1007/978-3-642-40415-3_5

Rickel, J., & Johnson, L. W. (2000). Task-Oriented Collaboration with Embodied Agents in Virtual Worlds. In J. Cassell, J. Sullivan, & S. Prevost (Eds.), *Embodied Conversational Agents*. Cambridge, MA: MIT Press.

Riedl, M. O., & Bulitko, V. (2013). Interactive narrative: An intelligent systems approach. *AI Magazine*, *34*(1), 67–77.

Riedl, M., Stern, A., Dini, D., & Alderman, J. (2008). Dynamic experience management in virtual worlds for entertainment, education, and training. *International Transactions on Systems Science and Applications*, *4*(2), 23–42.

Risi, S., Lehman, J., D'Ambrosio, D. B., Hall, R., & Stanley, K. O. (2012). Combining search-based procedural content generation and social gaming in the petalz video game. In *Proceedings of the 8th AAAI Conference on Artificial Intelligence and Interactive Digital Entertainment.*

Ritter, F., Kim, J. W., & Sanford, J. P. (2012). *Soar Frequently Asked Questions List.* Retrieved January 30, 2016, from http://acs.ist.psu.edu/projects/soar-faq/soar-faq.html#G3

Rizzo, A., Hartholt, A., Grimani, M., Leeds, A., & Liewer, M. (2014). Virtual Reality Exposure Therapy for Combat-Related Posttraumatic Stress Disorder. *Computer*, *47*(7), 31–37. doi:10.1109/MC.2014.199

Rizzo, P., Veloso, M., Miceli, M., & Cesta, A. (1997). Personality-driven social behaviors in believable agents. In *Proc. of the AAAI Fall Symposium on Socially Intelligent Agents.*

Robertson, G., & Watson, I. (2014). A review of real-time strategy game AI. *AI Magazine*, *35*(4), 75–204.

Robertson, J., & Young, R. M. (2014). Gameplay as on-line mediation search. In *Proceedings of the Experimental AI in Games Workshop.*

Rosenbloom, P. S. (2011b). From memory to problem solving: Mechanism reuse in a graphical cognitive architecture. In Artificial General Intelligence.

Rosenbloom, P. S. (2012). Deconstructing reinforcement learning in Sigma. In Artificial General Intelligence. doi:10.1007/978-3-642-35506-6_27

Rosenbloom, P. S. (2013). The Sigma cognitive architecture and system. *AISB Quarterly, 136.*

Rosenbloom, P., Newell, A., & Laird, J. (2014). Toward the knowledge level in SOAR: The role of the architecture in the use of knowledge. In K. van Lehn (Ed.), *Architectures for intelligence: The 22nd Carnegie Mellon Symposium on Cognition.* Psychology Press.

Rosenbloom, P. S. (2010). Combining procedural and declarative knowledge in a graphical architecture. In *Proceedings of the 10th International Conference on Cognitive Modeling.*

Rosenbloom, P. S. (2011a). Bridging dichotomies in cognitive architectures for virtual humans. In *Proceedings of the AAAI Fall Symposium on Advances in Cognitive Systems.*

Rosenbloom, P. S. (2012). Extending mental imagery in Sigma. In *Artificial General Intelligence* (pp. 272–281). Springer Berlin Heidelberg. doi:10.1007/978-3-642-35506-6_28

Rosenbloom, P. S. (2012). Towards a 50 msec cognitive cycle in a graphical architecture. In *Proceedings of the 11th international conference on cognitive modeling.*

Rosenbloom, P. S. (2013). The Sigma cognitive architecture and system. *AISB Quarterly*, *136*, 4–13.

Rosenbloom, P. S. (2014). Deconstructing episodic memory and learning in Sigma. In *Proceedings of the 36th Annual Conference of Cognitive Science Society.*

Rosenbloom, P. S., Demski, A., Han, T., & Ustun, V. (2013). Learning via gradient descent in Sigma. In *Proceedings of the 12th International Conference on Cognitive Modeling.*

Rosenbloom, P. S., Gratch, J., & Ustun, V. (2015). *Towards Emotion in Sigma: From Appraisal to Attention.* Artificial General Intelligence.

Ruijten, P., Midden, C., & Ham, J. (2013, April). I didn't know that virtual agent was angry at me: Investigating effects of gaze direction on emotion recognition and evaluation. In *Proceedings of the 8th International Conference on Persuasive Technology.* doi:10.1007/978-3-642-37157-8_23

Russell, S., & Norvig, P. (2010). Artificial Intelligence: A Modern Approach. Third Edition. Chps: Intro, 3, 12, 7-17. Prentice Hall.

Russell, J. A. (1980). A circumplex model of affect. *Journal of Personality and Social Psychology*, *39*(6), 1161–1178. doi:10.1037/h0077714

Russell, S., & Norvig, P. (2009). *Artificial intelligence: A modern approach* (3rd ed.). Upper Saddle River, NJ: Prentice Hall.

Russell, S., & Norvig, P. (2009). *Artificial Intelligence: A Modern Approach* (3rd ed.). Upper Saddle River, NJ: Prentice Hall.

Saberi, M., Bernardet, U., & DiPaola, S. (2014). An Architecture for Personality-based, Nonverbal Behaviour in Affective Virtual Humanoid Character. *Procedia Computer Science*, *41*, 204–211. doi:10.1016/j.procs.2014.11.104

Salen, K., & Zimmerman, E. (2003). *Rules of play: Game design fundamentals.* Cambridge, MA: MIT Press.

Salvucci, D. D. (2006). Modeling driver behavior in a cognitive architecture. *Human Factors*, *48*(2), 362–380. doi:10.1518/001872006777724417 PMID:16884055

Sandamirskaya, Y., Richter, M., & Schoner, G. (2011). A neural-dynamic architecture for behavioral organization of an embodied agent. In *2011 IEEE International Conference on Development and Learning (ICDL)* (*Vol. 2*, pp. 1–7). http://doi.org/ doi:10.1109/DEVLRN.2011.6037353

Schaeffer, J., Bulitki, V., & Buro, M. (2008). Bots Get Smart: Can video games breathe new life into AI research? *IEEE Spectrum*, *45*(12), 48–56. doi:10.1109/MSPEC.2008.4688952

Schank, R. C., & Abelson, R. P. (1975). *Scripts, plans, and knowledge.* Yale University.

Schaul, T. (2013). A video game description language for model-based or interactive learning. In *Proceedings of the IEEE Conference on Computational Intelligence and Games.* doi:10.1109/CIG.2013.6633610

Scherer, K. R. (1999). Appraisal theory. In T. Dalgleish & M. Power (Eds.), Handbook of cognition and emotion, (pp. 637–663). Chichester, UK: John Wiley & Sons. doi:10.1002/0470013494.ch30

Schmidhuber, J. (2015). *On Learning to Think: Algorithmic Information Theory for Novel Combinations of Reinforcement Learning Controllers and Recurrent Neural World Models*. Retrieved from http://arxiv.org/abs/1511.09249

Schreiber, G., Akkermans, H., Anjewierden, A., de Hoog, R., Shadbolt, N. R., Van de Velde, W., & Weilinga, B. (2000). *Knowledge Engineering and Management: The CommonKADS Methodology*. Cambridge, MA: MIT Press.

Segura, E. M., Kriegel, M., Aylett, R., Deshmukh, A., & Cramer, H. (2012). How Do You Like Me in This: User Embodiment Preferences for Companion Agents. In Y. Nakano, M. Neff, A. Paiva, & M. Walker (Eds.), *Intelligent Virtual Agents* (pp. 112–125). Springer Berlin Heidelberg. doi:10.1007/978-3-642-33197-8_12

Seif El-Nasr, M., Drachen, A., & Canossa, A. (Eds.). (2013). *Game analytics*. Springer London. doi:10.1007/978-1-4471-4769-5

Selic, B., Gullekson, G., & Ward, P. T. (1994). *Real-time object-oriented modeling* (Vol. 2). New York: John Wiley & Sons.

Shadbolt, N. R., & Smart, P. R. (2015). Knowledge Elicitation: Methods, Tools and Techniques. In J. R. Wilson & S. Sharples (Eds.), *Evaluation of Human Work* (4th ed.). Boca Raton, FL: CRC Press.

Shapiro, A. (2011). Building a character animation system. In Motion in Games. doi:10.1007/978-3-642-25090-3_9

Siegwart, R., & Nourbakhsh, I. R. (2004). *Introduction to Autonomous Mobile Robots* (1st ed.). Cambridge, MA: The MIT Press.

Siu, K., Zook, A., & Riedl, M. O. (2014). Collaboration versus competition: Design and evaluation of mechanics for games with a purpose. In *Proceedings of the 9th International Conference on the Foundations of Digital Games*.

Smart, P. R., & Sycara, K. (2015c). *Using a Cognitive Architecture to Control the Behaviour of Virtual Robots*. EuroAsianPacific Joint Conference on Cognitive Science, Turin, Italy.

Smart, P. R., Richardson, D. P., Sycara, K., & Tang, Y. (2014a). Towards a cognitively realistic computational model of team problem solving using ACT-R agents and the ELICIT experimentation framework.19th ICCRTS C2 Agility: Lessons Learned from Research and Operations. Southhampton University School of Electronics and Computer Science.

Smart, P. R., Sycara, K., & Tang, Y. (2014). Using Cognitive Architectures to Study Issues in Team Cognition in a Complex Task Environment. SPIE Defense, Security, and Sensing: Next Generation Analyst II, Baltimore, MD.

Smart, P. R., Sycara, K., Tang, Y., & Powell, G. (forthcoming). The ACT-R Unity Interface: Integrating ACT-R with the Unity Game Engine. Electronics and Computer Science, University of Southampton, Southampton, UK.

Smart, P. R., & Sycara, K. (2015, July) Situating Cognition within the Virtual World.*6th International Conference on Applied Human Factors and Ergonomics*, Las Vegas, NV.

Smart, P. R., & Sycara, K. (2015a). *Place Recognition and Topological Map Learning in a Virtual Cognitive Robot.17th International Conference on Artificial Intelligence*, Las Vegas, NV

Smart, P. R., & Sycara, K. (2015b). *Situating Cognition in the Virtual World.6th International Conference on Applied Human Factors and Ergonomics*, Las Vegas, NV.

Smart, P. R., Sycara, K., & Lebiere, C. (2014b, September). Cognitive Architectures and Virtual Worlds: Integrating ACT-R with the XNA Framework.*Annual Fall Meeting of the International Technology Alliance*, Cardiff, UK.

Smith, A. M., Andersen, E., Mateas, M., & Popović, Z. (2012). A case study of expressively constrainable level design automation tools for a puzzle game. In *Proceedings of the 7th International Conference on the Foundations of Digital Games*. doi:10.1145/2282338.2282370

Smith, A. M., Butler, E., & Popović, Z. (2013). Quantifying over play: Constraining undesirable solutions in puzzle design. In *Proceedings of the 8th International Conference on the Foundations of Digital Games*.

Smith, A. M., & Mateas, M. (2010). Variations Forever: Flexibly generating rulesets from a sculptable design space of mini-games. In *Proceedings of the IEEE Conference on Computational Intelligence and Games*. doi:10.1109/ITW.2010.5593343

Smith, A., Lewis, C., Hullett, K., Smith, G., & Sullivan, A. (2011). An inclusive view of player modeling. In *Proceedings of the 6th International Conference on the Foundations of Digital Games*.

Smith, B., Brown, B., Strong, W., & Rencher, A. (1975). Effects of Speech Rate on Personality Perceptions. *Language and Speech*, *18*, 145–152. PMID:1195957

Smith, G. (2014). The future of procedural content generation in games. In *Proceedings of the Experimental AI in Games Workshop*.

Smith, G., Othenin-Girard, A., Whitehead, J., & Wardrip-Fruin, N. (2012). PCG-based game design: creating Endless Web. In *Proceedings of the 7th International Conference on the Foundations of Digital Games*.

Smith, G., Whitehead, J., & Mateas, M. (2011). Tanagra: Reactive planning and constraint solving for mixed-initiative level design. *IEEE Transactions on Computational Intelligence and AI in Games*, *3*(3), 201–215. doi:10.1109/TCIAIG.2011.2159716

Smith, G., Whitehead, J., Mateas, M., Treanor, M., March, J., & Cha, M. (2011). Launchpad: A rhythm-based level generator for 2-D platformers. *IEEE Transactions on Computational Intelligence and AI in Games*, *3*(1), 1–16. doi:10.1109/TCIAIG.2010.2095855

Snodgrass, S., & Ontañón, S. (2014). A hierarchical approach to generating maps using markov chains. In *Proceedings of the 10th AAAI Conference on Artificial Intelligence and Interactive Digital Entertainment.*

Sorenson, N., Pasquier, P., & DiPaola, S. (2011). A generic approach to challenge modeling for the procedural creation of video game levels. *IEEE Transactions on Computational Intelligence and AI in Games, 3*(3), 229–244. doi:10.1109/TCIAIG.2011.2161310

Spinola, J., & Imbert, R. (2012). A Cognitive Social Agent Architecture for Cooperation in Social Simulations. In Y. Nakano, M. Neff, A. Paiva, & M. Walker (Eds.), *Intelligent Virtual Agents* (pp. 311–318). Springer Berlin Heidelberg. doi:10.1007/978-3-642-33197-8_32

Stanley, K. O., Bryant, B. D., & Miikkulainen, R. (2005). Real-time neuroevolution in the NERO video game. *IEEE Transactions on Evolutionary Computation, 9*(6), 653–668. doi:10.1109/TEVC.2005.856210

Starzyk, J. A., Graham, J., & Puzio, L. (2013). Simulation of a Motivated Learning Agent. In H. Papadopoulos, A. S. Andreou, L. Iliadis, & I. Maglogiannis (Eds.), *Artificial Intelligence Applications and Innovations* (pp. 205–214). Springer Berlin Heidelberg. doi:10.1007/978-3-642-41142-7_21

Stracuzzi, D. J., Fern, A., Ali, K., Hess, R., Pinto, J., Li, N., & Shapiro, D. G. et al. (2011). An application of transfer to american football: From observation of raw video to control in a simulated environment. *AI Magazine, 32*(2), 107–125.

Strannegård, C., von Haugwitz, R., Wessberg, J., & Balkenius, C. (2013). A cognitive architecture based on dual process theory. In *Artificial General Intelligence* (pp. 140–149). Springer Berlin Heidelberg. doi:10.1007/978-3-642-39521-5_15

Sullivan, K. (1974). *The teaching of elementary calculus: An approach using infinitesimals.* (PhD thesis). University of Wisconsin.

Sun, R. (2003). A tutorial on CLARION 5.0. In Cognitive Science Department Technical Reports. Rensselaer Polytechnic Institute. Retrieved from http://www.cogsci.rpi.edu/~rsun/sun.tutorial.pdf

Sun, R. (2006). The CLARION cognitive architecture: Extending cognitive modeling to social simulation. Cognition and multi-agent interaction, 79-99.

Sun, R. (2006). The CLARION cognitive architecture: Extending cognitive modeling to social simulation. In Cognition and multi-agent interaction, (pp. 79-99). New York, NY: Cambridge University Press.

Sun, R. (2006). The CLARION cognitive architecture: Extending cognitive modeling to social simulation. In Cognition and Multi-Agent Interaction. Cambridge University Press.

Sun, R. (1995). Robust reasoning: Integrating rule-based and similarity-based reasoning. *Artificial Intelligence, 75*(2), 241–295. doi:10.1016/0004-3702(94)00028-Y

Sun, R. (2001). *Duality of the mind.* Mahwah, NJ: Lawrence Erlbaum Associates.

Sun, R. (2002). *Duality of the Mind.* Mahwah, NJ: Lawrence Erlbaum Associates.

Sun, R. (2002). *Duality of the mind: A bottom up approach toward cognition.* Mahwah, NJ: Lawrence Erlbaum Associates.

Sun, R. (2004). Desiderata for cognitive architectures. *Philosophical Psychology, 17*(3), 341–373. doi:10.1080/0951508042000286721

Sun, R. (2006). *Cognition and multi-agent interaction: from cognitive modeling to social simulation.* Cambridge Univ Press.

Sun, R. (2006a). The CLARION Cognitive Architecture: Extending Cognitive Modeling to Social Simulation. In R. Sun (Ed.), *Cognition and Multi-Agent Interaction: From Cognitive Modeling to Social Interaction* (pp. 79–99). New York: Cambridge University Press.

Sun, R. (2007). Cognitive social simulation incorporating cognitive architectures. *Intelligent Systems, 22*(5), 33–39. doi:10.1109/MIS.2007.4338492

Sun, R. (2009). Motivational representations within a computational cognitive architecture. *Cognitive Computation, 1*(1), 91–103. doi:10.1007/s12559-009-9005-z

Sun, R. (Ed.). (2006b). *Cognition and Multi-Agent Interaction: From Cognitive Modeling to Social Interaction.* New York: Cambridge University Press.

Sun, R., & Zhang, X. (2004). Accounting for similarity-based reasoning within a cognitive architecture. In *Proceedings of the 26th Annual Conference of the Cognitive Science Society.* Mahwah, NJ: Lawrence Erlbaum Associates.

Sun, R., & Zhang, X. (2006). Accounting for a variety of reasoning data within a cognitive architecture. *Journal of Experimental & Theoretical Artificial Intelligence, 18*(2), 169–191. doi:10.1080/09528130600557713

Sutton, R., & Barto, A. (1998). *Reinforcement Learning: An Introduction.* MIT Press.

Su, W. P., Pham, B., & Wardhani, A. (2007). Personality and emotion-based high-level control of affective story characters. Visualization and Computer Graphics. *IEEE Transactions on, 13*(2), 281–293.

Swartout, W. (2010). Lessons learned from virtual humans. *AI Magazine, 31*(1), 9–20.

Taatgen, N., & Anderson, J. R. (2010). The Past, Present, and Future of Cognitive Architectures. *Topics in Cognitive Science, 2*(4), 693–704. doi:10.1111/j.1756-8765.2009.01063.x PMID:25164050

Tambe, M., Johnson, W. L., Jones, R. M., Koss, F., Laird, J. E., Rosenbloom, P. S., & Schwamb, K. (1995). Intelligent agents for interactive simulation environments. *AI Magazine, 16*(1), 15.

Tankard, J. W. Jr. (1970). Effects of eye position on person perception. *Perceptual and Motor Skills, 31*(3), 883–893. doi:10.2466/pms.1970.31.3.883 PMID:5498193

Taylor, G., & Sims, E. (2009). Developing Believable Interactive Cultural Characters for Cross-Cultural Training. In Online Communities and Social Computing (pp. 282-291). Springer Berlin Heidelberg.

Taylor, G., Quist, M., Furtwangler, S., & Knudsen, K. (2007, May). Toward a hybrid cultural cognitive architecture. *CogSci Workshop on Culture and Cognition.* SOARTech.

Taylor, S. J., Khan, A., Morse, K. L., Tolk, A., Yilmaz, L., Zander, J., & Mosterman, P. J. (2015). Grand challenges for modeling and simulation: simulation everywhere—from cyberinfrastructure to clouds to citizens. *Simulation: Transactions of the Society for Modeling and Simulation.*

Thagard, P. (2012). Cognitive architectures. In K. Frankish & W. M. Ramsey (Eds.), *The Cambridge Handbook of Cognitive Science* (pp. 50–70). Cambridge, UK: Cambridge University Press. doi:10.1017/CBO9781139033916.005

Thiebaux, M., Marsella, S., Marshall, A. N., & Kallmann, M. (2008, May). Smartbody: Behavior realization for embodied conversational agents. In *Proceedings of the 7th international joint conference on Autonomous agents and multiagent systems* (vol. 1, pp. 151-158). International Foundation for Autonomous Agents and Multiagent Systems.

Thompson, T., & Vinciguerra, M. (2014). Integrated pathfinding and player analysis for touch-driven games. In *Proceedings of the Experimental AI in Games Workshop.*

Thórisson, K. R. (1999). A Mind Model for Multimodal Communicative Creatures and Humanoids. *International Journal of Applied Artificial Intelligence*, *13*(4-5), 4. doi:10.1080/088395199117342

Togelius, J. (2011). A procedural critique of deontological reasoning. In *Proceedings of the Digital Games Research Association Conference.*

Togelius, J., & Schmidhuber, J. (2008). An experiment in automatic game design. In *Proceedings of the IEEE Symposium on Computational Intelligence and Games.*

Togelius, J., Yannakakis, G., Stanley, K., & Browne, C. (2011). Search-based procedural content generation: A taxonomy and survey. *IEEE Transactions on Computational Intelligence and AI in Games*, *3*(3), 172–186. doi:10.1109/TCIAIG.2011.2148116

Trafton, G., Hiatt, L., Harrison, A., Tamborello, F., Khemlani, S., & Schultz, A. (2013). ACT-R/E: An Embodied Cognitive Architecture for Human-Robot Interaction. *Journal of Human-Robot Interaction*, *2*(1), 30–54. doi:10.5898/JHRI.2.1.Trafton

Traum, D., Marsella, S. C., Gratch, J., Lee, J., & Hartholt, A. (2008). Multi-party, multi-issue, multi-strategy negotiation for multi-modal virtual agents. In *Intelligent Virtual Agents* (pp. 117–130). Springer Berlin Heidelberg. doi:10.1007/978-3-540-85483-8_12

Treanor, M., Schweizer, B., Bogost, I., & Mateas, M. (2012). The micro-rhetorics of Game-O-Matic. In *Proceedings of the 7th International Conference on the Foundations of Digital Games.*

Treanor, M., Zook, A., Eladhari, M. P., Togelius, J., Smith, G., Cook, M., & Smith, A. M. et al. (2015). AI-Based Game Design Patterns. In *Proceedings of the 10th International Conference on the Foundations of Digital Games.*

Tremblay, J., Borodovski, A., & Verbrugge, C. (2014). I can jump! exploring search algorithms for simulating platformer players. In *Proceedings of the Experimental AI in Games Workshop.*

Tremblay, J., Torres, P. A., & Verbrugge, C. (2014). An algorithmic approach to analyzing combat and stealth games. In *Proceedings of the IEEE Conference on Computational Intelligence and Games.* doi:10.1109/CIG.2014.6932898

Tupes, E., & Cristal, R. (1961). *Recurrent Personality Factors Based on Trait Ratings.* Technical Report ASD-TR-61-97. Lackland Air Force Base, TX: Personnel Laboratory, Air Force Systems Command.

Tversky, A. (1977). Features of similarity. *Psychological Review, 84*(4), 327–352. doi:10.1037/0033-295X.84.4.327

Ullmann, L. (2014). *The cold song.* New York, NY: Other Press.

Umarov, I., Mozgovoy, M., & Rogers, P. C. (2012). Believable and effective AI agents in virtual worlds: Current state and future perspectives. *International Journal of Gaming and Computer-Mediated Simulations, 4*(2), 37–59. doi:10.4018/jgcms.2012040103

Ustun, V., Rosenbloom, P. S., Sagae, K., & Demski, A. (2014). Distributed vector representations of words in the Sigma cognitive architecture. In Artificial General Intelligence. doi:10.1007/978-3-319-09274-4_19

Ustun, V., & Rosenbloom, P. S. (2015). Towards adaptive, interactive virtual humans in Sigma. In *Intelligent Virtual Agents* (pp. 98–108). Springer International Publishing. doi:10.1007/978-3-319-21996-7_10

Ustun, V., Rosenbloom, P. S., Kim, J., & Li, L. (2015). Building High Fidelity Human Behavior Model in the Sigma Cognitive Architecture. In *Proceedings of the 2015 Winter Simulation Conference.* Piscataway, NJ: Institute of Electrical and Electronics Engineers, Inc.

Ustun, V., Yilmaz, L., & Smith, J. S. (2006). A conceptual model for agent-based simulation of physical security systems. In *Proceedings of the 44th annual Southeast regional conference.* ACM. doi:10.1145/1185448.1185530

Vala, M., Ribeiro, T., & Paiva, A. (2012). A Model for Embodied Cognition in Autonomous Agents. In Y. Nakano, M. Neff, A. Paiva, & M. Walker (Eds.), *Intelligent Virtual Agents* (pp. 505–507). Springer Berlin Heidelberg. doi:10.1007/978-3-642-33197-8_60

van der Linden, R., Lopes, R., & Bidarra, R. (2013). Procedural generation of dungeons. *IEEE Transactions on Computational Intelligence and AI in Games, 6*(1), 78–89. doi:10.1109/TCIAIG.2013.2290371 van Lent, M., McAlinden, R., Probst, P., Silverman, B. G., O'Brien, K., & Cornwell, J. (2004). Enhancing the behaviorial fidelity of synthetic entities with human behavior models. Departmental Papers (ESE), 300.

van Oijen, J., Vanhée, L., & Dignum, F. (2012). CIGA: A Middleware for Intelligent Agents in Virtual Environments. In M. Beer, C. Brom, F. Dignum, & V.-W. Soo (Eds.), *Agents for Educational Games and Simulations* (pp. 22–37). Springer Berlin Heidelberg. doi:10.1007/978-3-642-32326-3_2

Vandromme, H., Hermans, D., & Spruyt, A. (2011). Indirectly Measured Self-esteem Predicts Gaze Avoidance. *Self and Identity*, *10*(1), 32–43. doi:10.1080/15298860903512149

Veale, T. (2012). From conceptual "mash-ups" to "bad-ass" blends: A robust computational model of conceptual blending. In *Proceedings of the International Conference on Computational Creativity*.

Veloso, M., Carbonell, J., Pérez, A., Borrajo, D., Fink, E., & Blythe, J.VELOSO. (1995). Integrating planning and learning: The PRODIGY architecture. *Journal of Experimental & Theoretical Artificial Intelligence*, *7*(1), 81–120. doi:10.1080/09528139508953801

Volpe, R., Nesnas, I., Estlin, T., Mutz, D., Petras, R., & Das, H. (2001). The CLARAty architecture for robotic autonomy. In *Aerospace Conference, 2001, IEEE Proceedings.* (*Vol. 1*, pp. 1/121–1/132). http://doi.org/ doi:10.1109/AERO.2001.931701

Wang, P., & Goertzel, B. (2007). Introduction: Aspects of Artificial General Intelligence. In Proceedings of the 2007 Conference on Advances in Artificial General Intelligence: Concepts, Architectures and Algorithms: Proceedings of the AGI Workshop 2006 (pp. 1–16). Amsterdam, The Netherlands, The Netherlands: IOS Press. Retrieved from http://dl.acm.org/citation.cfm?id=1565455.1565457

Wang, F., & Mckenzie, E. (1998). *Virtual life in virtual environments. Uni-versity of Edinburgh.* Computer Systems Group.

Wang, P., & Goertzel, B. (2007, June). Introduction: Aspects of artificial general intelligence. In *Proceedings of the 2007 conference on Advances in Artificial General Intelligence: Concepts, Architectures and Algorithms: Proceedings of the AGI Workshop 2006* (pp. 1-16). IOS Press.

Waxer, P. (1977). Nonverbal cues for anxiety: An examination of emotional leakage. *Journal of Abnormal Psychology, 86*(3), 306-314.

Weng, J. (2004). Developmental robotics: Theory and experiments. *International Journal of Humanoid Robotics*, *01*(02), 199–236. doi:10.1142/S0219843604000149

Whiten, A. (Ed.). (1991). *Natural Theories of Mind.* Oxford, UK: Basil Blackwell.

Wiggins, S. (2003). *Introduction to Applied Nonlinear Dynamical Systems and Chaos* (2nd ed.). New York: Springer-Verlag.

Wooldridge, M. (2002). Intelligent Agents: The Key Concepts. In V. Mařík, O. Štěpánková, H. Krautwurmová, & M. Luck (Eds.), *Multi-Agent Systems and Applications II* (pp. 3–43). Springer Berlin Heidelberg. doi:10.1007/3-540-45982-0_1

Yannakakis, G., & Togelius, J. (2011). Experience-driven procedural content generation. *IEEE Transactions on Affective Computing*, *2*(3), 147–161. doi:10.1109/T-AFFC.2011.6

Zammitto, V., DiPaola, S., & Arya, A. (2008). A methodology for incorpo-rating personality modeling in believable game characters. *Arya, 1*.

Zemla, J. C., Ustun, V., Byrne, M. D., Kirlik, A., Riddle, K., & Alexander, A. L. (2011). An ACT-R model of commercial jetliner taxiing. In *Proceedings of the Human Factors and Ergonomics Society Annual Meeting* (Vol. 55, No. 1, pp. 831-835). Sage Publications. doi:10.1177/1071181311551173

Zook, A., & Riedl, M. O. (2014). Automatic game design via mechanic generation. In *Proceedings of the 28th AAAI Conference on Artificial Intelligence*.

About the Contributors

Jeremy Owen Turner (b. 1974, Victoria, B.C., Canada) is currently a PhD Candidate at Simon Fraser University's School of Interactive Arts and Technology (Vancouver, Canada). Turner is also a sessional Professor at Simon Fraser University's Cognitive Science program. Turner's current research focus is on the subjects of: Artificial General Intelligence (AGI), cognitive science (cognitive architectures) and virtual character design. Since 1996, Turner has also developed an international portfolio as a performance artist, music composer, media-arts historian and art-critic within virtual worlds and video games. Turner's academic history includes an MA about avatar-design in Second Life and an interdisciplinary BA that focused on both Art-History and Music Composition. Turner's current PhD research explores developing cognitive architectural heuristics for virtual agents (automated characters/NPCs) in virtual worlds and video games.

Michael Nixon is a PhD candidate at the School of Interactive Arts & Technology at Simon Fraser University. He researches how to make virtual characters in digital environments more believable through the use of better cognitive models and non-verbal behavior in social contexts. His dissertation research focuses on the use of social signals as cues in the creation of unique identities for believable characters. Michael's M.Sc. thesis describes an investigation into the suitability of Delsarte's system of movement as a framework for the animation of believable characters.

Ulysses Bernardet is currently a postdoctoral fellow at the School of Interactive Arts and Technology of the Simon Fraser University, Vancouver, Canada. He holds a doctorate in psychology from the University of Zurich, and has a background in psychology, computer science and neurobiology. He is the main author of the large-scale neural systems simulator iqr, has developed models of insect cognition, and conceptualized and realized a number of complex, real-time interactive systems. Ulysses' research follows an interdisciplinary approach that brings together psychological and neurobiological models of behavior regulation, motivation, and emotion with mixed and virtual reality. At the core of his current research activity is the development of models of personality and nonverbal communication. These models are embodied in virtual humans and interact with biological humans in real-time. Ulysses likes to refer to this approach of "understanding humans by building them" as Synthetic Psychology.

Steve DiPaola, is a cognitive based AI computer scientist and former director of the Cognitive Science Program at Simon Fraser University. DiPaola also leads the iVizLab (ivizlab.sfu.ca), a research lab that strives to make computational systems bend more to the human experience by incorporating biological, cognitive and behavior knowledge models. Much of the lab's work is creating computation models of very human ideals such as expression, emotion, behavior and creativity. He is most known for his AI based computational creativity (darwinsgaze.com) and 3D facial expression systems. He came to SFU from Stanford University and before that NYIT Computer Graphics Lab, an early pioneering lab in high- end graphics techniques. He has held leadership positions at Electronic Arts, Saatchi Innovation and consulted for HP, Macromedia and the Institute for the Future. His computer based art has been exhibited internationally including the AIR and Tibor de Nagy galleries in NYC, Tenderpixel Gallery in London and Cambridge University's Kings Art Centre. The work has also been exhibited in major museums, including the Whitney Museum, the MIT Museum, and the Smithsonian.

***

Eugene Borovikov is an Image Processing (IP) Computer Vision (CV) scientist. Since 2010 he functioned as an independent IP/CV consultant specializing in smart technologies bridging the perception gaps in human-computer interactions on various computing platforms. In 2011, he joined the Communication Engineering Branch (CEB) at National Library of Medicine (NLM) and is actively contributing to the R&D efforts of the People Locator (PL) project by researching and developing face image retrieval (FIR) capabilities in web-scale real-world digital photo collections. During 2003-2010 he was an active member of the R&D team at CACI-Lanham primarily involved in multilingual OCR/ICR and NLP efforts in the document understanding area, authoring numerous papers and book chapters on Arabic OCR and handwriting recognition. In 2006, Dr. Borovikov directed an R&D team at ADF Solutions, advancing technology for content based image and video retrieval, collaborating with academic labs such as LAMP at UMCP. During 1997-2003, Eugene Borovikov worked as a Research Assistant at UMCP's Computer Vision Lab, performing research in the area of non-rigid shape recognition in applications to 3D human body shape estimation. Eugene Borovikov received his Ph.D. degree from University of Maryland, College Park in 2003, with the dissertation titled "High-performance visual computing in multi-perspective environments", advised by professors Larry Davis and Alan Sussman. Eugene Borovikov received the degree of Master of Arts in Applied Mathematics from University of Maryland, College Park in 1998 with the scholarly paper "Human Head Pose Estimation by Facial Features Location", advised by professors Larry Davis and David Harwood.

Selmer Bringsjord specializes in the logico-mathematical and philosophical foundations of articial intelligence (AI) and cognitive science (CogSci), and in collaboratively building AI systems on the basis of computational logic. Though he spends considerable engineering time in pursuit of ever-smarter computingmachines, he claims that \armchair" reasoning time has enabled him to deduce that the human mind will forever be superior to such machines. Bringsjord received the bachelor's degree from the University of Pennsylvania, and the PhD from Brown University, where he studied under Roderick Chisholm. Bringsjord is not unhappy about the apparent fact that he is through Chisholm an intellectual descendant of Leibniz, many of whose views to a high degree align with his own, and whose interest in a rather wide range of intellectual matters matches his own trans-disciplinary modus operandi. Bringsjord has long been on faculty at America's oldest technological university: Rensselaer Polytechnic Institute (RPI) in Troy NY; where he currently holds appointments in the Department of Cognitive Science, the Department of Computer Science, and the Lally School of Management, and where as a Full Professor he teaches AI, formal logic, formal human and machine reasoning and decision-making (and applications thereof, e.g. in nuclear strategy and micro-economics), and philosophy of AI and CogSci. In a break from things technical, he also teaches the intellectual history of New York City and the Hudson Valley, whose Occidental basis grounds out in Grotius. Funding for his r&d has come from the Luce Foundation, the National Science Foundation, the Templeton Foundation, AT&T, IBM, Apple, AFRL, ARDA/DTO/IARPA, ONR, DARPA, AFOSR, and other sponsors. Bringsjord has consulted to and advised many companies in the general realm of intelligent systems, and continues to do so. Bringsjord is author of What Robots Can & Can't Be (1992, Kluwer), concerned with the future of attempts to create robots that behave as humans, and also Superminds: People Harness Hypercomputation, and More (2003, Kluwer). Before the second of these books he wrote, with IBM's David Ferrucci, Artificial Intelligence and Literary Creativity: Inside the Mind of Brutus, A Storytelling Machine, published by Erlbaum. He is the author of Abortion: A Dialogue, published by Hackett. Bringsjord's first novel, Soft Wars, was published by Penguin USA. A forthcoming book, from Oxford University Press, is Gödel's Great Theorems, the current manuscript of which is in use in his pedagogy at RPI. Dr. Bringsjord is the author of papers and essays ranging in approach from the mathematical to the informal, and covering such areas as AI, logic, gaming, philosophy of mind, philosophy of religion, robotics, and ethics I and he has of late begun to move into the area of computational economics, for which he has invented a new paradigm, one based (unsurprisingly) on formal logic: logicist agent-based economics. A paper that sets out this paradigm is available here. An early paper in this paradigm, devoted to the modeling and simulation of bi-pay auctions, is available here. Most of Bringsjord's publications are unpublished; for example, he has written the play Calculi of Death. (Many of his writings, including some unpublished ones, are available directly through hotlinks in his vitae, available at http://www.rpi.edu/brings.) Bringsjord enjoys travel, and has lectured and interviewed in person across just about all of the 50 United States, and in many other countries, including England, Scotland, Norway, France, Finland, Ireland, Iceland, Italy, Belgium, Mexico, Australia, Germany, Romania, Poland, Denmark, China, Thailand, Japan, The Netherlands, Spain, Portugal, Egypt, and Canada. Bringsjord has interviewed and lectured on television, radio, and the internet in many additional countries.

Andrea Corradini has a PhD in Neuroscience and Cognitive Robotics from the Technical University of Ilmenau, Germany and a M.S. in Mathematics from the University of Trento, Italy. Currently, Andrea is a full professor in welfare design and technology at the Kolding School of Design, in Denmark. Andrea has a well-rounded background and research record in the field of multimodal user interfaces and human-computer interaction. He has 15+ years of expertise in designing and developing collaborative multimodal systems that support non-traditional input modality such as 3D gesture, speech and gaze for e.g. medical applications, virtual reality environments, embodied conversational agents and mobile applications.

Jacquelyne Forgette is a recent Masters graduate from the Department of Computer Science at Western University. Her research interests include artificial intelligence, machine learning, and issues in game design and implementation.

Kaveh Hassani holds a bachelor degree in software engineering and a master degree in Mechatronics engineering. He is currently a fourth year PhD candidate at University of Ottawa in computer and electrical engineering. His main research interests are artificial intelligence, machine learning, natural language processing, information retrieval, computational intelligence, and hybrid intelligence systems.

Michael James Katchabaw is an Associate Professor in the Department of Computer Science at Western University. His research focusses on various issues in game development and networking, with dozens of publications and numerous funded projects in the area, supported by various government and industry partners. At Western, Michael played a key role in establishing its program in game development, as well as the Digital Recreation, Entertainment, Art, and Media (DREAM) research group. With his efforts establishing Western as one of the first academic institutions with an interest in gaming, Michael is highly regarded in the gaming community.

Manish Mehta is a research scientist at Accenture Technology Labs where he works in the area of gamification, behavior shaping and crowdsourcing. Manish is a key subject matter expert on gamification and human motivation at Accenture that distinguishes Accenture's with a unique perspective on gamification placing it into a larger behavior change framework that has been translated into thought leadership with Accenture clients and guiding the business and research direction of Digital Workforce SII crowdsourcing projects. More recently, Manish was involved in designing behavior shaping strategies to incent active participation from fans in an online client community at a large technology firm Prior to joining Accenture, Manish worked as a Research Scientist at Disney Research where he worked in designing and conducting research on story based games for Disneyland theme park. Prior to Disney, he has worked on interactive game based installations that have been exhibited at Beall Center of Art and Technology, UCI Irvine. He has published over 40 scientific articles in top-tier international forums.

Paul S. Rosenbloom is Professor of Computer Science at the University of Southern California and Director for Cognitive Architecture Research at USC's Institute for Creative Technologies. He was a key member of USC's Information Sciences Institute for two decades, leading new directions activities over the second decade, and finishing his time there as Deputy Director. Earlier he was on the faculty at Carnegie Mellon University (where he had also received his MS and PhD in computer science) and Stanford University (where he had also received his BS in mathematical sciences, with distinction). His research concentrates on cognitive architectures – models of the fixed structure underlying minds, whether natural or artificial – and on understanding the nature, structure and stature of computing as a scientific domain. He is: a Fellow of both the Association for the Advancement of Artificial Intelligence (AAAI) and the Cognitive Science Society; the co-developer of Soar, one of the longest standing and most well developed cognitive architectures, during much of its early evolution; the primary developer of Sigma, which blends insights from earlier architectures such as Soar with ideas from graphical models; and the author of On Computing: The Fourth Great Scientific Domain (MIT Press, 2012).

Maryam Saberi is a PhD Candidate in Simon Fraser University, School of Interactive Arts and Technology.

Tom Scutt was a lecturer in AI at the University of Nottingham for seven years, before joining the video game industry in 1998. Since then he has worked in the games industry as programmer (including the AI programming for several of the Tomb Raider games), designer, and studio manager; while still regularly giving academic lectures. His research interests include MMO/social game AI & virtual economics; pervasive/locative games, and 'gamification' (using social/casual gaming reward systems to modify real-world behaviour).

Nigel Shadbolt is currently Principal of Jesus College, Oxford, and Professorial Research Fellow in the Department of Computer Science, University of Oxford. He is also a Chairman of the Open Data Institute, which he co-founded with Sir Tim Berners-Lee. Professor Sir Shadbolt is an interdisciplinary researcher, policy expert and commentator. He has studied and researched Psychology, Cognitive Science, Computational Neuroscience, Artificial Intelligence (AI), Computer Science and the emerging field of Web science. In 2013, Professor Shadbolt was knighted in the Queen's Birthday Honours "for services to science and engineering."

Paul Smart is a senior research fellow in the Web and Internet Science Research Group at the University of Southampton in the UK. His research interests include psychology, cognitive science, knowledge engineering, the Semantic Web, artificial intelligence, human factors, and the philosophy of mind. His current research activities seek to further our understanding of the effect of network-enabled technologies on human cognition at both the individual and collective (group) levels. He is particularly interested in the use of computer simulation techniques to advance our understanding of the factors that affect collective problem-solving performance in the context of complex socio-technical systems. He is also exploring some of the philosophical issues related to cognitive extension in contemporary and near-future network environments. Dr Smart has a DPhil in Experimental Psychology from the University of Sussex. Contact him at ps02v@ecs.soton.ac.uk.

Katia Sycara is a Research Professor in the School of Computer Science at Carnegie Mellon University and holds the Sixth Century Chair (part-time) in Computing Science at the University of Aberdeen in the U.K. She is also the Director of the Laboratory for Agents Technology and Semantic Web Technologies at Carnegie Mellon. Professor Sycara is a Fellow of the Institute of Electrical and Electronic Engineers (IEEE), Fellow of the American Association for Artificial Intelligence (AAAI) and the recipient of the 2002 ACM/SIGART Agents Research Award. She is also the recipient of the Outstanding Alumnus Award from the University of Wisconsin, which she received in 2005. She is a member of the Scientific Advisory

Board of France Telecom and a member of the Scientific Advisory Board of the Greek National Center of Scientific Research "Demokritos" Information Technology Division.

Volkan Ustun is an artificial intelligence researcher in the Cognitive Architecture group at USC Institute for Creative Technologies. His general research interests are computational cognitive models and simulation. He is currently involved in the various aspects of the development of the Sigma cognitive architecture. Volkan has B.S. and M.S. degrees from METU, Turkey and a Ph.D. degree from Auburn University in industrial and systems engineering. Papers out of his Ph.D. research are recipients of The Gene Newman Award of Excellence in Modeling and Simulation Research and the Winter Simulation Conference I-Sim/ACM-SIGSIM Best Computer Science Focused Student Paper Award. He is also been selected as the Most Outstanding Graduate International Student at Auburn University.

Sergey Yershov has over 14 years experience in secure client-server, mobile, cloud, and hardware engineering, as well as project management and requirements analysis. As a lead software developer for the Department of the Employment Services in Washington, DC he has created multiple enterprise and mobile applications, internal document management systems and supporting middleware, such as OCR and ESB. He also co-developed their internal rapid development framework "PeopleFirst", which uses visual tools to prototype and configure database, workflow, and form design. Sergey was a lead developer behind the award-winning OneApp program and Nutrition Services financial project for the DC State Superintendent of Education. As part of MACRO project for the Internal Revenue Service he has built a UML-XMI transformers and visual tools for code mapping and generation, and developed a COM bridge between MS Word and Java. Mr. Yershov is currently an expert consultant for Food and Drug Administration for their FURLS program. His other works include contributions to open-source projects, such as OpenFrameworks, and 3d computer vision for mobile platforms. He received a BA in Performing Arts from Odessa State Conservatory in Ukraine.

Ilya Zavorin has over 15 years of experience in applied research, large-scale, cloud and mobile software development, and big data analytics. As a Senior Principal Research Analyst at CACI International Inc., he has conducted research on applications of machine learning, pattern recognition and classifier combination to document image processing and natural language processing. As part of a project for Army Research Lab, Dr. Zavorin participated in development of a multi-evidence multi-engine system (MEMOE) that applied various types of fusion to combine outputs of multiple OCR engines to produce output of higher quality than individual engines or their combination based on majority vote. MEMOE was build using a flexible modular architecture allowing simple adding of components such as new OCR engines or pre/post-processing components and was applied to documents in several languages. As part of a large-scale software development project for a DoD client, participated in design and development of a mobile data collection, exploitation and dissemination tool for the Android platform. As part of the IARPA OSI EMBERS project, Dr. Zavorin participated in the design and development of a distributed highly modular real-time Amazon Web Services based system for social media predictive analytics. Performed evaluations of various third-party commercial, freeware and open-source tools in the areas of OCR, audio and video exploitation, content extraction and natural language processing. Collected test data corpora to support such evaluations. In some cases this required generating synthetic data designed to simulate real data when access to such real data was severely restricted due to its sensitivity. Before joining CACI Dr. Zavorin was a Research Staff Member at NASA Goddard Space Flight Center, where he developed efficient algorithms for automatic registration and fusion of multi- and hyper-spectral satellite imagery. He received his Ph.D. in Applied Mathematics from the University of Maryland, College Park.

Alexander Zook is a Ph.D. candidate in Human-Centered Computing at the Georgia Institute of Technology School of Interactive Computing. His research focuses on AI for game design and development, understanding how AI systems can augment or automate human design practices. The results of this research have implications for games used for entertainment, education, and training purposes. Alexander has worked with Blizzard Entertainment and Bioware to improve the design and ongoing development of analytic tools for major massively multiplayer online role-playing games.

Index

A

B

C

D

E

F

G

H

I

K

L

M

N

O

P

R

S

T

U

V

W

Z

Printed in the United States
By Bookmasters